Topics in Current Genetics

15

Series Editor: *Stefan Hohmann*

Per Sunnerhagen • Jure Piškur (Eds.)

Comparative Genomics

Using Fungi as Models

With 39 Figures, 14 in Color; and 20 Tables, 1 in Color

 Springer

Professor Dr. Per Sunnerhagen
Department of Cell and Molecular Biology
Lundberg Laboratory
Göteborg University
P.O. Box 462
405 30 Göteborg
Sweden

Professor Dr. Jure Piškur
Department of Cell and Organism Biology
Molecular Genetics, Hs 4
Sölvegatan 35
223 62 Lund
Sweden

The cover illustration depicts pseudohyphal filaments of the ascomycete *Saccharomyces cerevisiae* that enable this organism to forage for nutrients. Pseudohyphal filaments were induced here in a wild-type haploid MATa Σ1278b strain by an unknown readily diffusible factor provided by growth in confrontation with an isogenic petite yeast strain in a sealed petri dish for two weeks and photographed at 100X magnification (provided by Xuewen Pan and Joseph Heitman).

ISSN 1610-2096
ISBN-10 3-540-31480-6 Springer Berlin Heidelberg New York
ISBN-13 978-3-540-31480-6

Library of Congress Control Number: 2005938886

Springer-Verlag is a part of Springer Science + Business Media
springeronline.com

© Springer-Verlag Berlin Heidelberg 2006
Printed in Germany

Typesetting: Camera ready by editors
Data-conversion: PTP-Berlin, Stefan Sossna e.K.
Cover Design: Design & Production, Heidelberg
Printed on acid-free paper – 39/3152-YK – 5 4 3 2 1 0

Topics in Current Genetics publishes review articles of wide interest in volumes that centre around a specific topic in genetics, genomics as well as cell, molecular and developmental biology. Particular emphasis is placed on the comparison of several model organisms. Volume editors are invited by the series editor for special topics, but further suggestions for volume topics are highly welcomed.

Each volume is edited by one or several acknowledged leaders in the field, who ensure the highest standard of content and presentation. All contributions are peer-reviewed. All papers are published online prior to the print version. Individual DOIs (digital object identifiers) make each article fully citable from the moment of online publication.

Subscribers to the print version of *Topics in Current Genetics* receive free access to the online version. An online-only license is also available.

Editorial office:
Topics in Current Genetics
Series Editor: Stefan Hohmann
Cell and Molecular Biology
Göteborg University
Box 462
40530 Göteborg, Sweden
Phone: +46 733 547297
FAX: +46 31 7732595
E-mail: editor@topics-current-genetics.se
Website: http://www.topics-current-genetics.se

Preface

André Goffeau

For several millennia, Yeast has faithfully served mankind by producing carbon dioxide and alcohol. During the last three centuries, Yeast has emerged as a prominent scientific object. It has been seen under Antonie van Leeuwenhoek's optical microscope. It has been used by Antoine Lavoisier to demonstrate that chemical mass conservation applies to biological systems. It has been patented by Louis Pasteur in his attempt to upgrade the quality of French beer. Today about fifty thousand scientific papers describing properties of the model yeast *Saccharomyces cerevisiae* have been published and a community of over ten thousand scientists takes advantage of its remarkable genetic properties to study the basic mechanism of life that are conserved in all eukaryotic cells.

In 1996, an international consortium of 96 laboratories completed its genome sequence. Today, the genome of the laboratory strain *Saccharomyces cerevisiae* SC288c is the most verified and the best annotated eukaryotic genome sequence available. It comprises the full centromeric and telomeric regions, which are repetitive and difficult to sequence and therefore are often omitted in the sequencing of other eukaryotic genomes.

It took about six other years to complete a second yeast genome sequence; that of the fission yeast *Schizosaccharomyces pombe* which over 500 millions years ago shared a common ancestor with the budding yeast *Saccharomyces cerevisiae*.

This book marks the development of "fungal comparative genomics" which took off in the year 2000 by the partial genome sequencing and comparison of thirteen yeast species by a consortium of French scientists. During these last five years, over 20 fungal genome sequences have become available, providing a unique material to explore the mechanisms of genome evolution. Even though the evolution time of these species is estimated to span nearly one billion years (more than the divergence time from fish to man) the fungal species have conserved similar morphology and related life style. This conservation simplifies the tracing of genes and chromosomes during evolution as well as that of protein domains or that of metabolic and regulatory pathways.

The comparison of the fungal genomes has provided evidence for an important evolution mechanism predicted 35 years ago by Suzumi Ohno who stated that "speciation requires prior gene duplication". It has unravelled the existence of different gene duplication mechanisms operating to different extent in different yeast species. Many specific or general mechanisms of evolution have already emerged from fungal comparative genomics and many of those are expected to be extrapolated to the evolution of higher species.

This book describes the tools of "comparative fungal genomics" which progressively intermix those used by "phylogenetic analysis". It gives a general overview of the evolution of the fungal world. It scrutinises and compares the genomes of a variety of Hemiascomycete species such as those from brewer, baker, and laboratory *Saccharomyces cerevisiae* strains, the plant pathogen *Ashbya gossypii*, the hydrocarbon-loving *Yarrowia lipolytica*, the industrial yeast *Kluyveromyces lactis*, and the human pathogen *Candida glabrata* as well as that of the Archiascomycete *Schizosaccharomyces pombe.*

This book provides the "starting kit" of a new research area that investigates basic or specific evolution mechanisms and vanguards the molecular exploration of many yeast species that were poorly known up to now. It may announce a progressive drift of interest from the "biblical" yeast *Saccharomyces cerevisiae* to the genomic and molecular scrutiny of a multitude of other fungal species.

Goffeau, André
    Institut des Sciences de la Vie, Université catholique de Louvain, Croix du Sud, 5/15, 1348 Louvain-la-Neuve, Belgium
    goffeau@fysa.ucl.ac.be

Table of contents

List of contributors

Aigle, Michel
 IBGC. CNRS/Université Victor Segalen Bordeaux 2, rue Camille Saint Saens, 233077 Bordeaux Cedex, France

Axelson-Fisk, Marina
 Fraunhofer-Chalmers Research Center for Industrial Mathematics, Chalmers Science Park, SE-412 88 Göteborg, Sweden

Barton, David B. H.
 Institute of Genetics, University of Nottingham, Queen's Medical Centre, Nottingham NG7 2UH, UK

Bolotin-Fukuhara, Monique
 Institut de Génétique et Microbiologie, Université Paris Sud/CNRS UMR 8621, 91405 Orsay Cedex, France
 bolotin@igmors.u-psud.fr

Brachat, Sophie
 Biozentrum Universität Basel, Klingelbergstrasse 50, CH-4056 Basel, Switzerland and Novartis-Pharma (Basel)

Casaregola, Serge
 Microbiologie et Génétique Moléculaire, INRA UMR 1238/ CNRS UMR 2585, INA-PG; 78850 Thiverval-Grignon, France

Cohn, Marita
 Department of Cell and Organism Biology, Molecular Genetics, Lund University, Sölvegatan 35, S-223 62 Lund, Sweden
 marita.cohn@cob.lu.se

Dietrich, Fred
 Biozentrum Universität Basel, Klingelbergstrasse 50, CH-4056 Basel, Switzerland and Department of Molecular Genetics and Microbiology, Duke University, Medical Center, Research Drive, Durham, NC 27710, USA

Gaffney, Tom
 Syngenta Biotechnology Inc., 3054 Cornwallis Road, Research Triangle Park, NC 27709, USA

Gouveia-Oliveira, Rodrigo
 Center for Biological Sequence Analysis, BioCentrum-DTU, Building 208, Technical University of Denmark, DK-2800 Kgs. Lyngby, Denmark

Hallin, Peter F.
  Center for Biological Sequence Analysis, BioCentrum-DTU, Building 208, Technical University of Denmark, DK-2800 Kgs. Lyngby, Denmark

Hansen, Jørgen
  Poalis A/S, Bülowsvej 25, DK-1870 Frederiksberg C, Denmark

Hohmann, Stefan
  Department for Cell and Molecular Biology, Göteborg University, Box 462, S-40530 Göteborg, Sweden
  hohmann@gmm.gu.se

Kielland-Brandt, Morten C.
  Carlsberg Laboratory, Gamle Carlsberg Vej 10, DK-2500 Copenhagen Valby, Denmark
  mkb@crc.dk

Kodama, Yukiko
  Suntory Research Center 1-1-1, Wakayamadai, Shimamoto-cho, Mishima-gun, Osaka 618-8503, Japan
  Yukiko_Kodama@suntory.co.jp

Krantz, Marcus
  Department for Cell and Molecular Biology, Göteborg University, Box 462, S-40530 Göteborg, Sweden. Present address: The Systems Biology Institute, 6-31-15 Jingumae M31 6A, Shibuya, Tokyo 150-0001, Japan

Kurtzman, Cletus P.
  Microbial Genomics and Bioprocessing Research Unit, National Center for Agricultural Utilization Research, Agricultural Research Service, U.S. Department of Agriculture, Peoria, Illinois, USA
  kurtzman@ncaur.usda.gov

Liti, Gianni
  Institute of Genetics, University of Nottingham, Queen's Medical Centre, Nottingham NG7 2UH, UK

Philippsen, Peter
  Biozentrum Universität Basel, Klingelbergstrasse 50, CH-4056 Basel, Switzerland
  Peter.Philippsen@unibas.ch

Piškur, Jure
  BioCentrum-DTU, Technical University of Denmark and Cell and Organism Biology, University of Lund, Sweden
  jure.piskur@biocentrum.dtu.dk

Sunnerhagen, Per
    Department of Cell and Molecular Biology, Lundberg Laboratory, Göteborg
    University, P.O. Box 462, SE-405 30 Göteborg, Sweden
    per.sunnerhagen@molbio.gu.se

Ussery, David
    Center for Biological Sequence Analysis, BioCentrum-DTU, Building 208,
    Technical University of Denmark, DK-2800 Kgs. Lyngby, Denmark
    dave@cbs.dtu.dk

Voegeli, Sylvia
    Biozentrum Universität Basel, Klingelbergstrasse 50, CH-4056 Basel, Switzer-
    land

Wanchanthuek, Phatthanaphong
    Center for Biological Sequence Analysis, BioCentrum-DTU, Building 208,
    Technical University of Denmark, DK-2800 Kgs. Lyngby, Denmark

Wolfe, Kenneth H.
    Department of Genetics, Smurfit Institute, University of Dublin, Trinity Col-
    lege, Dublin 2, Ireland
    khwolfe@tcd.ie

Wong, Simon
    Department of Genetics, Smurfit Institute, University of Dublin, Trinity Col-
    lege, Dublin 2, Ireland

Wood, Valerie
    Wellcome Trust Sanger Institute, Wellcome Trust Genome Campus, Hinxton,
    CB10 1SA, UK
    val@sanger.ac.uk

Comparative genomics and gene finding in fungi

Marina Axelson-Fisk and Per Sunnerhagen

Abstract

In the spring of 2005, we had access to 18 fully sequenced fungal genomes, and more are coming rapidly. New approaches and methods are being developed to harvest this information source to derive functional predictions and understanding of genome anatomy. Comparative genomics also tells us stories about the evolution of yeasts and filamentous fungi, and the genome rearrangements that marked their history. For example, several genes encoding proteins required for heterochromatin formation and RNA interference have been lost uniformly throughout the *Hemiascomycetes*, although some genes remain in a few species in a scattered pattern.

Being the first eukaryote to have its genome fully sequenced, *Saccharomyces cerevisiae* was the forerunner for *in silico* methods of genome annotation in general, and gene finding in particular. Lessons learned from the comparatively simple genome of this budding yeast have paved the way for efficient genome analysis in other fungi as well as eukaryotes in general.

Several fungal species are of important applied interest for mankind, and so it is essential to utilise comparative genomics to derive functional information about them. The set of fungal genomes: simple, related in evolution, and with a high density of functional information, can serve as a highly efficient test bed for the further development of comparative genomics.

1 Comparative genomics

Comparative genomics is on the rise as a potent tool in molecular biology. Comparisons of single sequences, protein or nucleic acid, preceded comparisons of whole genomes by two decades. Classical similarity searches of amino acid sequences identified orthologues and paralogues of proteins from widely divergent species, and comparison of ribosomal RNA sequence was used to determine phylogenetic relationships. Since these are among the most highly conserved features that can be derived directly from genomes, comparisons over long evolutionary distances are possible and desirable. More recently, the availability of many fully sequenced genomes has made possible a broad collection of comparative exercises. For instance, study of closely related species allow identification of syntenic blocks in chromosomes, conservation of *cis*-regulatory sequences, spreading of repetitive sequence elements, development of pseudogenes etc.

Topics in Current Genetics, Vol. 15
P. Sunnerhagen, J. Piškur (Eds.): Comparative Genomics
DOI 10.1007/4735_111 / Published online: 11 November 2005
© Springer-Verlag Berlin Heidelberg 2005

Comparative genomics attains its full power only when experimental genetic and molecular biology data are available from at least one of the species. Prominent cases are mammalian genomes (mouse, rat, and human), where functional data from mouse and human can be drawn upon, and the nematodes *Caenorhabditis elegans vs. C. briggsae*. Among plants, full genome sequences are available from *Arabidopsis thaliana* and rice (*Oryza sativa*), and more genome sequences are underway. Studies of genomes from higher plants face the obstacle of quite differing sizes, ranging from 1.2×10^8 bp (*A. thaliana*) to over 1.5×10^{10} bp (*e.g.* some *Allium* species).

1.1 Comparative genomics of fungi

The publicly available genomes from yeasts and filamentous fungi, 18 at the time of writing, represent a unique resource for comparative genomics, by two arguments. First, a wide range of evolutionary distances is represented, from separation times between 5 and 20 MYr (within the "*Saccharomyces sensu stricto*" group; Kellis et al. 2003) to 600 – 1200 MYr (between basidiomycetes and ascomycetes; Heckman et al. 2001; Douzery et al. 2004). Second, among the 18 species, many are genetically tractable experimental organisms. Thus, it is possible to directly verify inferences from genome comparisons using molecular genetics, opening up a multitude of interesting possibilities. Further, analysis of pathway conservation can reveal if whole signalling or metabolic pathways, or branches thereof, are missing or differently wired in some species (see Chapter 6 by Krantz and Hohmann in this volume).

Comparative genomics of yeasts has been reviewed with emphasis on the protein-coding complements of the different species (Herrero et al. 2003). The potential of comparative genomics of closely related *Saccharomyces* species for identification of regulatory elements has recently been highlighted (Kellis et al. 2004b), and the usefulness of genome sequencing for shedding light on phylogenetic relationships among yeasts has also been emphasised (Piškur and Langkjaer 2004). The purpose of the present volume is to draw attention to the considerable potential of a combination of bioinformatics and experimental approaches utilising information from the many fungal genomes on hand, representing yeast and filamentous fungi.

A highly useful tool for comparative genomics of fungi, FungalBlast, has recently been developed at the Saccharomyces Genome Database (SGD; www.yeastgenome.org/) (Balakrishnan et al. 2005). This takes advantage of all completely or partially sequenced fungal genomes, representing at the time of writing 38 species, and allows parallel searches in these for protein or DNA sequences similar to the query. Other tools, such as the Fungal Alignment (displaying amino acid sequence homologies) and the Synteny Viewer (displaying the gene arrangement around corresponding gene loci) exploit the genome sequences of closely related *Saccharomyces* species. These and other bioinformatics devices developed explicitly for comparisons between fungal genomes have quickly become popular among molecular biologists.

1.2 Relationships between sequenced fungal genomes

From a *Saccharomyces cerevisiae*-centric perspective, the presently sequenced fungal genomes represent a sliding scale from sibling species to quite distant relatives. There is first a set of closely related *Saccharomyces* species (*S. paradoxus, S. mikatae, S. bayanus, S. kudriavzevii*). These are estimated to have diverged between 5 and 20 Myr ago. Extensive studies have been invested into the *Hemiascomycetes* as a whole, comprising the vast majority of known ascomycetous yeast species. Thus, genome sequences are available from *S. castellii, Candida albicans, C. glabrata, Yarrowia lipolytica, Debaryomyces hansenii, Kluyveromyces lactis, K. waltii, Hansenula polymorpha,* and *Ashbya gossypii*. By virtue of its relatedness on the sequence level, *A. gossypii* is classified with the *Hemiascomycetes* despite its predominantly filamentous mode of growth. A summary of what has been observed from genome comparisons within the *Hemiascomycetes* is found in this volume, Chapter 8 by Bolotin-Fukuhara et al.

Small gene families, most often consisting of two to three members, are quite common in many hemiascomycetes. The *S. cerevisiae* genome sequence revealed that the organisation of such duplications was such that blocks of genes often could be mapped to corresponding blocks of seemingly duplicated genes elsewhere in the genome. This prompted the suggestion that a series of large duplication and recombination events were key in shaping of the budding yeast genome (Philippsen et al. 1997; Wolfe and Shields 1997). Direct confirmation of this prediction came recently with the sequences of genomes from fungi that split off from the *Saccharomyces* branch before these duplications took place, namely *Ashbya gossypii* (Dietrich et al. 2004), *Kluyveromyces lactis* (Dujon et al. 2004), and *K. waltii* (Kellis et al. 2004a). Here, it is possible to find relationships between long syntenic blocks of genes in *Saccharomyces* vs. these other non-duplicated species on a 2:1 basis (see Chapter 4 by Wong and Wolfe in this volume). Beside the basic rule that extensive gene duplications are a distinctive feature of the *Saccharomyces sensu lato* group, there are cases where a gene has been duplicated independently in two branches of the fungal tree. Thus, an investigation comparing *S. cerevisiae* and *Sz. pombe* revealed 56 such duplications (Hughes and Friedman 2003). Other examples are pyruvate decarboxylase genes, which have been independently duplicated in *S. cerevisiae* and *S. kluyveri* (Møller et al. 2004), and genes encoding mitochondrial ADP/ATP carriers in *S. cerevisiae* and *Y. lipolytica* (Mentel et al. 2005). Obviously, assignment of orthologous relationships is often ambiguous in such cases (see section on orthologue mapping in Chapter 10 by Wood).

Representatives of two more subclasses of *Ascomycetes* have been fully sequenced. The fission yeast *Schizosaccharomyces pombe* (see Chapter 10 by Wood), a widely used experimental organism, belongs to the *Archiascomycetes*. The genus *Schizosaccharomyces* has only three characterised species, and no other close relatives are known. The fission yeasts are thought to lack many of the special evolutionary adaptations of the *Hemiascomycetes*. Several ascomycetous filamentous fungi, classified in *Euascomycetes*, (*Aspergillus nidulans, Giberella zeae* [a.k.a. *Fusarium graminearum*], *Magnaporthe grisea, Neurospora crassa*) have been fully sequenced. Some of these (*A. nidulans*, *N. crassa*) are important genetic model organisms with a long scientific history. The larger complexity of the filamentous lifestyle is reflected in a gene number about twice as high as in the typical yeast (Table 1). In contrast to the recently diverged *Saccharomyces* species, the split between these three branches of *Ascomycetes* (*Hemiascomycetes*, *Archiascomycetes*, and *Euascomycetes*) took place as long as 0.3 – 1 GYr ago (Maddison 1997; Sipiczki 2000; Heckman et al. 2001; Douzery et al. 2004), thus comparable to the distance separating vertebrates and arthropods.

Even further away on the evolutionary scale are the basidiomycetes. Genome sequences are available from *Phanerochaete chrysosporium* (a filamentous fungus causing white-rot of wood) and *Ustilago maydis* (a maize pathogen with a multicellular as well as a unicellular, yeast-like, life phase). This is also the basidiomycete with the best-studied genetics. The full genome sequence is available from *Cryptococcus neoformans*, a yeast pathogenic for humans. It should be noted that the concept of yeasts is operational, since unicellular fungi occur both among ascomycetes and basidiomycetes. The predominant theory for the evolution of ascomycetous yeasts is by evolution from filamentous ancestors (Liu and Hall 2004); for the basidiomycetous yeasts, such a tracing of evolutionary history is less apparent. *Coprinopsis cinerea*, a free-living mushroom that can be cultivated in defined medium and for this reason has permitted genetic analysis, has also been extensively sequenced.

Finally, there is one sequenced representative of *Microsporidia*, for which the relationships to other major classes of fungi have long remained unresolved, that of the intracellular parasite *Encephalitozoon cuniculi*. The impact of genomic sequencing on the phylogeny of fungi is treated in Chapter 2 in this volume by Piškur and Kurtzman. On the other hand, phylogenetic advances can impact genome sequencing by suggesting new species to be sequenced. For instance, can we map more closely the point where a whole-genome duplication event took place within the *Hemiascomycetes*?

1.3 Properties of sequenced fungal genomes

Compared to those of higher plants and animals, the presently sequenced fungal genomes are compact; the gene density exceeds 0.26 per kb (Table 1). Consequently, intergenic regions are short. Introns, which are a rare commodity in budding yeast genomes, are more frequent in other fungal genomes. However, fungal introns tend to be short even where they are numerous (Table 1). Repetitive sequences, which make up a major fraction of vertebrate DNA, are low in abundance. Transposable elements, both DNA transposons and retroelements, are found in all branches of the fungal kingdom (Daboussi 1997). The genome of the fully sequenced basidiomycetes, *Cryptococcus neoformans*, reveals a considerably higher abundance of both introns and transposable elements than seen in ascomycetes (Loftus et al. 2005).

Table 1. Key features of some sequenced and annotated fungal genomes

	Genome size (Mb)	Gene number	Genes per kb	Average gene length	Fraction coding DNA	Average intergenic length	Introns per gene	Average length introns
Saccharomyces cerevisiae	12.5	5777	0.46	1460	0.71	515	0.05	216
Kluyveromyces waltii	10.6	5230	0.49					
Candida albicans	14.9	6419	0.43		0.73		0.42	
Debaryomyces hansenii	12.2	6906	0.57		0.79			
Ashbya gossypii	8.7	4718	0.54		0.80	341	0.05	
Schizosaccharomyces pombe	14.1	4973	0.35	1430	0.57	952	0.95	78
Gibberella zeae	35.6	11640	0.33	1744	0.50	1301	2.22	93
Magnaporthe grisea	38.8	11108	0.29	1683	0.41	1503	1.89	143
Neurospora crassa	38.9	10082	0.26	1673	0.38	1953	1.70	135
Aspergillus nidulans	30.1	9457	0.31	1882	0.51	1266	2.69	101
Cryptococcus neoformans	19.1	6572	0.34	2195	0.51	916	5.30	67
Ustilago maydis	19.7	6522	0.33	1935	0.61	1040	0.75	127
Encephalitozoon cuniculi	2.5	1996	0.80		0.86			

Numbers were derived from the following sources: *Saccharomyces cerevisiae*, Chapter 10 by Wood (this volume); *Kluyveromyces waltii*, Kellis et al. (2004b); *Candida albicans*, Jones et al. (2004); *Debaryomyces hansenii*, www.ebi.ac.uk/2can/genomes/eukaryotes/Debaryomyces_hansenii.html; *Ashbya gossypii*, Dietrich et al. (2004); *Schizosaccharomyces pombe*, Chapter 10 by Wood (this volume); *Gibberella zeae* (=*Fusarium graminearum*) http://www.broad.mit.edu/annotation/fungi/fusarium/; *Magnaporthe grisea*, www.broad.mit.edu/annotation/fungi/magnaporthe/; *Neurospora crassa*, www.broad.mit.edu/annotation/fungi/neurospora_crassa_7/index.html; *Aspergillus nidulans*, www.broad.mit.edu/annotation/fungi/aspergillus/; *Cryptococcus neoformans*, Loftus et al. (2005); *Ustilago maydis* www.broad.mit.edu/annotation/fungi/ustilago_maydis/; *Encephalitozoon cuniculi*, Katinka et al. (2001)

Throughout, there is a clear trend that unicellular organisms have smaller and more compact genomes than multicellular organisms. Among the presently sequenced fungal genomes, there are representatives for both free-living unicellular and filamentous species. Also here, the obvious tendency is for the unicellular organisms (the yeasts) to have the more compact genomes; the average genome size for the filamentous fungi is about 2.5-fold larger than for the yeasts (30 vs. 12 Mb). There is one interesting exception to this rule of thumb. The ascomycete *Ashbya gossypii* (see Chapter 9 by Brachat et al. in this volume), which has predominantly been observed in a hyphal form, surprisingly has a genome smaller than all the unicellular yeast species the collection (9 Mb), in fact the smallest genome reported so far for a free-living eukaryote. It also has a gene number in the lower end of the spectrum (4718). Another species that differs distinctly from the rest of the collection is *Encephalitozoon cuniculi*. This atypical fungus possesses the smallest eukaryotic genome reported to date, only 2.5 Mb containing 1996 genes (Katinka et al. 2001). This is considerably less than most free-living prokaryotes. Being an intracellular parasite, *E. cuniculi* has lost not only much of the metabolic and transport capacity common to most eukaryotes, but also lack organelles including mitochondria and peroxisomes. A broad overview of information organisation, coding capacity, and DNA sequence properties of sequenced fungal genomes is given in Chapter 3 by Wanchanthuek et al. in this volume.

Beside genes, chromosomes carry elements necessary for their own maintenance and stability – origins of replication, telomeres, and centromeres. The overall organisation of telomeres in fungi is similar to that in other eukaryotes. Experiments in yeasts (*S. cerevisiae* and *Sz. pombe*) have laid much of the foundation for our present understanding of telomere biology in general, including the mechanism of telomere replication and length maintenance, telomere-binding proteins, and silencing. However, the sequence of the telomere repeats in fungi, especially the yeasts, are more divergent between species than what has so far been observed in other organism groups, where the repeat unit TTAGGG is predominant. Also, fungal telomeres are considerably shorter (several hundred nucleotides) than telomeres in vertebrates or plants, which are typically thousands to tens of thousand nucleotides long (McKnight et al. 1997). It should be noted that telomeric sequences are underrepresented in genome sequencing project because of cloning problems, and that the telomeric regions are not necessarily fully mapped in all the "completed" genomes; targeted measures are usually required to resolve these problematic regions. Telomeres are treated by Cohn et al. in Chapter 5 of this volume.

Centromeres constitute another region of fundamental importance for chromosome function. Even though the entire centromeric regions have been sequenced in most fungal genome projects, most of our understanding of the function of centromeres is limited to the most experimentally tractable species, *S. cerevisiae* and *Sz. pombe*. The currently limited information indicates a considerable variation between fungal species in the size of the functional entity. Thus, while a functional centromere in the budding yeast *S. cerevisiae* covers little more than a single nucleosome, centromeres in the fission yeast *Sz. pombe* comprise 40 kb, and the size of the centromeric region in *Neurospora crassa*, at 400 kb, approaches the size of human centromeres (Centola and Carbon 1994). No functional information is yet available about the size of centromeres in basidiomycetes. However, clustering of transposons in single blocks comprising 40 – 100 kb on each chromosome from *Cryptococcus neoformans* is an indication that the size of centromeres in this organism are similar to those of *Sz. pombe* (Loftus et al. 2005).

We are only beginning to find the overall architecture of domain organisation in chromosomes. In metazoans, there is an accumulation of highly expressed genes in early replicating regions of the genome. However, such a correlation was not seen in *S. cerevisiae* (Raghuraman et al. 2001; Gilbert 2002). Global expression studies and mapping of replication origins in other fungal species should reveal if such an organisation is found or not in the fungal kingdom at large. In order to fully benefit from the wealth of fungal genome data to resolve these and other issues, there may be a need to expand the range of experimental species. For instance, the extent of non-homologous recombination varies greatly between yeast species such as *S. cerevisiae* (very low), *Sz. pombe* (intermediate), and *K. lactis* (high). Systematic correlations of this property with genomic and other properties of several other yeast species may be quite efficient tools to find out the fundamental reasons for this.

1.4 Objectives of comparative genomics using different evolutionary distances

The different features in a genome evolve at quite different rates and so different evolutionary distances are required for comparative approaches to capture them efficiently. Protein sequences are more conserved than DNA sequences in general, and functionally important amino acid residues stand out above the rest. Consequently, comparative genomics among rather distantly related fungal species has been utilised successfully to validate predicted phosphorylation sites in proteins based on their evolutionary conservation (Escote et al. 2004).

Conservation of gene order, synteny, can be used as an additional tool for identification of genes. Synteny between the human, mouse, and rat genomes has been used for gene verification and prediction (Gibbs et al. 2004). Synteny is extensive within closely related species such as within the *Saccharomyces sensu stricto* group, but is gradually broken up as one moves further away among the *Hemiascomycetes* species. At longer evolutionary distances between fungi, synteny is completely lost. Knowing the gene order from closely related species can be used *e.g.* to distinguish orthologues from paralogues. It is also possible to utilise synteny for *de novo* gene finding, as demonstrated by Kellis et al. (2003).

Using evolutionary conservation as the criterion, it is harder to identify regulatory regions than coding sequences. This is because the sequences of *cis*-regulatory elements evolve faster than proteins in general, including the regulatory proteins that bind to them. Consequently, the logical set of genomes to use for identification of *cis*-regulatory elements is derived of several closely related species, such as the *Saccharomyces* species used for identification of conserved transcription factor binding sites (Cliften et al. 2003). The present set of fungal genome sequences holds the promise to allow finding also other types of *cis*-regulatory elements. The sequences that define regulation of mRNA stability and translation are considerably less well explored than promoter elements. Thus, we do not know their degree of conservation, and so it is valuable to have access to a wide range of evolutionary distances between genomes in order to be able to discriminate these different elements against the noise background. It may be necessary to apply quite different criteria for the various types of post-transcriptional regulatory elements. For example, upstream ORFs (uORFs) in the 5' untranslated region of an mRNA in some cases serve to attenuate translation of the main ORF. Functional uORFs could be characterised by their DNA sequence, their distance from the start codon of the main ORF, by the amino acid sequence of the encoded peptide, by the length of the uORF, or by other criteria. Once the important criteria have been established, comparative genomics can be used to identify functional uORFs on a genome-wide scale by evolutionary conservation. Thus, using conservation between hemiascomycetous species of uORF sequence and position as criteria, Zhang and Dietrich (2005) found 38 new candidate uORF-containing genes, of which 15 were shown to carry the uORF within the 5' UTR. Mutation of about half of these uORFs affected apparent translational efficiency and/or mRNA stability. Other sequence elements involved in post-transcriptional regulation include internal ribosome entry sites (IREs), A/U-rich elements (AREs) near the 3' end of the transcript, determining stability of the mRNA molecule, and binding sites for small regulatory RNAs. In most of these cases, more information about the defining properties of the elements will be needed, and this will have to come from experiments. Ultimately, however, comparative genomics of fungi is a very promising tool for elucidation of these regulatory elements.

2 Gene finding in fungal genomes

The detection and analysis of functional elements in fungal genomes is, due to their compactness, much more straightforward than in higher organisms. In *S. cerevisiae*, in particular, the coding sequence covers over 70% of the genome, only a fraction of the genes ($\sim$ 5%) contain introns, the intergenic regions are short, repetitive sequences infrequent and transposons relatively few.

However, although being the most analysed genome of all organisms, the *S. cerevisiae* genome is still far from fully characterised. As of January 27, 2005, SGD reports 5888 annotated ORFs in *S. cerevisiae*, including 4952 experimentally verified ORFs, and 936 hypothetical ORFs with only non-experimental evidence (such as sequence homology to known residues) and with most or all GO-attributes unknown. The Munich Information Center for Protein Sequences (MIPS) database (mips.gsf.de/genre/proj/yeast/) includes 6335 annotated ORFs where 3535 either have known protein products or show strong similarity to known proteins. Thus, almost a decade after the sequencing of the *S. cerevisiae* genome, we still do not know the exact number of genes. The estimates reported in the literature have ranged from 4800 to 6000 (see Table 2), and the various sources available show high discrepancies. Some of the differences can be accounted for by the differing opinions in whether to include ORFs overlapping Ty elements, and whether the truncated and frameshifted copies of genes that tend to

Table 2. Estimated number of genes in *S. cerevisiae*

Goffeau et al. (1996)	5885
Cebrat et al. (1997); Mackiewicz et al. (1999)	~4800
Kowalczuk et al. (1999)	>4800
Blandin et al. (2000); Malpertuy et al. (2000)	5651
Zhang and Wang (2000)	5645
Wood et al. (2001)	<5570
Mackiewicz et al. (2002)	5322
Kumar et al. (2002)	~6000
Kellis et al. (2003)	5726
SGD (January 2005)	5888
MIPS (January 2005)	6335

Publications more recent than those included in this table do not include estimated gene numbers.

appear near the telomeres should be counted or not. However, some of the differences are due to the wide variety of prediction methods used, combining both experimental and computational approaches in different ways.

One of the primary tasks when deciphering a newly sequenced genome is the identification of its protein coding genes. Existing computational methods for gene prediction include *de novo* (or *ab initio*) approaches, which use the intrinsic statistical patterns in the DNA sequence itself as the only source of information, comparative methods using multiple genomic sequences or sources, and similarity based methods (or homology searches) which employ the sequence similarity between evolutionary related organisms and proteins. Comparative methods aim at identifying the gene structure using multiple sources of information, including the statistical properties of a protein coding gene, as well as the sequence similarities between them. Homology searches look for sequence similarity between a query sequence and a database of known genes, using linguistic pattern matching methods.

While homology searches are a powerful tool for validating potential genes, the success of a search depends on the existence of a homologue in the database, and they often fail to resolve the complete gene structure and protein sequence. Comparative gene finding improves the identification of new genes, and facilitates the specification of the entire gene structure, including the detection of translation start and translation stop, as well as internal exon-intron boundaries.

In this section, we give an overview of the gene finding methods that have been used in the yeast genome from the early estimates in Goffeau et al. (1996), and through the updated and adjusted predictions in the years to come.

2.1 Gene finding in *S. cerevisiae*

2.1.1 At first there were 6,000...

When the sequence of chromosome III in *S. cerevisiae* was published in 1992 (Oliver et al. 1992), it was the first eukaryotic chromosome ever to be sequenced, and up to this point perhaps some 1000 yeast genes had been defined using classical methods (Mortimer et al. 1992). Being the third smallest chromosome in the *S. cerevisiae* genome (~ 0.32 Mb), it contained 182 putative protein coding ORFs of at least 100 amino acids in length. The cut-off 100 was chosen in order to limit the number of actual genes omitted, while at the same time minimizing the probability of random ORFs occurring in the set (less than 0.2% for this cutoff; Sharp and Cowe 1991). Out of the 182 ORFs listed, only 34 appeared on the *S. cerevisiae* genetic map, and more surprisingly, approximately half showed no clear sequence homology to known genes in any organism, including yeast itself. Thus, it seemed that even in a genome as small and as extensively studied as that of yeast, only a fraction of the genes had previously been identified, a conclusion that was reinforced as the sequencing project progressed.

Using the same cut-off as for chromosome III, the entire *S. cerevisiae* genome sequence revealed 7472 ORFs of at least 100 codons (Goffeau et al. 1996). However, about 3000 of these ORFs overlapped, and since overlaps were known to be rare, only the longer ORF in an overlapping pair was deemed as coding, causing the number to drop to 6275. Using various measures of coding capacity, such as the Codon Bias Index (CBI) (Bennetzen and Hall 1982), and the Codon Adaptation Index (CAI) (Sharp and Li 1987), an additional 390 genes, that were thought not to be translated into protein, were omitted. For instance, since there seemed to be an overrepresentation of ORFs in the 100-150 codon range, Dujon et al. (1994, 1997) suggested that ORFs shorter than 150 codons and with CAI < 0.11 would be labelled as questionable. This criterion was not used on all chromosomes, however. The final estimate reported in Goffeau et al. (1996) was 5885 putative protein-encoding ORFs in addition to some 450 RNA-encoding genes.

Now a peculiar phenomenon appeared, named 'the mystery of orphans'. Out of the 5885 predicted, over half were 'orphan' genes with unknown function or homology. In comparison, out of the yeast genes previously identified using traditional methods, only a quarter turned out as orphans. In fact, as the sequencing project progressed, the number of orphans grew faster than the number of homologues. This is a paradox, because as the list of genes grows, so should the proportion of homologues matching that list. Cebrat et al. (1997) provided an explanation that assumed the number of published protein-coding genes to be too high (see next section). By using an asymmetry of purines and pyrimidines in the first and second codon positions to measure coding capacity of an ORF, they arrived at an estimate closer to 4800 genes. Out of these 4800, only about 200 – 300 (or 5%) were orphans, and thus the 'mystery of orphans' disappeared. Using a similar measure of asymmetry, Kowalczuk et al. (1999) later reinforced the claim that the number of genes was in fact closer to 4800 than the previous estimate of around 6000. This, however, did not explain the high number of relatively long, non-functional ORFs in the genome. The probability of generating an ORF of 100 codons or longer by chance is fairly small, and thus, the number of such sequences should be much lower than observed. Mackiewicz et al. (1999) suggested that these sequences were the results of duplication mechanisms inside coding sequences in non-coding frames. By their estimates, many of the ORFs annotated in MIPS at the time had very low coding capacity, and shared properties with antisense sequences of protein coding ORFs. It had earlier been noted that long protein coding sequences generate ORFs within the coding sequence, and in a specific phase of the antisense strand, to a much higher extent than random sequences (Cebrat 1998). But while protein-coding sequences reveal a prevalence of purines in the first codon positions, these antisense ORFs are richer in pyrimidines, and thus have a much lower coding capacity. Mackiewicz et al. looked for homology between known proteins and antisense ORFs translated in all six frames, and found paralogues and orthologues to the antisense ORFs in phases that differed from the annotated phase in MIPS.

It should be noted that the reports by Cebrat et al. (1997, 1998), Kowalczuk et al. (1999), and Mackiewicz et al. (1999) all came from the same group, and while the number of genes has varied over the years, the estimates now seem to have stabilised around 5600 – 5700.

2.1.2 Adjusting the estimates

Since the early estimates (Goffeau et al. 1996) the list of potential protein-coding genes in *S. cerevisiae* has been adjusted several times (see Table 2), using a variety of different methods, ranging from various geometrical representations of the DNA sequence to *de novo* approaches to cross-species comparisons. Kowalczuk et al. (1999) arrived at a new estimate of the number of protein-coding ORFs by devising a method that compares the set of all ORFs longer than 100 codons to the set of ORFs with known function. The method utilises an asymmetry in base composition between codon positions in protein-coding ORFs, and is based on the measure arctan(G-C)/(A-T). By comparing the distributions of the two ORF sets, Kowalczuk et al. (1999) predicted that there were at most 4800 protein-coding ORFs in the yeast genome.

Another geometrical representation of DNA sequences, called the Z-curve, was introduced by Zhang and Wang (2000). In their method, the DNA sequence of a given ORF was represented by a three-dimensional space curve, with the dimensions corresponding to measures for purine versus pyrimidine, amino versus keto, and weak versus strong hydrogen bond, respectively. By calculating a phase-specific version of the Z-curve the authors arrived at an estimate of at most 5645 protein-coding genes.

Kumar et al. (2002) brought the estimated gene count in *S. cerevisiae* back up to about 6000 genes. By combining experimental and computational methods, using gene-trapping and expression analysis methods, they identified 137 previously overlooked genes, out of which 104 were shorter than 100 codons. Since these data were drawn from the analysis of only 40% of the *S. cerevisiae* genome, Kumar et al. extrapolated to an additional 150 genes that were to be identified using the same method. By comparing the *S. cerevisiae* genome to that of *Ashbya gossypii*, Brachat et al. (2003) found that nearly half of the 2000 ORFs in *S. cerevisiae* annotated as hypothetical at that time had homologues in *Ashbya*. Furthermore, the over 400 ORFs overlapping other ORFs in *S. cerevisiae* and lacking homologues in *Ashbya* were suggested to be spurious.

The initial annotation of *S. cerevisiae* in Goffeau et al. (1996) did not include many ORFs shorter than 100 amino acids. Since then, a number of short protein coding ORFs have been identified. In the framework of the Génolevures project (reviewed in Chapter 8 by Bolotin-Fukuhara et al.), Blandin et al. (2000) managed to identify previously overlooked small genes by comparing *S. cerevisiae* inter-genic regions to other hemiascomycetous yeasts. In an independent effort, similar work was performed by Cliften et al. (2001). As an extension of this work, Kessler et al. (2003) detected nearly 100 new small *S. cerevisiae* genes that were tran-scribed from the predicted DNA strand, by scanning all *S. cerevisiae* ORFs of at least 18 amino acids toward a special fungal database and performing RT-PCR tests of transcription. The database was composed by all NCBI entries listed under "fungi" (excluding *S. cerevisiae* sequences) in conjunction with sequences avail-able from various fungal projects.

2.1.3 Gene finding in Schizosaccharomyces pombe

The availability of another yeast genome, *Schizosaccharomyces pombe*, facilitated whole genome comparisons of two well-studied unicellular eukaryotes. Wood et al. (2001) started out by analysing the 6282 *S. cerevisiae* ORFs listed in SGD at that time, using standard search procedures in various public databases. Over 300 new genes had just recently been identified in *Sz. pombe* that had homologues in other organisms, but were absent from the *S. cerevisiae* dataset (Wood et al. 2001). Wood et al. compared these to *S. cerevisiae* to determine whether they had been wrongfully omitted from the predictions. Furthermore, FASTA alignments were performed within the Artemis tool (Rutherford et al. 2000) on existing gene predictions to assess their accuracy. ORFs were removed from the final estimate if they lacked convincing similarity to known proteins, absence of functional data, overlapping another functional feature (especially in the N-terminus), or had an extreme GC-content. Out of the initial 6282 ORFs, 370 were disregarded using the above procedure, 42 represented pseudogenes or frameshifted sequences, and 193 were very hypothetical proteins, arriving at a final upper limit of 5570 protein cod-ing genes in *S. cerevisiae*.

When the entire *Sz. pombe* genome was sequenced (Wood et al. 2002), a more thorough sequence analysis was undertaken. Gene finding in *Sz. pombe* is some-what harder than in *S. cerevisiae*. Due to the lower gene density and higher occur-rence of introns, looking for ORFs longer than a certain threshold is not sufficient. The gene prediction in Wood et al. (2002) was carried out by using GENEFINDER (Wilson et al., unpublished; http://ftp.genome.washington.edu/ cgi/bin/Genefinder) and HMMER (Eddy et al., unpublished; http://hmmer.wustl.edu/) trained on experimentally verified *Sz. pombe* coding and intronic sequences. In addition, homology searches against SWISS-PROT and TrEMBL (Bairoch and Apweiler 1999; Bairoch et al. 2005), the EMBL database (Stoesser et al. 1999; Kanz et al. 2005) and Pfam (Bateman et al. 1999, 2004) were performed using BLAST (Altschul et al. 1990), FASTA (Pearson and Lip-man 1988), MSPcrunch (Sonnhammer and Durbin 1994), and Genewise (Birney et al. 1996). The predictions were refined further using the Artemis tool (Rutherford et al. 2000).

After including all ORFs longer than 100 amino acids, and excluding those deemed questionable based on their length, coding potential and lack of homologues, Wood et al. (2002) predicted 4824 genes, out of which 43% shared a total of 4370 confirmed introns (compared to 5% of the genes sharing 272 introns in *S. cerevisiae*). Moreover, the large-scale genome duplications observed in *S. cerevisiae* did not seem to appear in *Sz. pombe*. The conclusion was that *Sz. pombe* had significantly fewer protein coding genes and less gene redundancy than *S. cerevisiae*, but considerably more complex gene structure with more introns and longer intergenic regions.

2.2 Comparative gene finding in yeast

2.2.1 Multiple comparisons

At the time the *S. cerevisiae* genome was sequenced, genome annotation posed a huge challenge, because of the limited amount of eukaryotic sequence data available. The genomic sequence and functional genomics data has since exploded, resulting in an unprecedented opportunity for multiple comparisons of related species at a wide variety of evolutionary distances.

The next eukaryote to be sequenced after *S. cerevisiae* was the roundworm *C. elegans* (The *C. elegans* sequencing consortium, 1998). Comparing it to *S. cerevisiae* revealed a substantial fraction of genes with one-to-one orthologous relationships in the two species. By comparing the protein sets of two such highly diverged eukaryotes also allowed the identification of genes responsible for the core biological processes shared by the two, such as various metabolisms, protein folding and degradation (Chervitz et al. 1998). The authors suggested that such core processes were carried out by a similar number of proteins, and therefore it would be sufficient to study these processes in the simpler organism. Moreover, the proteins involved in biological processes characteristic of multicellular life seem to be different from those involved in the core processes, even though they might share common domains. Chervitz et al. inferred that model organisms can provide reliable functional annotation of human genes.

Brachat et al. (2003) proved the power of comparative genomics by comparing the *S. cerevisiae* genome to that of *A. gossypii*. Comparisons of the two genomes revealed an unexpectedly high degree of homology and gene order conservation. Of the 4700 protein coding genes annotated in *Ashbya*, 95% showed homology in *S. cerevisiae*, and 90% of these mapped at syntenic positions. By comparing the amino acid sequences of the *Ashbya* genome to annotation-free regions in the *S. cerevisiae* genome, Brachat et al. identified 23 novel ORFs in *S. cerevisiae* and 69 regions in need of sequence corrections either by ORF extensions or fusions.

In the Génolevures project, *S. cerevisiae* was compared to 13 other hemiascomycetous yeast species. Since there were inconsistencies in the criteria and methods used between the chromosome sequencing projects, Blandin et al. (2000) reviewed the entire yeast genome using consistent criteria for all 16 chromosomes. By comparing the intergenes of *S. cerevisiae* to the 13 other species, 50 novel genes were found, and 26 possible extensions to previously annotated genes. The revisited *S. cerevisiae* genome was estimated to contain at least 5651 protein-coding genes.

As part of the same project, Malpertuy et al. (2000) compared the set of 6213 *S. cerevisiae* protein products, described in Blandin et al. (2000) and Tekaia et al. (2000) (including the 50 novel genes), to the protein collections of a number of non-Ascomycetous organisms (including *C. elegans*, human and rat). The *S. cerevisiae* genes were divided into a 'common' set of genes having homologues in this comparison, and a 'maverick' set of genes lacking convincing homologues among non-Ascomycetes. Of the 6213 genes analyzed, 3759 (or 60%) fell into the common set, and among the resulting maverick genes, 728 (12%) were known beforehand to have homologues in other *Ascomycetes* (mostly *Sz. pombe*).

In order to estimate the number of true genes in the maverick set, Souciet et al. (2000) assigned a Poisson distribution to the number of species among the hemi-ascomycetes having a homologue to a given *S. cerevisiae* gene. The common set appeared to follow this distribution fairly well, while the maverick genes were highly biased around zero, probably due to a significant number of questionable ORFs in the set. Using the same Poisson distribution for the maverick set, the *S. cerevisiae* genome was estimated to contain 5651 protein coding genes, out of which 1892 were *Ascomycetes*-specific. The authors further proposed the removal of 612 false predictions from the databases.

The initial annotation of *S. cerevisiae* did not include many small ORFs (smORFs), simply because the set of ORFs of < 100 codons contains an exceedingly high fraction of non-coding sequences. Nevertheless, this set includes numerous important gene classes, such as mating pheromones, proteins involved in energy metabolism, transcriptional regulators, ribosomal proteins, etc. (Basrai et al. 1997). Several methods have been used, both experimental and computational, to identify such smORFs, and with all the sequences at hand a natural approach is the comparative one. In an attempt to estimate which species were best suited for the identification of smORFs as well as regulatory regions, Cliften et al. (2001) compared *S. cerevisiae* sequences to a number of other *Saccharomyces* species. According to their analysis, at least three sequences of various degree of similarity need to be aligned to *S. cerevisiae* in order to identify conserved regulatory sequences. Preferably, two would come from the *sensu stricto* group and the third from the *sensu lato* or petite-negative groups. In the *sensu stricto* group, being the farthest within the group, *S. mikatae*, *S. kudriavzevii*, and *S. bayanus* all seemed to be at a suitable evolutionary distance from *S. cerevisiae* with an estimated 70% identity in non-coding sequences. In the *sensu lato* and petite-negative groups, having an estimated 40% identity in non-coding regions, *S. castelli* and *S. kluyveri* were favoured for comparisons.

Using phylogenetic footprinting, Cliften et al. (2003) revised the *S. cerevisiae* gene catalogue by comparing the genomes of six *Saccharomyces* species. The sequence comparisons performed affected more than 10% of the existing gene annotations and based on the sequence conservation found 43 novel genes were predicted, all with fewer than 100 codons. Moreover, by identifying nonsense codons and frameshift mutations in the orthologues, 515 annotated *S. cerevisiae* genes were predicted to be false.

Kellis et al. (2003) devised a comparative analysis test to conduct systematic whole genome comparisons of the *S. cerevisiae* genome and three other *sensu stricto* species (*S. paradoxus*, *S. mikatae*, and *S. bayanus*). For each *S. cerevisiae* ORF analysed, the other three were set to 'vote' on its validity, based on the conservation in reading frame. Applying the test to all ORFs longer than 50 amino acids resulted in a revised gene catalogue of 5726 genes, where 188 were ORFs shorter than 100 amino acids.

To explore the mechanisms of eukaryotic genome evolution among the hemiascomycetous yeasts, Dujon et al. (2004) compared *S. cerevisiae* to four other yeast species (*Candida glabrata*, *Kluyveromyces lactis*, *Debaryomyces hansenii*, *Yarrowia lipolytica*). Although with similar genome sizes, lifestyles, and physiological properties, these four yeasts were chosen, among other things, because they display different reproductive mechanisms and span an evolutionary range as large as the entire phylum of chordates. The study revealed a variety of evolutionary events and mechanisms that should allow useful comparisons with other phyla. More importantly in this context, approximately 24,200 novel genes were identified, classified into 4700 *S. cerevisiae* protein families, and forming an excellent basis for future interspecific comparisons.

What is presented here is a historical summary of different approaches and different estimates produced over the years. By no means do we want give the impression that the most recent estimate is the most accurate one, but rather give the reader a flavour of what methods have been used and how difficult the gene finding problem is, even in such a compact and well characterised genome as that of *S. cerevisiae*. The question is what have we learned? Has all this activity made us better at predicting genes, or are we only going round in circles? We know now that predicting ORFs based on their length alone is not enough. Spurious long ORFs seem to be rather frequent in *S. cerevisiae*, while a significant amount of small ORFs have been and probably still are overlooked. Adding measures of coding capacity, although increasing the accuracy, is not sufficient either. Comparative approaches show great promise, and the speed at which sequence data is produced already allows us to make comparisons we could only dream of a decade ago. But we already have difficulty to keep up, and there is a crying need for more efficient, more powerful methods and tools, to be able to perform the multiple analyses our data allow us.

However, as the genome of an organism is full of intricacies and exceptions, there will never be one method that solves it all. We will have to continue to combine experimental methods with computational, one-dimensional analyses with multiple. The only thing we know is that we will have fun while we are doing it.

16 Marina Axelson-Fisk and Per Sunnerhagen

Table 3. Evolutionary conservation of proteins involved in RNA interference and establishment and maintenance of heterochromatin

| | | RNAi | | | | Heterochromatin | | | | | |
| | | | | | | Structural | Histone methylases[a] | | | DNA methylases | |
Group	Species	Dicer	Rde-1	Argonaute	RdR	HP1	SuVar3-9	Suv4-20h2	Smyd3	Dnmt1	Dnmt2
		1,4E-33	7,3E-13	4,6E-38	1,9E-23	2,7E-04	2,0E-18	5,0E-19	8,6E-14	2,2E-18	4,5E-29
Hemiasco	*Saccharomyces cerevisiae*	-	-	-	-	-	-	-	-	-	-
	Saccharomyces mikatae	-	-	-	-	-	-	-	-	-	-
	Saccharomyces paradoxus	-	-	-	-	-	-	-	-	-	-
	Saccharomyces kudriavzevii	-	-	-	-	-	-	-	-	-	-
	Saccharomyces bayanus	-	-	-	-	-	-	-	-	-	-
	Saccharomyces castellii	-	-	+	-	-	-	-	-	-	-
	Candida glabrata	-	-	-	-	-	-	-	-	-	-
	Saccharomyces kluyveri	-	-	-	-	-	-	-	-	-	-
	Kluyveromyces waltii	-	-	-	-	-	-	-	-	-	-
	Kluyveromyces lactis	-	-	-	-	-	-	-	-	-	-
	Candida albicans	-	-	+	-	-	-	-	-	-	-
	Yarrowia lipolytica	-	-	-	-	-	-	-	-	-	-
	Debaryomyces hansenii	-	-	-	-	-	-	-	-	-	-
	Ashbya gossypii	-	-	-	-	-	-	-	-	-	-
Archi	*Schizosaccharomyces pombe*	+	+	+	+	+	+	+	+	-	+
Euasco	*Gibberella zeae*	3	+	3	3	+	2	2	+	+	-
	Magnaporthe grisea	4	3	4	+	+	2	+	2	2	-
	Neurospora crassa	3	2	3	4	+	2	+	+	2	-
	Aspergillus nidulans	2	2	+	3	+	+	+	2	-	-
Basidio[b]	*Cryptococcus neoformans*			+							
	Coprinopsis cinerea	2	+	2		+				+	+
	Ustilago maydis					+	+	+	+		
Micro	*Encephalitozoon cuniculi*	-	-	-	-	+	+	-	-	-	-

Footnote to Table 3 overleaf. Protein sequences were used for BLAST searches against conceptually translated DNA sequence databases representing the fungal organisms indicated. To avoid species bias, non-fungal orthologues were chosen as queries: Dicer, *H. sapiens* Dicer1 (NP_803187); RdR, *C. elegans* putative RNA-directed RNA polymerase (CAA88315); Argonaute, *M. musculus* Argonaute 1 (AY135687.1); Rde-1, *C. elegans* Rde-1 (NM_171525.1); HP1, *H. sapiens* Heterochromatin protein 1 (AF136630); Su-Var3-9, *H. sapiens* suppressor of variegation 3-9 homolog 1 (NM_003173.1); Suv4-20h2, *M. musculus* suppressor of variegation 4-20 homolog 2 (AY555193.1); Smyd3, *M. musculus* SET and MYND domain containing 3 (NM_027188.2); Dnmt1, *Ovis aries* Dnmt1 (NP_001009473); Dnmt2, *H. sapiens* Dnmt2 (1G55_A).
The numbers below the queries are the cut-off E-values below which Blast hits are reported. The cut-off value for HP1 is considerably higher than the others since this protein is quite small.
Species are organised roughly according to evolutionary distance from *S. cerevisiae*, and in phylogenetic groups: "Hemiasco" = *Hemiascomycetes;* "Archi" = *Archiascomycetes;* "Euasco" = *Euascomycetes;* "Basidio" = *Basidiomycetes;* "Micro" = *Microsporidia*
"-" signs indicate no hits below the cut-off; "+" signs, one presumed orthologue; numbers indicate the number of different presumed orthologues in that organism.
[a] For the histone methylases, there is a continuous range of hits above the cut-off value within the *Hemiascomycetes* due to matches with other SET-domain proteins.
[b] For the *Basidiomycetes,* the lack of a hit has not been marked with a "-" sign, since the full genome sequences are not available.
Individual BLAST E-values and sequence accession numbers for hits in fungal genomes can be found in the web supplement.

3 Extending the applications of fungal comparative genomics

3.1 Conservation of pathways

Evolutionary conservation, loss, or gain, of groups of functionally related genes is expected to follow the life-styles of organisms or viruses carrying them (Pellegrini et al. 1999). Modularity, in the sense of physical proximity of functionally related genes, which has been observed in *S. cerevisiae*, accentuates the tendency of co-conservation of such gene groups (Overbeek et al. 1999; Ettema et al. 2001). This phenomenon has been studied among the *Hemiascomycetes*. There, several entire pathways have been lost after the whole genome duplication; genes involved in degradation of pyrimidines and purines, metabolism of pentoses, and pathways involved in metabolism of aromatic compounds (reviewed in Piškur and Langkjaer 2004).

Based on a comparison between *S. cerevisiae* and *Sz. pombe*, it was pointed out that genes involved in RNA interference (RNAi) were absent in the *S. cerevisiae* genome (Aravind et al. 2000). Likewise, it has long been realised that fundamental differences exist between budding and fission yeast in the extent of heterochromatin, primarily in centromeric regions. This is reflected in the absence in the *S. cerevisiae* genome of genes encoding several components of structural and catalytic proteins required for formation of heterochromatin (Huang 2002).

Extending this analysis to a wider range of fungal species, it becomes apparent that the whole hemiascomycetes branch has lost these heterochromatin-related genes (Table 3). Thus, this includes homologues of Heterochromatin Protein 1 (orthologue of *Sz. pombe* Swi6), a component of centromeric heterochromatin, and SuVar3-9 (orthologue of *Sz. pombe* Clr4), a histone 3 Lys9 methyl transferase. It has been shown that loss of proteins required for establishment of RNAi results in loss of silencing at centromeres, aberrant RNAs derived from centromeric repeats, and loss of cohesion from centromeres (Volpe et al. 2002; Hall et al. 2003). In accordance with this, the ribonuclease Dicer and RNA-mediated RNA polymerase, both required for RNAi, are absent in all hemiascomycetes, but are generally found in all other fungi. In many cases, there are several homologues of each protein in the filamentous *Euascomycetes*, while the unicellular *Sz. pombe* has only one homologue. Whatever triggered the collective loss of genes in these two interrelated functions, it must have happened before the large duplications that are common to the *Saccharomyces sensu lato* group, since *Ashbya gossypii*, *Kluyveromyces lactis*, and *K. waltii* all have lost these genes.

There are some intriguing exceptions to this monotonous loss of genes implicated in heterochromatin and RNAi within *Hemiascomycetes*. The Argonaute protein, member of the RISC complex, is present in all fungal genomes outside *Hemiascomycetes*, but oddly also in *Candida albicans* and *Saccharomyces castellii*. There is no simple explanation why this protein appears in these two seemingly isolated species within *Hemiascomycetes*, and in only one species of the *Saccharomyces sensu stricto* group. Possibly Argonaute performs additional functions outside its role in RNAi. The phylogenetic distribution of Dnmt2, a predicted DNA cytosine methyl transferase (Dong et al. 2001), is also worth commenting. Well-conserved homologues of this protein are found throughout metazoans and vascular plants, and also in protists and some bacteria. In fungi, however, Dnmt2 homologues are restricted to *Sz. pombe*, *Coprinopsis cinerea*, and *Phanerochaete chrysosporium*. As for Argonaute, there is no obvious explanation for this scattered distribution within fungi.

For the basidiomycetes, the picture is less clear, since the three investigated species display homologues to some, but not all genes for these functions. For example, *C. neoformans* lacks an obvious Dicer orthologue but does have an Argonaute candidate orthologue. Despite these findings, there is experimental evidence that RNAi works in *C. neoformans* (Liu et al. 2002).

A pertinent question regarding the fungal species presently sequenced is to what extent a particular function is associated with a certain life style (*e.g.* unicellular vs. filamentous), or with the taxonomic group. For instance, are genes encoding proteins required for apoptosis retained in genomes of filamentous fungi and selectively lost in unicellular yeasts, irrespective of whether these are ascomycetes or basidiomycetes? Since all four classes of organisms exist, such questions can in principle be answered using a similar analysis as in Table 3.

3.2 Fungal genomes as models for other genomes

Comparative genomics in fungi is likely to become a focal point in molecular biology at large, both for experimentalists and theoreticians. The collection of fungal genome sequences also holds the promise to serve as a resource for identification of genes in other organisms with more expanded and complex genomes, for three reasons. First, it is much harder to discern coding sequences against the background noise solely by computer searches in such genomes simply because of their sheer size and lower gene density. Second, having a relatively large set of related sequences with different degrees of relatedness allows robust identification of coding sequences based on conservation of amino acid sequence, and in many cases also of synteny. This comparative genomics approach was used to find about 100 new small ORFs in the *S. cerevisiae* genome, and further to functionally verify a conserved human homologue of one of them (Kessler et al. 2003). de Groot et al. (2003) found putative GPI-anchored proteins in *S. cerevisiae* and *C. albicans* by screening four fungal genomes for homologous genes using a consensus sequence for the GPI attachment site. Third, the genetic tractability of several fungal species allows phenotypic testing of candidate genes found in *e.g.* the human genome by knockout of the fungal homologue. If one takes the other perspective and compares to bacterial genomes, at least the yeast genomes appear to have experienced less horizontal gene transfer (although there are reports of its occurrence; see Hall et al. 2005) and hybrid stages, leading to purer lineages.

Comparative genomics in fungi also has potential for applied science. Some fungi are pathogenic for humans and animals. Well-known examples are the human pathogens *C. albicans* and *Cryptococcus neoformans*. Further, opportunistic infections are caused by *Aspergillus* species; *Histoplasma* causes systemic infections; and *Trichophyton* and other genera are dermatopathogens. For plants, fungal pathogens are plentiful and cause significant damage to major crops. *Magnaporthe grisea* (rice blast), *Ustilago maydis* (corn smut), and *Gibberella zeae* (head blight of wheat and ear rot of corn) are important examples. Knowledge about the well-studied model organisms (*S. cerevisiae*, *Sz. pombe*, and others) can be used to predict *e.g.* essential genes in fungi pathogenic for plants or animals; the products of those are promising drug targets (Haselbeck et al. 2002).

Fungi can kill bacteria and other fungi, too, and in this capacity they can be put to use by man. Comparisons between fungal species again will prove helpful. Penicillin is traditionally harvested from *Penicillium chrysogenum*, but it is also produced by *Aspergillus* species. Insights into the regulation of penicillin production have been gained from the genetically tractable *A. nidulans* (Kato et al. 2003). Polyketides are a diverse class of secondary metabolites derived from coenzyme A, which includes many toxic compounds and antibiotics. Polyketide synthase genes have been isolated from a range of fungal species using degenerate PCR (Lee et al. 2001). Additionally, fungi are used to produce chemicals such as vitamins, polysaccharides, pigments etc. Biodegradation by fungi (degradation of wood by filamentous fungi being a prime example) is another area of biotechnology where comparative genomics holds promise to improve efficiency.

Also the traditional uses of yeast, baking and brewing, may ultimately benefit from comparative genomics. The genome of lager-brewing yeast (*S. pastorianus/S. carlsbergensis*) has been sequenced and its hybrid nature clarified, with contributions from *S. cerevisiae* and *S. bayanus* (see Chapter 7 by Kodoma et al.).

In some respects, fungi have genetic properties that set them apart from other eukaryotes; a relevant example for this text is their meiotic recombination frequency, which is several orders of magnitude higher than in most plants and animals. This obviously leads to more rapid fragmentation of syntenic blocks. Using the appropriate evolutionary distances however, the history of genome rearrangements can be traced, as exemplified in Chapters 4, 7, and 9 in this volume. Another case is repeat-induced point mutations (RIP), where repeated sequences selectively accumulate G:C to A:T transversions through an active process. This phenomenon was first discovered in *Neurospora crassa* (Cambareri et al. 1989), where it is linked to DNA methylation. More recently, similar events have been reported to occur in the filamentous fungi *Podospora anserina* (Graia et al. 2001), *Magnaporthe grisea* (Ikeda et al. 2002), and *A. nidulans* (Clutterbuck 2004), however DNA methylation is less obviously involved in the process in those species. Conceivably, genomic processes akin to RIP could occur also in eukaryotes outside the fungal kingdom.

In addition to the uses of fungal genomes for wet scientists to obtain direct predictions for experimental work, this genome collection is an essential resource for the development of methods in comparative genomics. Just like sequencing of the *S. cerevisiae* genome preceded that of other eukaryotes, and its annotation process therefore pioneered the methods for gene finding, the set of fungal genomes now at hand can serve as a starting point for the development of more sophisticated methods in comparative genomics. We have already mentioned the importance in this regard of the varying evolutionary distances between the fungal species in question, and the abundance of experimental information on genes and gene products. Moreover, the relative simplicity of gene and genome organisation (scarcity of introns, pseudogenes, and repetitive elements, as well as the high overall gene density) makes gene identification a less complicated task than in most plant or animal genomes. Thus, fungi are an ideal test-bed for novel concepts in gene finding algorithms.

Fungal genomes and their exploration will prove useful as models to biology in general, both for basic and applied science, both for experimentalists and theoreticians, as this volume will hopefully underscore.

Acknowledgements

M.A. acknowledges financial support from the Swedish Foundation for Strategic Research, and P.S. from the Swedish Research Council and the Swedish Cancer Fund.

References

Altschul SF, Gish W, Miller W, Myers EW, Lipman DJ (1990) Basic local alignment search tool. J Mol Biol 215:403-410

Aravind L, Watanabe H, Lipman DJ, Koonin EV (2000) Lineage-specific loss and divergence of functionally linked genes in eukaryotes. Proc Natl Acad Sci USA 97:11319-11324

Bairoch A, Apweiler R (1999) The SWISS-PROT protein sequence data bank and its supplement TrEMBL in 1999. Nucleic Acids Res 27:49-54

Bairoch A, Apweiler R, Wu CH, Barker WC, Boeckmann B, Ferro S, Gasteiger E, Huang H, Lopez R, Magrane M, Martin MJ, Natale DA, O'Donovan C, Redaschi N, Yeh LS (2005) The Universal Protein Resource (UniProt). Nucleic Acids Res 33 (Database Issue):D154-159

Balakrishnan R, Christie KR, Costanzo MC, Dolinski K, Dwight SS, Engel SR, Fisk DG, Hirschman JE, Hong EL, Nash R, Oughtred R, Skrzypek M, Theesfeld CL, Binkley G, Dong Q, Lane C, Sethuraman A, Weng S, Botstein D, Cherry JM (2005) Fungal BLAST and Model Organism BLASTP Best Hits: new comparison resources at the Saccharomyces Genome Database (SGD). Nucleic Acids Res 33 (Database Issue):D374-377

Basrai MA, Hieter P, Boeke JD (1997) Small open reading frames: beautiful needles in the haystack. Genome Res 7:768-771

Bateman A, Birney E, Durbin R, Eddy SR, Finn RD, Sonnhammer EL (1999) Pfam 3.1: 1313 multiple alignments and profile HMMs match the majority of proteins. Nucleic Acids Res 27:260-262

Bateman A, Coin L, Durbin R, Finn RD, Hollich V, Griffiths-Jones S, Khanna A, Marshall M, Moxon S, Sonnhammer EL, Studholme DJ, Yeats C, Eddy SR (2004) The Pfam protein families database. Nucleic Acids Res 32:D138-141

Bennetzen JL, Hall BD (1982) Codon selection in yeast. J Biol Chem 257:3026-3031

Birney E, Thompson JD, Gibson TJ (1996) PairWise and SearchWise: finding the optimal alignment in a simultaneous comparison of a protein profile against all DNA translation frames. Nucleic Acids Res 24:2730-2739

Blandin G, Durrens P, Tekaia F, Aigle M, Bolotin-Fukuhara M, Bon E, Casaregola S, de Montigny J, Gaillardin C, Lepingle A, Llorente B, Malpertuy A, Neuveglise C, Ozier-Kalogeropoulos O, Perrin A, Potier S, Souciet J, Talla E, Toffano-Nioche C, Wesolowski-Louvel M, Marck C, Dujon B (2000) Genomic exploration of the hemiascomycetous yeasts: 4. The genome of *Saccharomyces cerevisiae* revisited. FEBS Lett 487:31-36

Brachat S, Dietrich FS, Voegeli S, Zhang Z, Stuart L, Lerch A, Gates K, Gaffney T, Philippsen P (2003) Reinvestigation of the *Saccharomyces cerevisiae* genome annotation by comparison to the genome of a related fungus: *Ashbya gossypii*. Genome Biol 4:R45

Cambareri EB, Jensen BC, Schabtach E, Selker EU (1989) Repeat-induced G-C to A-T mutations in *Neurospora*. Science 244:1571-1575

Cebrat S, Dudek MR, Mackiewicz P, Kowalczuk M, Fita M (1997) Asymmetry of coding versus noncoding strand in coding sequences of different genomes. Microb Comp Genomics 2:259-268

Cebrat S, Mackiewicz P, Dudek MR (1998) The role of the genetic code in generating new coding sequences inside existing genes. Biosystems 45:165-176

Centola M, Carbon J (1994) Cloning and characterization of centromeric DNA from *Neurospora crassa*. Mol Cell Biol 14:1510-1519

Chervitz SA, Aravind L, Sherlock G, Ball CA, Koonin EV, Dwight SS, Harris MA, Dolinski K, Mohr S, Smith T, Weng S, Cherry JM, Botstein D (1998) Comparison of the complete protein sets of worm and yeast: orthology and divergence. Science 282:2022-2028

Cliften P, Sudarsanam P, Desikan A, Fulton L, Fulton B, Majors J, Waterston R, Cohen BA, Johnston M (2003) Finding functional features in *Saccharomyces* genomes by phylogenetic footprinting. Science 301:71-76

Cliften PF, Hillier LW, Fulton L, Graves T, Miner T, Gish WR, Waterston RH, Johnston M (2001) Surveying *Saccharomyces* genomes to identify functional elements by comparative DNA sequence analysis. Genome Res 11:1175-1186

Clutterbuck AJ (2004) MATE transposable elements in *Aspergillus nidulans*: evidence of repeat-induced point mutation. Fungal Genet Biol 41:308-316

The *C. elegans* sequencing consortium (1998) Genome sequence of the nematode *C. elegans*: a platform for investigating biology. Science 282:2012-2018

Daboussi MJ (1997) Fungal transposable elements and genome evolution. Genetica 100:253-260

de Groot PW, Hellingwerf KJ, Klis FM (2003) Genome-wide identification of fungal GPI proteins. Yeast 20:781-796

Dietrich FS, Voegeli S, Brachat S, Lerch A, Gates K, Steiner S, Mohr C, Pohlmann R, Luedi P, Choi S, Wing RA, Flavier A, Gaffney TD, Philippsen P (2004) The *Ashbya gossypii* genome as a tool for mapping the ancient *Saccharomyces cerevisiae* genome. Science 304:304-307

Dong A, Yoder JA, Zhang X, Zhou L, Bestor TH, Cheng X (2001) Structure of human DNMT2, an enigmatic DNA methyltransferase homolog that displays denaturant-resistant binding to DNA. Nucleic Acids Res 29:439-448

Douzery EJ, Snell EA, Bapteste E, Delsuc F, Philippe H (2004) The timing of eukaryotic evolution: does a relaxed molecular clock reconcile proteins and fossils? Proc Natl Acad Sci USA 101:15386-15391

Dujon B, Albermann K, Aldea M, Alexandraki D, Ansorge W, Arino J, Benes V, Bohn C, Bolotin-Fukuhara M, Bordonné R, Boyer J, Camasses A, Casamayor A, Casas C, Chéret G, Cziepluch C, Daignan-Fornier B, Dang DV, de Haan M, Delius H, Durand P, Fairhead C, Feldmann H, Gaillon L, Galisson F, Gamo FJ, Gancedo C, Goffeau A, Goulding SE, Grivell LA, Habbig B, Hand NJ, Hani J, Hattenhorst U, Hebling U, Hernando Y, Herrero E, Heumann K, Hiesel R, Hilger F, Hofmann B, Hollenberg CP, Hughes B, Jauniaux JC, Kalogeropoulos A, Katsoulou C, Kordes E, Lafuente MJ, Landt O, Louis EJ, Maarse AC, Madania A, Mannhaupt G, Marck C, Martin RP, Mewes HW, Michaux G, Paces V, Parle-McDermott AG, Pearson BM, Perrin A, Pettersson B, Poch O, Pohl TM, Poirey R, Portetelle D, Pujol A, Purnelle B, Ramezani Rad M, Rechmann S, Schwager C, Schweizer M, Sor F, Sterky F, Tarassov IA, Teodoru C, Tettelin H, Thierry A, Tobiasch E, Tzermia M, Uhlen M, Unseld M, Valens M, Vandenbol M, Vetter I, Vicek C, Voet M, Volckaert G, Voss H, Wambutt R, Wedler H, Wiemann S, Winsor B, Wolfe KH, Zollner A, Zumstein E, Kleine K (1997) The nucleotide sequence of *Saccharomyces cerevisiae* chromosome XV. Nature 387:98-102

Dujon B, Alexandraki D, Andre B, Ansorge W, Baladron V, Ballesta JP, Banrevi A, Bolle PA, Bolotin-Fukuhara M, Bossier P, Bou G, Boyer J, Buitrago MJ, Cheret G, Colleaux L, Dalgan-Fornier B, del Rey F, Dion C, Domdey H, Düsterhöft A, Düsterhus S, Entian KD, Erfle H, Esteban PF, Feldmann H, Fernandes L, Fobo GM, Fritz C, Fukuhara H, Gabel C, Gaillon L, Garcia-Cantalejo JM, Garcia-Ramirez JJ, Gent ME, Ghazvini M, Goffeau A, Gonzalez A, Grouthes D, Guerreiro P, Hegemann J, Hewitt N, Hilger F, Hollenberg CP, Horaitis O, Indge KJ, Jacquier A, James CM, Jauniauz JC, Jiminez A, Keuchel H, Kirchrath L, Kleine K, Kötter P, Legrain P, Liebl S, Louis EJ, Maia e Silva A, Marck C, Monnier A-L, Möstl D, Müller S, Obermaier B, Oliver SG, Pallier C, Pascolo S, Pfeiffer F, Philippsen P, Planta RJ, Pohl FM, Pohl TM, Pöhlmann R, Portetelle D, Purnelle B, Puzos V, Ramezani Rad M, Rasmussen SW, Remacha M, Revuelta JL, Richard GF, Rieger M, Rodrigues-Pousada C, Rose M, Rupp T, Santos MA, Schwager C, Sensen C, Skala J, Soares H, Sor F, Stegemann J, Tettelin H, Thierry A, Tzermia M, Urrestarazu LA, van Dyck L, van Vliet-Reedijk JC, Valens M, Vandenbol M, Vilela C, Vissers S, von Wettstein D, Voss H, Wiemann S, Xu G, Zimermann J, Haasemann M, Becker I, Mewes HW (1994) Complete DNA sequence of yeast chromosome XI. Nature 369:371-378

Dujon B, Sherman D, Fischer G, Durrens P, Casaregola S, Lafontaine I, De Montigny J, Marck C, Neuveglise C, Talla E, Goffard N, Frangeul L, Aigle M, Anthouard V, Babour A, Barbe V, Barnay S, Blanchin S, Beckerich JM, Beyne E, Bleykasten C, Boisrame A, Boyer J, Cattolico L, Confanioleri F, De Daruvar A, Despons L, Fabre E, Fairhead C, Ferry-Dumazet H, Groppi A, Hantraye F, Hennequin C, Jauniaux N, Joyet P, Kachouri R, Kerrest A, Koszul R, Lemaire M, Lesur I, Ma L, Muller H, Nicaud JM, Nikolski M, Oztas S, Ozier-Kalogeropoulos O, Pellenz S, Potier S, Richard GF, Straub ML, Suleau A, Swennen D, Tekaia F, Wesolowski-Louvel M, Westhof E, Wirth B, Zeniou-Meyer M, Zivanovic I, Bolotin-Fukuhara M, Thierry A, Bouchier C, Caudron B, Scarpelli C, Gaillardin C, Weissenbach J, Wincker P, Souciet JL (2004) Genome evolution in yeasts. Nature 430:35-44

Escote X, Zapater M, Clotet J, Posas F (2004) Hog1 mediates cell-cycle arrest in G1 phase by the dual targeting of Sic1. Nat Cell Biol 6:997-1002

Ettema T, van der Oost J, Huynen M (2001) Modularity in the gain and loss of genes: applications for function prediction. Trends Genet 17:485-487

Gibbs RA, Weinstock GM, Metzker ML, Muzny DM, Sodergren EJ, Scherer S, Scott G, Steffen D, Worley KC, Burch PE, Okwuonu G, Hines S, Lewis L, DeRamo C, Delgado O, Dugan-Rocha S, Miner G, Morgan M, Hawes A, Gill R, Celera, Holt RA, Adams MD, Amanatides PG, Baden-Tillson H, Barnstead M, Chin S, Evans CA, Ferriera S, Fosler C, Glodek A, Gu Z, Jennings D, Kraft CL, Nguyen T, Pfannkoch CM, Sitter C, Sutton GG, Venter JC, Woodage T, Smith D, Lee HM, Gustafson E, Cahill P, Kana A, Doucette-Stamm L, Weinstock K, Fechtel K, Weiss RB, Dunn DM, Green ED, Blakesley RW, Bouffard GG, De Jong PJ, Osoegawa K, Zhu B, Marra M, Schein J, Bosdet I, Fjell C, Jones S, Krzywinski M, Mathewson C, Siddiqui A, Wye N, McPherson J, Zhao S, Fraser CM, Shetty J, Shatsman S, Geer K, Chen Y, Abramzon S, Nierman WC, Havlak PH, Chen R, Durbin KJ, Egan A, Ren Y, Song XZ, Li B, Liu Y, Qin X, Cawley S, Cooney AJ, D'Souza LM, Martin K, Wu JQ, Gonzalez-Garay ML, Jackson AR, Kalafus KJ, McLeod MP, Milosavljevic A, Virk D, Volkov A, Wheeler DA, Zhang Z, Bailey JA, Eichler EE, Tuzun E, Birney E, Mongin E, Ureta-Vidal A, Woodwark C, Zdobnov E, Bork P, Suyama M, Torrents D, Alexandersson M, Trask BJ, Young JM, Huang H, Wang H, Xing H, Daniels S, Gietzen D, Schmidt J,

Stevens K, Vitt U, Wingrove J, Camara F, Mar Alba M, Abril JF, Guigo R, Smit A, Dubchak I, Rubin EM, Couronne O, Poliakov A, Hubner N, Ganten D, Goesele C, Hummel O, Kreitler T, Lee YA, Monti J, Schulz H, Zimdahl H, Himmelbauer H, Lehrach H, Jacob HJ, Bromberg S, Gullings-Handley J, Jensen-Seaman MI, Kwitek AE, Lazar J, Pasko D, Tonellato PJ, Twigger S, Ponting CP, Duarte JM, Rice S, Goodstadt L, Beatson SA, Emes RD, Winter EE, Webber C, Brandt P, Nyakatura G, Adetobi M, Chiaromonte F, Elnitski L, Eswara P, Hardison RC, Hou M, Kolbe D, Makova K, Miller W, Nekrutenko A, Riemer C, Schwartz S, Taylor J, Yang S, Zhang Y, Lindpaintner K, Andrews TD, Caccamo M, Clamp M, Clarke L, Curwen V, Durbin R, Eyras E, Searle SM, Cooper GM, Batzoglou S, Brudno M, Sidow A, Stone EA, Payseur BA, Bourque G, Lopez-Otin C, Puente XS, Chakrabarti K, Chatterji S, Dewey C, Pachter L, Bray N, Yap VB, Caspi A, Tesler G, Pevzner PA, Haussler D, Roskin KM, Baertsch R, Clawson H, Furey TS, Hinrichs AS, Karolchik D, Kent WJ, Rosenbloom KR, Trumbower H, Weirauch M, Cooper DN, Stenson PD, Ma B, Brent M, Arumugam M, Shteynberg D, Copley RR, Taylor MS, Riethman H, Mudunuri U, Peterson J, Guyer M, Felsenfeld A, Old S, Mockrin S, Collins F (2004) Genome sequence of the Brown Norway rat yields insights into mammalian evolution. Nature 428:493-521
Gilbert DM (2002) Replication timing and metazoan evolution. Nat Genet 32:336-337
Goffeau A, Barrell BG, Bussey H, Davis RW, Dujon B, Feldmann H, Galibert F, Hoheisel JD, Jacq C, Johnston M, Louis EJ, Mewes HW, Murakami Y, Philippsen P, Tettelin H, Oliver SG (1996) Life with 6000 genes. Science 274:546, 563-567
Graia F, Lespinet O, Rimbault B, Dequard-Chablat M, Coppin E, Picard M (2001) Genome quality control: RIP (repeat-induced point mutation) comes to *Podospora*. Mol Microbiol 40:586-595
Hall C, Brachat S, Dietrich FS (2005) Contribution of horizontal gene transfer to the evolution of *Saccharomyces cerevisiae*. Eukaryot Cell 4:1102-1115
Hall IM, Noma K, Grewal SI (2003) RNA interference machinery regulates chromosome dynamics during mitosis and meiosis in fission yeast. Proc Natl Acad Sci USA 100:193-198
Haselbeck R, Wall D, Jiang B, Ketela T, Zyskind J, Bussey H, Foulkes JG, Roemer T (2002) Comprehensive essential gene identification as a platform for novel anti-infective drug discovery. Curr Pharm Des 8:1155-1172
Heckman DS, Geiser DM, Eidell BR, Stauffer RL, Kardos NL, Hedges SB (2001) Molecular evidence for the early colonization of land by fungi and plants. Science 293:1129-1133
Herrero E, de la Torre MA, Valentin E (2003) Comparative genomics of yeast species: new insights into their biology. Int Microbiol 6:183-190
Huang Y (2002) Transcriptional silencing in *Saccharomyces cerevisiae* and *Schizosaccharomyces pombe*. Nucleic Acids Res 30:1465-1482
Hughes AL, Friedman R (2003) Parallel evolution by gene duplication in the genomes of two unicellular fungi. Genome Res 13:1259-1264
Ikeda K, Nakayashiki H, Kataoka T, Tamba H, Hashimoto Y, Tosa Y, Mayama S (2002) Repeat-induced point mutation (RIP) in *Magnaporthe grisea*: implications for its sexual cycle in the natural field context. Mol Microbiol 45:1355-1364
Kanz C, Aldebert P, Althorpe N, Baker W, Baldwin A, Bates K, Browne P, van den Broek A, Castro M, Cochrane G, Duggan K, Eberhardt R, Faruque N, Gamble J, Diez FG, Harte N, Kulikova T, Lin Q, Lombard V, Lopez R, Mancuso R, McHale M, Nardone F, Silventoinen V, Sobhany S, Stoehr P, Tuli MA, Tzouvara K, Vaughan R, Wu D,

Zhu W, Apweiler R (2005) The EMBL Nucleotide Sequence Database. Nucleic Acids Res 33 (Database Issue):D29-33

Katinka MD, Duprat S, Cornillot E, Metenier G, Thomarat F, Prensier G, Barbe V, Peyretaillade E, Brottier P, Wincker P, Delbac F, El Alaoui H, Peyret P, Saurin W, Gouy M, Weissenbach J, Vivares CP (2001) Genome sequence and gene compaction of the eukaryote parasite *Encephalitozoon cuniculi*. Nature 414:450-453

Kato N, Brooks W, Calvo AM (2003) The expression of sterigmatocystin and penicillin genes in *Aspergillus nidulans* is controlled by *veA*, a gene required for sexual development. Eukaryot Cell 2:1178-1186

Kellis M, Birren BW, Lander ES (2004a) Proof and evolutionary analysis of ancient genome duplication in the yeast *Saccharomyces cerevisiae*. Nature 428:617-624

Kellis M, Patterson N, Birren B, Berger B, Lander ES (2004b) Methods in comparative genomics: genome correspondence, gene identification and regulatory motif discovery. J Comput Biol 11:319-355

Kellis M, Patterson N, Endrizzi M, Birren B, Lander ES (2003) Sequencing and comparison of yeast species to identify genes and regulatory elements. Nature 423:241-254

Kessler MM, Zeng Q, Hogan S, Cook R, Morales AJ, Cottarel G (2003) Systematic discovery of new genes in the *Saccharomyces cerevisiae* genome. Genome Res 13:264-271

Kowalczuk M, Mackiewicz P, Gierlik A, Dudek MR, Cebrat S (1999) Total number of coding open reading frames in the yeast genome. Yeast 15:1031-1034

Kumar A, Harrison PM, Cheung KH, Lan N, Echols N, Bertone P, Miller P, Gerstein MB, Snyder M (2002) An integrated approach for finding overlooked genes in yeast. Nat Biotechnol 20:58-63

Lee T, Yun SH, Hodge KT, Humber RA, Krasnoff SB, Turgeon GB, Yoder OC, Gibson DM (2001) Polyketide synthase genes in insect- and nematode-associated fungi. Appl Microbiol Biotechnol 56:181-187

Liu H, Cottrell TR, Pierini LM, Goldman WE, Doering TL (2002) RNA interference in the pathogenic fungus *Cryptococcus neoformans*. Genetics 160:463-470

Liu YJ, Hall BD (2004) Body plan evolution of ascomycetes, as inferred from an RNA polymerase II phylogeny. Proc Natl Acad Sci USA 101:4507-4512

Loftus BJ, Fung E, Roncaglia P, Rowley D, Amedeo P, Bruno D, Vamathevan J, Miranda M, Anderson IJ, Fraser JA, Allen JE, Bosdet IE, Brent MR, Chiu R, Doering TL, Donlin MJ, D'Souza CA, Fox DS, Grinberg V, Fu J, Fukushima M, Haas BJ, Huang JC, Janbon G, Jones SJ, Koo HL, Krzywinski MI, Kwon-Chung JK, Lengeler KB, Maiti R, Marra MA, Marra RE, Mathewson CA, Mitchell TG, Pertea M, Riggs FR, Salzberg SL, Schein JE, Shvartsbeyn A, Shin H, Shumway M, Specht CA, Suh BB, Tenney A, Utterback TR, Wickes BL, Wortman JR, Wye NH, Kronstad JW, Lodge JK, Heitman J, Davis RW, Fraser CM, Hyman RW (2005) The genome of the basidiomycetous yeast and human pathogen *Cryptococcus neoformans*. Science 307:1321-1324

Mackiewicz P, Kowalczuk M, Gierlik A, Dudek MR, Cebrat S (1999) Origin and properties of non-coding ORFs in the yeast genome. Nucleic Acids Res 27:3503-3509

Maddison WP (1997) Gene tree and species trees. Systematic Biology 46:523-536

Malpertuy A, Tekaia F, Casaregola S, Aigle M, Artiguenave F, Blandin G, Bolotin-Fukuhara M, Bon E, Brottier P, de Montigny J, Durrens P, Gaillardin C, Lepingle A, Llorente B, Neuveglise C, Ozier-Kalogeropoulos O, Potier S, Saurin W, Toffano-Nioche C, Wesolowski-Louvel M, Wincker P, Weissenbach J, Souciet J, Dujon B

(2000) Genomic exploration of the hemiascomycetous yeasts: 19. *Ascomycetes*-specific genes. FEBS Lett. 487:113-121

McKnight TD, Fitzgerald MS, Shippen DE (1997) Plant telomeres and telomerases. A review. Biochemistry (Mosc) 62:1224-1231

Mentel M, Piškur J, Neuveglise C, Rycovska A, Cellengova G, Kolarov J (2005) Triplicate genes for mitochondrial ADP/ATP carriers in the aerobic yeast *Yarrowia lipolytica* are regulated differentially in the absence of oxygen. Mol Genet Genomics 273:84-91

Mortimer RK, Contopoulou CR, King JS (1992) Genetic and physical maps of *Saccharomyces cerevisiae*, Edition 11. Yeast 8:817-902

Møller K, Langkjaer RB, Nielsen J, Piškur J, Olsson L (2004) Pyruvate decarboxylases from the petite-negative yeast *Saccharomyces kluyveri*. Mol Genet Genomics 270:558-568

Oliver SG, van der Aart QJM, Agostoni-Carbone ML, Aigle M, Alberghina L, Alexandraki D, Antoine G, Anwar R, Ballesta JPG, Benit P, Berben G, Bergantino E, Biteau N, Bolle PA, Bolotin-Fukuhara M, Brown A, Brown AJP, Buhler JM, Carcano C, Carignani G, Cederberg H, Chanet R, Contreras R, Crouzet M, Daignan-Fornier B, Defoor E, Delgado M, Demolder J, Doira C, Dubois E, Dujon B, Dusterhoft A, Erdmann D, Esteban M, Fabre F, Fairhead C, Faye G, Feldmann H, Fiers W, Francingues-Gaillard MC, Franco L, Frontali L, Fukuhara H, Fuller LJ, Galland P, Gent ME, Gigot D, Gilliquet V, Glansdorff N, Goffeau A, Grenson M, Grisanti P, Grivell LA, de Haan M, Haasemann M, Hatat D, Hoenicka J, Hegemann J, Herbert CJ, Hilger F, Hohmann S, Hollenberg CP, Huse K, Iborra F, Indge KJ, Isono K, Jacq C, Jacquet M, James CM, Jauniaux JC, Jia Y, Jimenez A, Kelly A, Kleinhans U, Kreisl P, Lanfranchi G, Lewis C, van der Linden CG, Lucchini G, Lutzenkirchen K, Maat MJ, Mallet L, Mannhaupt G, Martegani E, Mathieu A, Maurer CTC, McConnell D, McKee RA, Messenguy F, Mewes HW, Molemans F, Montague MA, Muzi Falconi M, Navas L, Newlon CS, Noone D, Pallier C, Panzeri L, Pearson BM, Perea J, Philippsen P, Pierard A, Planta RJ, Plevani P, Poetsch B, Pohl F, Purnelle B, Ramezani Rad M, Rasmussen SW, Raynal A, Remacha M, Richterich P, Roberts AB, Rodriguez F, Sanz E, Schaaff-Gerstenschlager I, Scherens B, Schweitzer B, Shu Y, Skala J, Slonimski PP, Sor F, Soustelle C, Spiegelberg R, Stateva LI, Steensma HY, Steiner S, Thierry A, Thireos G, Tzermia M, Urrestarazu LA, Valle G, Vetter I, van Vliet-Reedijk JC, Voet M, Volckaert G, Vreken P, Wang H, Warmington JR, von Wettstein D, Wicksteed BL, Wilson C, Wurst H, Xu G, Yoshikawa A, Zimmermann FK, Sgouros JG (1992) The complete DNA sequence of yeast chromosome III. Nature 357:38-46

Overbeek R, Fonstein M, D'Souza M, Pusch GD, Maltsev N (1999) The use of gene clusters to infer functional coupling. Proc Natl Acad Sci USA 96:2896-2901

Pearson WR, Lipman DJ (1988) Improved tools for biological sequence comparison. Proc Natl Acad Sci USA 85:2444-2448

Pellegrini M, Marcotte EM, Thompson MJ, Eisenberg D, Yeates TO (1999) Assigning protein functions by comparative genome analysis: protein phylogenetic profiles. Proc Natl Acad Sci USA 96:4285-4288

Philippsen P, Kleine K, Pohlmann R, Dusterhoft A, Hamberg K, Hegemann JH, Obermaier B, Urrestarazu LA, Aert R, Albermann K, Altmann R, André B, Baladron V, Ballesta JPG, Becam AM, Beinhauer J, Boskovic J, Buitrago MJ, Bussereau F, Coster F, Crouzet M, D'Angelo M, Dal Pero F, De Antoni A, Del Rey F, Doignon F, Domdey H, Dubois E, Fiedler T, Fleig U, Floeth M, Fritz C, Gaillardin C, Garcia-Cantalejo JM, Glansdorff NN, Goffeau A, Güldener U, Herbert C, Heumann K, Heuss-Neitzel D,

Hilbert H, Hinni K, Iraqui Houssaini I, Jacquet M, Jimenez A, Jonniaux JL, Karpfinger L, Lanfranchi G, Lepingle A, Levesque H, Lyck R, Maftahi M, Mallet L, Maurer KCT, Messenguy F, Mewes HW, Mostl D, Nasr F, Nicaud JM, Niedenthal RK, Pandolfo D, Pierard A, Piravandi E, Planta RJ, Pöhl TM, Purnelle B, Rebischung C, Remacha M, Revuelta JL, Rinke M, Saiz JE, Sartorello F, Scherens B, Sen-Gupta M, Soler-Mira A, Urbanus JHM, Valle G, Van Dyck L, Verhasselt P, Vierendeels F, Vissers S, Voet M, Volckaert G, Wach A, Wambutt R, Wedler H, Zollner A, Hani J (1997) The nucleotide sequence of *Saccharomyces cerevisiae* chromosome XIV and its evolutionary implications. Nature 387:93-98

Piškur J, Langkjaer RB (2004) Yeast genome sequencing: the power of comparative genomics. Mol Microbiol 53:381-389

Raghuraman MK, Winzeler EA, Collingwood D, Hunt S, Wodicka L, Conway A, Lockhart DJ, Davis RW, Brewer BJ, Fangman WL (2001) Replication dynamics of the yeast genome. Science 294:115-121

Rutherford K, Parkhill J, Crook J, Horsnell T, Rice P, Rajandream MA, Barrell B (2000) Artemis: sequence visualization and annotation. Bioinformatics 16:944-945

Sharp PM, Cowe E (1991) Synonymous codon usage in *Saccharomyces cerevisiae*. Yeast 7:657-678

Sharp PM, Li WH (1987) The codon Adaptation Index - a measure of directional synonymous codon usage bias, and its potential applications. Nucleic Acids Res 15:1281-1295

Sipiczki M (2000) Where does fission yeast sit on the tree of life? Genome Biol 1:reviews1011.1011-1014

Sonnhammer EL, Durbin R (1994) A workbench for large-scale sequence homology analysis. Comput Appl Biosci 10:301-307

Souciet J, Aigle M, Artiguenave F, Blandin G, Bolotin-Fukuhara M, Bon E, Brottier P, Casaregola S, de Montigny J, Dujon B, Durrens P, Gaillardin C, Lepingle A, Llorente B, Malpertuy A, Neuveglise C, Ozier-Kalogeropoulos O, Potier S, Saurin W, Tekaia F, Toffano-Nioche C, Wesolowski-Louvel M, Wincker P, Weissenbach J (2000) Genomic exploration of the hemiascomycetous yeasts: 1. A set of yeast species for molecular evolution studies. FEBS Lett 487:3-12

Stoesser G, Tuli MA, Lopez R, Sterk P (1999) The EMBL Nucleotide Sequence Database. Nucleic Acids Res 27:18-24

Tekaia F, Blandin G, Malpertuy A, Llorente B, Durrens P, Toffano-Nioche C, Ozier-Kalogeropoulos O, Bon E, Gaillardin C, Aigle M, Bolotin-Fukuhara M, Casaregola S, de Montigny J, Lepingle A, Neuveglise C, Potier S, Souciet J, Wesolowski-Louvel M, Dujon B (2000) Genomic exploration of the hemiascomycetous yeasts: 3. Methods and strategies used for sequence analysis and annotation. FEBS Lett 487:17-30

Wolfe KH, Shields DC (1997) Molecular evidence for an ancient duplication of the entire yeast genome. Nature 387:708-713

Volpe T, Kidner C, Hall IM, Teng G, Grewal SI, Martienssen R (2002) Regulation of heterochromatic silencing and histone H3 lysine-9 methylation by RNAi. Science 297:1833-1837

Wood V, Gwilliam R, Rajandream MA, Lyne M, Lyne R, Stewart A, Sgouros J, Peat N, Hayles J, Baker S, Basham D, Bowman S, Brooks K, Brown D, Brown S, Chillingworth T, Churcher C, Collins M, Connor R, Cronin A, Davis P, Feltwell T, Fraser A, Gentles S, Goble A, Hamlin N, Harris D, Hidalgo J, Hodgson G, Holroyd S, Hornsby T, Howarth S, Huckle EJ, Hunt S, Jagels K, James K, Jones L, Jones M, Leather S,

McDonald S, McLean J, Mooney P, Moule S, Mungall K, Murphy L, Niblett D, Odell C, Oliver K, O'Neil S, Pearson D, Quail MA, Rabbinowitsch E, Rutherford K, Rutter S, Saunders D, Seeger K, Sharp S, Skelton J, Simmonds M, Squares R, Squares S, Stevens K, Taylor K, Taylor RG, Tivey A, Walsh S, Warren T, Whitehead S, Woodward J, Volckaert G, Aert R, Robben J, Grymonprez B, Weltjens I, Vanstreels E, Rieger M, Schafer M, Muller-Auer S, Gabel C, Fuchs M, Fritzc C, Holzer E, Moestl D, Hilbert H, Borzym K, Langer I, Beck A, Lehrach H, Reinhardt R, Pohl TM, Eger P, Zimmermann W, Wedler H, Wambutt R, Purnelle B, Goffeau A, Cadieu E, Dreano S, Gloux S, Lelaure V, Mottier S, Galibert F, Aves SJ, Xiang Z, Hunt C, Moore K, Hurst SM, Lucas M, Rochet M, Gaillardin C, Tallada VA, Garzon A, Thode G, Daga RR, Cruzado L, Jimenez J, Sanchez M, del Rey F, Benito J, Dominguez A, Revuelta JL, Moreno S, Armstrong J, Forsburg SL, Cerrutti L, Lowe T, McCombie WR, Paulsen I, Potashkin J, Shpakovski GV, Ussery D, Barrell BG, Nurse P (2002) The genome sequence of *Schizosaccharomyces pombe*. Nature 415:871-880

Wood V, Rutherford KM, Ivens A, Rajandream MA, Barrell BG (2001) A re-annotation of the *Saccharomyces cerevisiae* genome. Comp Funct Genom 2:143-154

Zhang CT, Wang J (2000) Recognition of protein coding genes in the yeast genome at better than 95% accuracy based on the Z curve. Nucleic Acids Res 28:2804-2814

Zhang Z, Dietrich FS (2005) Identification and characterization of upstream open reading frames (uORF) in the 5' untranslated regions (UTR) of genes in *Saccharomyces cerevisiae*. Curr Genet 48:77-87

Electronic Supplementary Material:
Supplementary material is available for this article at http://dx.doi.org/10.1007/4735_117.

Axelson-Fisk, Marina
Fraunhofer-Chalmers Research Center for Industrial Mathematics, Chalmers Science Park, SE-412 88 Göteborg, Sweden

Sunnerhagen, Per
Department of Cell and Molecular Biology, Lundberg Laboratory, Göteborg University, P.O. Box 462, SE-405 30 Göteborg, Sweden
per.sunnerhagen@molbio.gu.se

Taxonomy and phylogenetic diversity among the yeasts

Cletus P. Kurtzman and Jure Piškur

Abstract

Yeasts are among the economically and scientifically most important eukaryotic microorganisms known. At present, there are 1,500 recognized species, which are distributed between the ascomycetes and the basidiomycetes, but only a small fraction of these species have undergone extensive genetic analyses. In this chapter, we discuss application of molecular methods for identification of species and for their classification from phylogenetic analysis of gene sequences. The resulting phylogeny is considered in the context of comparative genomics and evolution, and provides a useful background for selection of additional species for whole genome sequencing as well as for new biotechnological applications.

1 Introduction

Yeasts are fungi that predominantly exist as unicellular organisms. However, some yeasts can become multicellular through formation of strands of elongated buds known as pseudohyphae, or through the formation of true hyphae that have well developed crosswalls like those seen in typical filamentous fungi. *Candida albicans* not only buds profusely, but forms pseudohyphae and occasional true hyphae. However, a few species, such as *Eremothecium* (*Ashbya*) *gossypii*, grow exclusively by formation of true hyphae. So far, about 1,500 yeast species have been described, classified, and are available from culture collections worldwide (for further details, see the references in Kurtzman and Fell 1998). Other yeasts maintained in culture collections include industrially important strains of *Saccharomyces cerevisiae*, *C. albicans*, and other common pathogenic species, and popular laboratory organisms, such as genetically characterized strains of *S. cerevisiae* and *Schizosaccharomyces pombe*, which serve as general models to understand the eukaryotic cell. *Saccharomyces cerevisiae* has for decades been the best characterized eukaryotic organism, especially from the perspective of genetics and physiology. Following elucidation of its genome sequence (Goffeau et al. 1996), *S. cerevisiae* is now one of the primary organisms used to study comparative genomics.

For several decades, molecular biologists have focused on a small number of yeasts, mainly because of the limitations of available genetic and biochemical tools. However, during the last few years, with the onset of a variety of novel mo-

Topics in Current Genetics, Vol. 15
P. Sunnerhagen, J. Piškur (Eds.): Comparative Genomics
DOI 10.1007/b106654 / Published online: 20 January 2005

lecular biology approaches, additional species have undergone a thorough molecular analysis. This has recently culminated with the sequencing of over a dozen different yeast genomes (for details, see Piškur and Langkjaer 2004). With this new information, yeasts are becoming increasingly interesting to molecular biologists because understanding diversity and phylogenetic relationships is essential for comparative genomics studies.

Yeasts must be correctly identified and phylogenetically classified if the great amount of genomic data being developed from various species is to be effectively analyzed. The identification and classification of yeasts is now being profoundly affected by advances in molecular biology. Initially, identification of yeasts and other fungi was from use of phenotypic characters, such as fermentation of sugars, growth on various organic compounds, and presence of diagnostic morphological features. This is often an uncertain process, and to remedy this problem, taxonomists began to use molecular comparisons. Early studies employed nuclear DNA reassociation to measure relatedness between strains, which demonstrated that many commonly used phenotypic characters were inadequate to correctly identify species. With the development of rapid gene sequencing technologies, it became possible to more easily compare species and to assess their relatedness through phylogenetic analysis of sequences. These latter studies brought many surprises, among them that the budding ascomycetous yeasts represent a distinct clade that is sister to the euascomycetes, the so-called filamentous ascomycete fungi such as *Aspergillus* and *Neurospora*, and that the fission yeasts are basal to both groups. Basidiomycetous yeasts were found to be distributed among the three major clades, those related to species with a mushroom-type sexual state, those related to the smuts, and those related to the rusts. For taxonomists, this new information raised the possibility that the classification of yeasts and other fungi can actually be based on evolutionary relationships. For molecular biologists and biotechnologists, a classification system derived from genetic relatedness provides names with a predictive value that convey genetic and biotechnological similarities. In this chapter, we will discuss some of the early molecular comparisons that have been used as well as the work now being done and where it may lead us in future studies.

2 Whole genome comparisons from measurements of DNA reassociation

Nuclear DNA reassociation or hybridization represents the first quantitative molecular method used to assess species relatedness. DNA from the species pair of interest is sheared, mixed, made single-stranded, and the degree of relatedness determined from quantitation of the extent of reassociation. Many different methods are used to measure this process, which can be done spectrophotometrically or through use of radioisotopes or other markers (Kurtzman 1993a). Bak and Stenderup (1969) appear to be the first to have made extensive use of nucleic acid reassociation as a means to determine relatedness among yeasts. They showed from

Table 1. DNA relatedness between conspecific strains that were previously considered separate yeast species because of differences in traditional taxonomic characteristics

Species	Characteristic (+ or -)	DNA relatedness (%)
	True Hyphae	
Hansenula wingei	+	78[a]
Hansenula canadensis	-	
	Glucose fermentation	
Debaryomyces formicarius	+	96[b]
Debaryomyces vanriji	-	
	Lactose assimilation	
Schwanniomyces castellii	+	97[b]
Schwanniomyces occiden-talis	-	
	Nitrate assimilation	
Hansenula minuta	+	75[c]
Pichia lindneri	-	
	Nitrate assimilation	
Sterigmatomyces halophilus	+	100[d]
Sterigmatomyces indicus	-	

[a] Fuson et al. (1979).
[b] Price et al. (1978).
[c] Kurtzman (1984b).
[d] Kurtzman (1990b).

high DNA relatedness that the pathogens *Candida albicans*, *C. stellatoidea*, and *C. claussenii* represent the same species. Following this work, Meyer and Phaff (1972) demonstrated high DNA homology between *Candida lusitaniae* and *C. obtusa* and between *C. salmonicola* and *C. sake*. The extensive DNA comparisons of Price et al. (1978) clarified a number of taxonomic questions in the genera *Saccharomyces*, *Pichia*, *Debaryomyces*, and *Torulaspora*. Other DNA relatedness studies showed that the fodder yeast *Candida utilis* represents the asexual form of *Pichia jadinii*, an ascosporic species known only from human infections (Kurtzman et al. 1979), and that many species placed in *Saccharomyces* are conspecific with *S. cerevisiae* and *S. bayanus* (Vaughan-Martini and Kurtzman 1985). Consequently, DNA relatedness studies have shown that many of the phenotypic characters commonly used for separation of species and genera are often only strain-specific (Table 1).

A major question has been how to interpret DNA reassociation data. Measurements of DNA complementarity are commonly expressed as percent relatedness. This usage can be misleading because DNA strands must show at least 75-80% base sequence similarity before duplexing can occur and a reading is registered on the scale of percent relatedness (Bonner et al. 1973; Britten et al. 1974). Experimental conditions can greatly influence extent of duplex formation, but when measured under optimum conditions, different methods of assessing DNA relatedness do give essentially the same result (Kurtzman 1993a). Percent DNA relatedness provides an approximation of overall genome similarity between two organisms, but the technique does not detect single gene differences or exact multiples of ploidy, although aneuploidy can sometimes be detected (Vaughan-Martini and Kurtzman 1985).

On the basis of shared phenotype, strains that show 80% or greater nuclear DNA relatedness were believed to represent members of the same yeast species (Martini and Phaff 1973; Price et al. 1978). This issue was also examined on the basis of the biological species concept (e. g. Dobzhansky 1976), asking what is the fertility between strains showing varying degrees of DNA relatedness (Kurtzman 1984a, 1984b, 1987; Kurtzman et al. 1980a, 1980b). In one of these studies, the heterothallic species *Pichia amylophila* and *P. mississippiensis*, which show 25% DNA relatedness, gave abundant mating, but ascus formation was limited and no ascospores were formed. Similar results were found for crosses between *Pichia americana* and *P. bimundalis* (21% DNA relatedness) and between *Pichia alni* and *P. canadensis* (*Hansenula wingei*), the latter pair showing just 6% DNA relatedness. The varieties of *Issatchenkia scutulata*, which show 25% DNA relatedness, behaved somewhat differently. Crosses between var. *scutulata* and var. *exigua* gave extent of mating and ascospore formation comparable to intravarietal crosses. Ascospore viability from these intervarietal crosses was about 5%, but sib-matings of these progeny gave 17% ascospore viability. However, backcrosses to the parentals gave poor ascosporulation and very low viability. *Williopsis saturnus* is a homothallic species with five varieties that range in DNA relatedness from 37-79 % (Kurtzman 1987). Intervarietal fertility is reduced and varies depending on the strains crossed. Consequently, the preceding studies show that mating among heterothallic as well as homothallic taxa can occur over a wide range of DNA relatedness values, but that highly fertile crosses, which demonstrate conspecificity, seem to require 70-80% or greater DNA relatedness.

Recognizing biological species from genetic crosses requires not only high viability of progeny, but high viability from crosses between progeny, and that backcrosses to the parental mating types also produce highly viable progeny that are interfertile. Most biologically defined yeast species have only undergone the first genetic cross in this series of tests. Chromosomal mutations such as deletions, duplications, inversions, translocations, fusions, or fission may markedly impact fertility, but may not noticeably diminish DNA relatedness and would therefore go undetected by measurements from DNA reassociation experiments. This may be the case for *Saccharomyces bayanus* and *S. uvarum*. Vaughan-Martini and Kurtzman (1985) reported 98% DNA reassociation between the two taxa, but Naumov (2000) found low fertility when the two were crossed.

Fischer et al. (2000) examined chromosomal translocations among the closely related species of the *Saccharomyces cerevisiae* clade. These species show low DNA relatedness, correspondingly divergent gene sequences, and produce infertile progeny when crossed (Naumov et al. 2000). Despite widespread translocations in some species, Fischer et al. (2000) concluded that there was no correlation between gene-based species phylogenies and the presence of translocations. In a continuation of this approach, Delneri et al. (2003) reengineered chromosomes of *S. cerevisiae* to be collinear with those of *S. mikatae*. The improved synteny resulted in a higher number of viable ascospores from interspecific crosses, but there was extensive aneuploidy among these progeny. Consequently, species barriers are complex and involve a number of factors. As a result of the preceding genetic comparisons, the numerical range of 70-100 % DNA relatedness as indicative of conspecificity, a range that is also applied to species with no known sexual state, should be viewed as a prediction because many factors impact fertility and the assessment of biological species.

3 Identification of species from analysis of gene sequence divergence

As described in the previous section, nuclear DNA reassociation studies have had a marked impact on recognizing yeast species. Although quite useful, the method is time consuming and the extent of genetic resolution goes no further than that of closely related species. Gene sequencing offers a rapid method for recognizing species and resolution is not limited to closely related taxa. Peterson and Kurtzman (1991) determined that domain 2 of large subunit (26S) ribosomal RNA (rRNA) was sufficiently variable to resolve individual species when tested with the closely related reference species listed in the previous section. This work further confirmed the report of Vaughan-Martini and Kurtzman (1985) that *Saccharomyces pastorianus* is a natural hybrid of *S. cerevisiae* and *S. bayanus* by showing that the rRNA sequence of *S. pastorianus* is identical to that of *S. bayanus*. At the same time, this finding added the cautionary note that not all species can be resolved from a single gene sequence and that hybrids may go undetected, which is important to recognize because several natural and industrial yeast strains are hybrids (Nilsson-Tillgren et al. 1981). Kurtzman and Robnett (1998) expanded the proceeding work by sequencing both domains 1 and 2 (ca. 600 nucleotides) of 26S ribosomal DNA (rDNA) for all known ascomycetous yeasts, thus, providing a universally available database for rapid identification of known species and the detection of new species and for their initial phylogenetic placement. Fell et al. (2000) published the D1/D2 sequences for all known basidiomycetous yeasts, thus, completing the database for all known yeasts. Resolution provided by the D1/D2 domain was estimated from comparisons of taxa determined to be closely related from genetic crosses and from DNA reassociations (Table 2). In general,

Table 2. Correlation of nuclear DNA relatedness and sequence divergence in large subunit rDNA domains D1/D2 among ascomycetous yeastsa

Strain pairs	% nDNA relatedness	D1/D2 nucleotide differences
70 Conspecific pairs	70 - 100	0 - 3
Ca. 200 unrelated pairs	0 - 20	6 ~ 250

[a] Data from Kurtzman and Robnett (1998).

strains of a species show no more than 0-3 nucleotide differences (0-0.3 %), and strains showing 6 or more noncontiguous substitutions (1 %) are separate species. Strains with intermediate nucleotide substitutions are also likely to be separate species. One impact of the D1/D2 database has been to permit detection of a large number of new species causing a near doubling of known species since publication of the most recent edition (4th) of *The Yeasts, A Taxonomic Study* (Kurtzman and Fell 1998). Another use is that the non-taxonomist can now quickly and accurately identify most known species, as well as recognize new species, by sequencing circa 600 nucleotides and doing a BLAST Search in GenBank.

The focus of our discussion on species identification from gene sequences has been on rDNA. A major advantage of rDNA is that it is present in all living organisms, has a common origin, occurs as multiple copies and is easy to sequence because primers pairs for conserved regions can generally be used for all organisms. Small subunit (18S) rDNA is generally too conserved to allow separation of individual species (James et al. 1996; Kurtzman and Robnett 2003). Domains 1 and 2 of large subunit rDNA generally are sufficiently substituted to allow recognition of most individual species. Some exceptions to resolution of closely related species are given in Table 3 and include the hybrid species *Saccharomyces pastorianus* as well as varieties of *Candida shehatae* (Kurtzman 1990a). Additionally, there is one well documented exception that 1% nucleotide divergence indicates strains to be separate species. Lachance et al. (2003) found interfertile strains of *Clavispora lusitaniae* that are highly polymorphic in the D1/D2 domain. Undoubtedly, other exceptions exist and additional gene sequences will be necessary to recognize them. In this context, Groth et al. (1999) discovered a natural chimeric isolate of *Saccharomyces* with genetic material from three species.

The internal transcribed spacer regions ITS 1 and ITS 2, which are separated by the 5.8S gene of rDNA, are also highly substituted and have been used for species identification, but for many species, ITS sequences give no greater resolution than that obtained from 26S domains D1/D2 (James et al. 1996; Kurtzman and Robnett 2003). However, Fell and Blatt (1999) were able to resolve cryptic species in the *Xanthophyllomyces dendrorhous* species complex that had been unresolved from D1/D2 sequence analysis, and Scorzetti et al. (2002) reported ITS sequences to provide somewhat greater resolution among many basidiomycetous species than was found for D1/D2, although, a few species were less well resolved by ITS than by D1/D2. The intergenic spacer region (IGS) of rDNA tends to be highly substituted and sequences of this region have been used with good success to separate closely related lineages of *Cryptococcus* (Fan et al. 1995), *Xanthophyllomyces* (Fell and Blatt 1999), and *Saccharomyces* (Kurtzman et al. unpublished). Because of the occurrence of repetitive sequences and homopolymeric regions, IGS tends to be difficult to sequence.

Separation of species using gene sequences other than those of the rDNA repeat has been successful for a variety of fungi (Geiser et al. 1998; O'Donnell et al. 2000), including the yeasts. Belloch et al. (2000) have demonstrated the utility of cytochrome oxidase II for resolution of *Kluyveromyces* species, Daniel et al. (2001) have successfully used actin-1 for species of *Candida*, and Kurtzman and

Table 3. Extent of nuclear DNA complementarity and nucleotide divergence in large subunit rDNA D1/D2 domains between certain closely related yeasts[a]

Taxa	% nDNA reassociation	D1/D2 divergence
Candida shehatae var. she-hatae		
var. *insectosa*	49	1
var. *lignosa*	46	0
Pichia amylophila		
Pichia mississippiensis	27	4
Pichia toletana		
Pichia xylosa	29	1
Saccharomyces bayanus		
Saccharomyces pastorianus	72	0
Torulaspora delbrueckii		
Torulaspora pretoriensis	13	5

[a] Data summarized from Kurtzman and Robnett (1998).

Robnett (2003) have shown the usefulness of elongation factor 1-α and RNA polymerase II for resolution of *Saccharomyces* species. At present, the main impediment to widespread use of other gene sequences is developing sequencing primers that are effective for essentially all species, and construction of databases that include sequences from all known species. Daniel et al. (2001) and Daniel and Meyer (2003) have made considerable progress in development of an actin sequence database for species identification, although there are no primer sets that are effective for all species, thus requiring extra work to obtain these sequences. Resolution of taxa from actin is somewhat greater than from D1/D2, but not surprisingly, clear separation of closely related species is not always certain. These comparisons point out that a definitive separation of species from single genes is not possible in all cases, but use of single gene sequence databases for rapid species identification is extremely useful even with the risk that hybrid species and other closely related taxa may be unresolved.

Other molecular-based methods commonly used for species identification include species-specific primer pairs and probes, randomly amplified polymorphic DNA (RAPD), amplified fragment length polymorphisms (AFLP), restriction fragment length polymorphisms (RFLP) and karyotyping. Species-specific primers are effective when used for PCR-based identifications involving a small number of species or when a particular species is the subject of the search (Fell 1993; Mannarelli and Kurtzman 1998). Otherwise, there is the likelihood that mixtures with large numbers of primer pairs will lead to uncertain banding patterns. Microsatellite-primed RAPDs (Gadanho et al. 2003) and AFLP fingerprint (de Barrios Lopes et al. 1999) have been effectively used in some laboratories. The main criticism of these techniques is reproducibility between laboratories because small differences in PCR conditions may impact the species-specific patterns that serve as reference. Karyotyping with pulse field electrophoresis and RAPD on mito-chondrial DNA can serve in the initial characterization and identification of yeast species. However, the interpretation of the chromosome band patterns and mito-chondrial restriction fragments for taxonomic purposes is complicated by the high degree of polymorphism, such as chromosomal rearrangements within some yeast taxa (Spirek et al. 2003).

4 Relationships of fungi from phylogenetic analysis of gene sequence divergence

Gene sequence analyses have brought many surprises to our understanding of rela-tionships among the fungi. The distinction between yeasts and dimorphic filamen-tous fungi has often been uncertain. Some have viewed the yeasts as primitive fungi, whereas others perceived them to be reduced forms of more evolved taxa (Cain 1972; Redhead and Malloch 1977). Phylogenetic analyses of rDNA se-quences demonstrated the ascomycetous yeasts, as well as yeast-like genera such as *Ascoidea* and *Cephaloascus*, to comprise a clade that is a sister group to the "filamentous" ascomycetes (euascomycetes). *Schizosaccharomyces*, *Taphrina*, *Protomyces*, *Saitoella*, *Pneumocystis*, and *Neolecta*, a mushroom-like fungus, form a divergent clade basal to the yeast-euascomycete branch (Hausner et al. 1992; Hendriks et al. 1992; Kurtzman 1993b, Kurtzman and Robnett 1994, 1995, 1998; Kurtzman and Sugiyama 2001; Landvik 1996; Nishida and Sugiyama 1993; Sjamsuridzal et al. 1997; Sugiyama 1998; Wilmotte et al. 1993). Nishida and Su-giyama (1994) have termed the basal ascomycete clade the 'archiascomycetes'. Some members of the yeast clade, such as certain species of *Ascoidea* and *Eremo-thecium*, show no typical budding, whereas dimorphism occurs among the so-called black yeasts in the genera *Aureobasidium* and *Phialophora* as well as in certain other euascomycete genera. Similarly, vegetative reproduction by fission is shared by *Dipodascus* and *Galactomyces*, members of the yeast clade, as well as by the distantly related genus *Schizosaccharomyces*. Consequently, yeasts cannot be recognized solely on the basis of presence or absence of budding. However, with a few exceptions, ascomycetous yeasts can be separated phenotypically from euascomycetes by the presence of budding or fission and the formation of sexual states unenclosed in a fruiting body.

Basidiomycetous yeasts are usually characterized by budding, but some, such as *Trichosporon*, also reproduce by fission. Yeast forms are found in all three ma-jor basidiomycete lineages, i.e., among certain mushrooms (Tremellales), as well as the smut and rust clades (Swann and Taylor 1993; Fell et al. 2000). As with as-comycetous species, the basidiomycetous yeasts, with few exceptions, can be rec-ognized morphologically by the presence of budding or fission, and the formation of sexual states that are unenclosed in a fruiting body.

Phylogenetic relationships among the yeasts and other fungi have been resolved from analysis of gene sequence divergence as cited in the preceding references.

These studies presume that horizontal transfer among different lineages has been limited, which can be tested by comparing the congruence of phylogenies derived from individual genes. Most of the analyses have used rDNA sequences, but there are generally no major differences in tree topologies whether analyses are from rDNA sequences or from those of other genes (Kurtzman and Robnett 2003; Liu et al. 1999). Although phylogenetic trees derived from analyses of various genes are generally congruent, support for basal lineages from single gene analyses is often quite weak (Kurtzman and Robnett 2003; Rokas et al. 2003). Because of this, branching order is uncertain as is the taxonomy that is synthesized from these trees. Use of multiple genes in the analysis generally strengthens support for basal lineages. Kurtzman and Robnett (2003) examined relationships among the circa 80 species of the 'Saccharomyces complex'. Combined analysis of 18S, 26S and mitochondrial small subunit rDNAs with elongation factor-1α and cytochrome oxidase II gave high bootstrap support to moderately deep lineages, but more basal lineages will require a greater number of genes for strong support. From this study, Kurtzman (2003) taxonomically circumscribed as genera the 14 clades defined by the preceding analysis (Fig. 1), but deeper lineages (e.g. families) were not strongly supported. A major question is how many gene sequences are required for an accurate phylogenetic reconstruction? The six gene sequences used by Kurtzman and Robnett (2003) as a concatenated dataset gave strong bootstrap support for a majority of the 14 clades that were resolved. However, the *Lachancea* clade (Clade 10) was weakly supported at 58%, primarily because of the inclusion of the divergent species *Lachancea* (*Saccharomyces*) *kluyveri*.

Rokas et al. (2003) screened the published genome sequences from seven *Saccharomyces* species and that of *Candida albicans* and selected 106 widely distributed orthologous genes for phylogenetic analysis. The resulting analyses showed that a dataset comprised of a concatenation of a minimum of nearly any 20 genes gave well supported trees that were comparable to a dataset comprised of 106 genes. This work clearly illustrates that a much larger number of genes is required for reconstructing phylogenies than is currently being sequenced in most laboratories. Whether 20 gene sequences will strongly resolve species clades larger than *Saccharomyces* needs to be determined. However, partial genome sequencing may be sufficient to resolve phylogenetic relationships within different groups of yeasts. Therefore, intensified partial genome sequencing covering a large number of yeast taxa should be continued to provide sufficient datasets to determine the phylogenetic relationships among the major yeast taxa. Another factor that impacts resolution, as well as circumscription of genera, is the issue of missing taxa. It seems likely that fewer than 1% of extant species are known, which can be inferred from the high frequency of single-species branches in phylogenetic trees. In other words, this means that a majority of yeasts are yet to be discovered and isolated from nature and then characterized in the laboratory. In coming years, more resources and effort will need to be spent on one of the most neglected aspects of yeast biology, their ecology and biodiversity, to provide the necessary support for yeast molecular biology and especially comparative genomics studies.

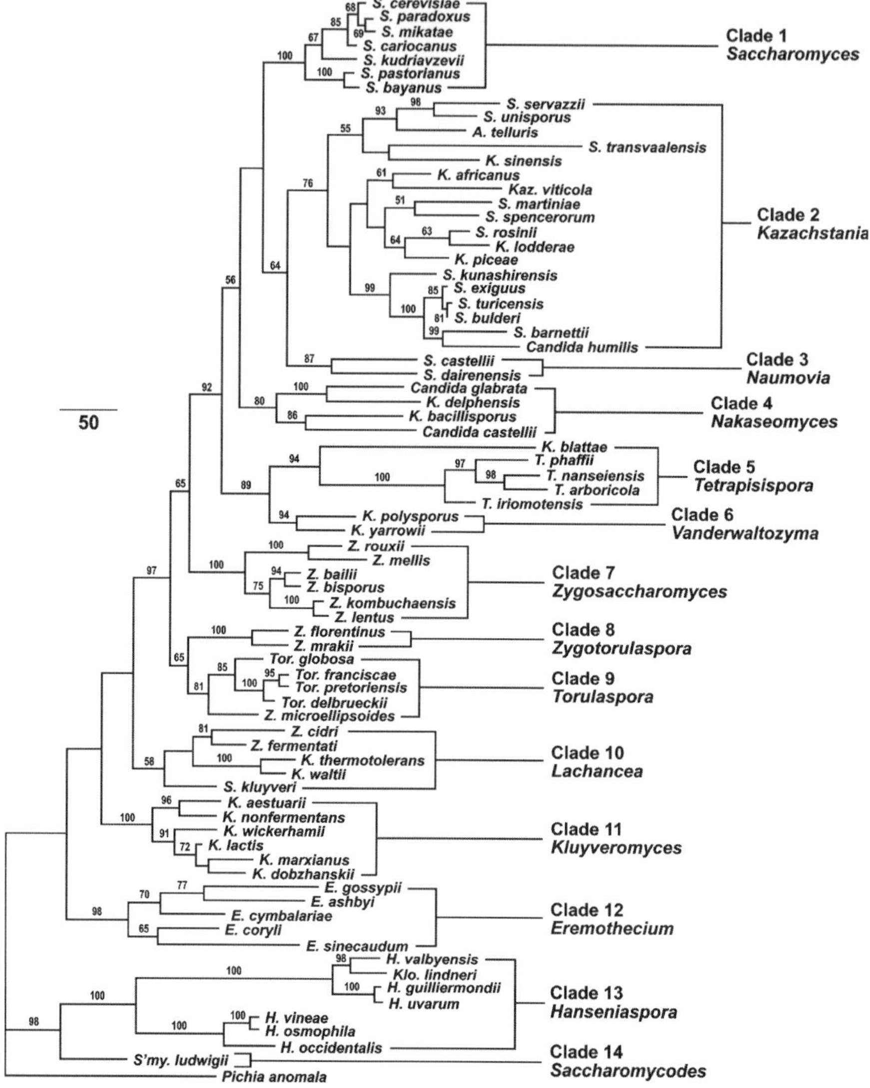

Fig. 1. Phylogenetic tree resolving species of the 'Saccharomyces complex' into 14 clades, which are interpreted as phylogenetically circumscribed genera. One of three most parsimonius trees derived from maximum parsimony analysis of a dataset comprised of nucleotide sequences from 18S, 5.8S/alignable ITS, and 26S (three regions) rDNAs, EF-1α, mitochondrial small subunit rDNA and COXII. Branch lengths based on nucleotide substitutions are indicated by the bar. Bootstrap values ≥ 50% are given. *Pichia anomala* is the outgroup species in the analysis. Modified from Kurtzman (2003) and Kurtzman and Robnett (2003).

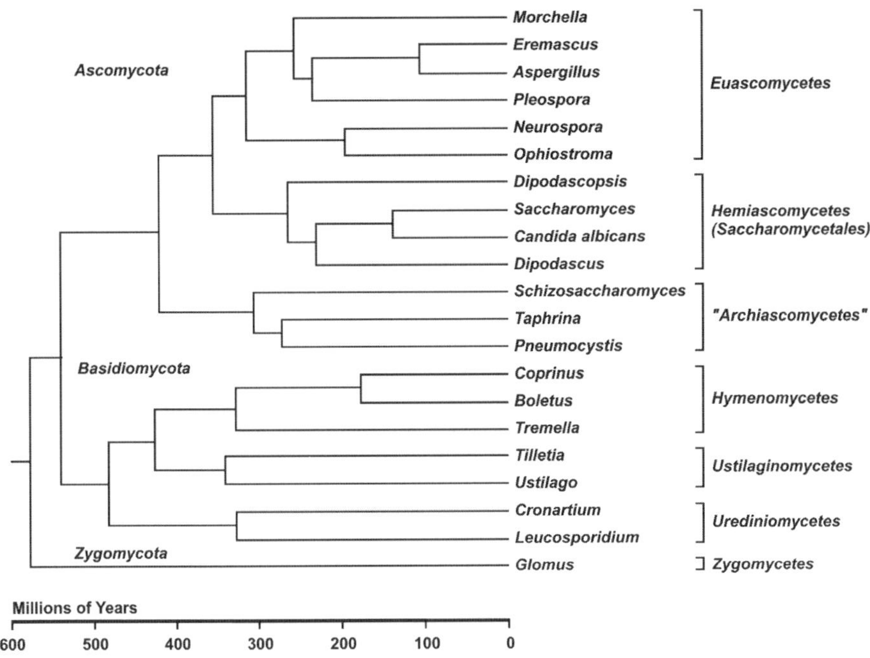

Fig. 2. Phylogenetic relationships among the major groups of fungi and their evolutionary divergence times as estimated from 18S rDNA. This tree is based on the analysis of Berbee and Taylor (2001) and is useful to determine the approximate timing of different evolutionary events (Dujon et al. 2004; Piškur and Langkjaer 2004; Wolfe 2004). The *Saccharomyces* branch represents the site of placement for all species shown in Figure 1.

5 Evolution and timing of modern yeast traits

Phylogenetic analyses have revealed that the ascomycetes and basidiomycetes are sister taxa and that the zygomycetes represent a closely related basal lineage. Berbee and Taylor (2001) presented a timeline of fungal evolution based on 18S gene sequences and suggested that the Glomales (zygomycetes) diverged from the ascomycete/basidiomycete clade prior to 400 mya, and that among ascomycetous yeasts, the *Saccharomyces* lineage may have separated from other yeasts around 150 mya (Fig. 2). Divergence times are based on many assumptions and Graur and Martin (2004) have rather pointedly discussed some of the problems encountered when estimating molecular timescales. For example, rates of substitution in specific rDNA regions vary among lineages, which can significantly influence estimates of time of divergence (Petersen et al. 2000). Nonetheless, an approximate time-scale of divergence times among different yeast lineages is very important to understand the origin of modern yeast traits. The main rationale to understand the evolutionary history of yeasts is to compare a similar molecular or cellular trait in different species and then deduce how the original progenitor of such a trait looked. On the basis of phylogenetic relationships among species compared, the predicted progenitor and the accompanying molecular events can then be set into an estimated evolutionary time frame.

The extensive sequencing within the *Saccharomyces* complex and the initial molecular studies of several *Saccharomyces* and *Kluveromyces* species other than *S. cerevisiae*, and the inclusion of *Eremothecium (Ashbya) gossypii* and *Candida glabrata*, have recently provided the phylogenetic framework to understand the origin and timing of several modern properties unique for the *Saccharomyces*-complex yeasts (Kellis et al. 2004; Dietrich et al. 2004; Dujon et al. 2004; Piškur and Langkjaer 2004). For example, *Saccharomyces* sensu stricto yeasts, as well as other *Saccharomyces* members, primarily degrade hexoses only to pyruvate and ethanol, even in the presence of oxygen. This phenomenon relies on a "glucose repression" circuit that represses the respiratory pathway in the presence of glucose (Johnston 1999). Other unique traits are the ability to survive without oxygen (Andreasen and Stier 1953; Møller et al. 2001) as well as the complexity of the life cycle. Sexually reproducing yeasts can either undergo mating between heterothallic lines, which are self-sterile, or homothallic lines, which are self-fertile. In *S. cerevisiae*, homothallism occurs because cells can switch mating type; specifically a maternal cell can, upon budding, switch its mating type and then mate with its daughter (Strathern and Herskowitz 1979). A site-specific nuclease, encoded by *HO*, represents a central element in this high frequency mating type switching (Haber 1998).

At least two molecular mechanisms: (i) whole-genome duplication and (ii) horizontal gene transfer have been proposed to play a major role in the evolutionary history of the *Saccharomyces*-complex yeasts. The segmental duplication, which occurred circa 100–150 million years ago (Wolfe and Shields 1997), apparently provided new genes, which were the basis for major remodeling of metabolism, including the development of an efficient glucose repression pathway and oxygen independence, in *Saccharomyces*-complex yeasts. For example, some of the duplicated genes have "remodeled" their expression to become dependent on the presence/absence of oxygen and glucose (Kwast et al. 2002). The developing yeast lineages apparently had a competitive advantage; they could grow rapidly with or without oxygen and they preferentially produced ethanol, which is toxic to some competitors. The whole issue of genome duplication will be discussed in detail in some of the following chapters.

One of the crucial requirements for facultative anaerobiosis in yeasts is the independence of the fourth pyrimidine *de novo* biosynthesis of enzyme activity catalysed by dihydroorotate dehydrogenase (DHODase), from the active respiratory chain (Nagy et al. 1992). Apparently, the gene for the respiratory chain-independent enzyme, closely related to some modern bacterial DHODases, was adopted by the *S. cerevisiae* lineage before the *S. cerevisiae* and *Lachancea (Saccharomyces) kluyveri* lineages separated (Gojković et al. 2004) and promoted gradual independence from the presence of oxygen. Homothallism, as known in the modern *Saccharomyces* sensu stricto yeasts, apparently originated in a similar way, through acquisition of an intein-like sequence, a selfish mobile element, which can be inserted in-frame into host genes (Keeling and Roger 1995). After separation of the *S. cerevisiae* and *L. kluyveri* lineages, an intein invaded the yeast *VMA1* gene from an unknown source and subsequently duplicated and gave rise to *HO* (Butler et al. 2004). Therefore, this event is much younger than the adoption of a bacterial-like DHODase (Piškur and Langkjaer 2004; Fig. 1). Efficient homothallic switching as seen in *S. cerevisiae* is the background for a predominantly diploid lifestyle found among modern *Saccharomyces* yeasts and increases the level of robustness of these yeasts. The precise timing of the preceding events is still unclear (Piškur and Langkjaer 2004), but further analyses of species within the *Saccharomyces*-complex will facilitate resolution of temporal uncertainties. Analysis of species assigned to *Zygosaccharomyces, Torulaspora,* and *Zygotorulaspora* should be particularly helpful because these three genera appear basal to *Saccharomyces* and some species within these genera are homothallic whereas others are heterothallic. Increased coordination within the scientific community regarding which species to focus on for further work will be highly beneficial for future progress.

6 Future prospects

The basic outline of phylogenetic relationships among the fungi has been determined from gene sequence analyses. Multigene comparisons are now needed to verify and strengthen these observations, and as basal branches become better resolved, a robust taxonomy can be developed that has its foundation on evolutionary relationships. This new taxonomy should have a predictiveness that will be useful to molecular geneticists, biotechnologists, evolutionary biologists and ecologists. The extensive gene sequence databases that develop from this work will have a profound effect on the way in which fungi are identified. The sequences can be used to detect individual species or populations through their use in probes, chips, and other diagnostic methods. However, one of the crucial immediate problems to be solved is which yeasts (on the species and isolate level) to select for further sequencing and laboratory work, and further, on the availability of the data and how to organize the yeast genome sequences into a single database or at least databases that can successfully communicate with each other.

References

Andreasen AA, Stier TJB (1953) Anaerobic nutrition of *Saccharomyces cerevisiae*. I. Ergosterol requirement for the growth in a defined medium. J Cell Comp Physiol 41:23-36

Bak AL, Stenderup A (1969) Deoxyribonucleic acid homology in yeasts. Genetic relatedness within the genus *Candida*. J Gen Microbiol 59:21-30

Belloch C, Querol A, Garcia MD, Barrio E (2000) Phylogeny of the genus *Kluyveromyces* inferred from the mitochondrial cytochrome-*c* oxidase II gene. Int J Syst Evol Microbiol 50:405-416

Berbee ML, Taylor JW (2001) Fungal molecular evolution: gene trees and geologic time. In: McLaughlin DJ, McLaughlin E, Lemke P (eds) The Mycota VII, Part B. Springer-Verlag, Berlin, pp 231-245

Bonner TI, Brenner DJ, Neufeld BR, Britten RJ (1973) Reduction in the rate of DNA reassociation by sequence divergence J Mol Biol 81:123-135

Britten RJ, Graham DE, Neufeld BR (1974) Analysis of repeating DNA sequences by reassociation. In: Grossman L, Moldave K (eds) Methods in Enzymology, Vol 29. Academic Press, NY, pp 363-418

Butler G, Kenny C, Fagan A, Kurischko C, Gaillardin C, Wolfe KH (2004) Evolution of the *MAT* locus and its Ho endonuclease in yeast species. Proc Natl Acad Sci USA 101:1632-1637

Cain RF (1972) Evolution of the fungi. Mycologia 64:1-14

Daniel H-M, Meyer W (2003) Evaluation of ribosomal RNA and actin gene sequences for the identification of ascomycetous yeasts. Int J Food Microbiol 86:61-78

Daniel H-M, Sorrell TC, Meyer W (2001) Partial sequence analysis of the actin gene and its potential for studying the phylogeny of *Candida* species and their teleomorphs. Int J Syst Evol Microbiol 51:1593-1606

De Barros Lopes M, Rainiere S, Henschje PA, Langridge P (1999) AFLP fingerprinting for analysis of yeast genetic variation. Int J Syst Bacteriol 49:915-924

Delneri D, Colson I, Grammenoudi S, Roberts IN, Louis, EJ, Oliver SG (2003) Engineering evolution to study speciation in yeasts. Nature 422:68-72

Dietrich FS, Voegeli S, Brachat S, Lerch A, Gates K, Steiner S, Mohr C, Pohlmann R, Luedi P, Choi S, Wing RA, Flavier A, Gaffney TD, Philippsen P (2004) The *Ashbya gossypii* genome as a tool for mapping the ancient *Saccharomyces cerevisiae* genome. Science 304:304-307

Dobzhansky T (1976) Organismic and molecular aspects of species formation. In: Molecular Evolution, Ayala FJ (ed), Sinauer Assoc. Sunderland, Massachusetts, pp 95-105

Dujon B, Sherman D, Fischer G, Durrens P, Casaregola S, Lafontaine I, De Montigny J, Marck C, Neuveglise C, Talla E, Goffard N, Frangeul L, Aigle M, Anthouard V, Babour A, Barbe V, Barnay S, Blanchin S, Beckerich JM, Beyne E, Bleykasten C, Boisrame A, Boyer J, Cattolico L, Confanioleri F, De Daruvar A, Despons L, Fabre E, Fairhead C, Ferry-Dumazet H, Groppi A, Hantraye F, Hennequin C, Jauniaux N, Joyet P, Kachouri R, Kerrest A, Koszul R, Lemaire M, Lesur I, Ma L, Muller H, Nicaud JM, Nikolski M, Oztas S, Ozier-Kalogeropoulos O, Pellenz S, Potier S, Richard GF, Straub ML, Suleau A, Swennen D, Tekaia F, Wesolowski-Louvel M, Westhof E, Wirth B, Zeniou-Meyer M, Zivanovic I, Bolotin-Fukuhara M, Thierry A, Bouchier C, Caudron B, Scarpelli C, Gaillardin C, Weissenbach J, Wincker P, Souciet JL (2004) Genome Evolution in Yeast. Nature 430:35-44

Fan M, Chen LC, Ragan MA, Gutell R, Warner JR, Currie BP, Casadevall, A (1995) The 5S rRNA and the RNA intergenic spacer of the two varieties of *Cryptococcus neoformans*. J Med Vet Mycol 33:215-221

Fell JW (1993) Rapid identification of yeast species using three primers in a polymerase chain reaction. Mol Mar Biol Biotechnol 1:175-186

Fell JW, Blatt G (1999) Separation of strains of the yeasts *Xanthophyllomyces dendrorhous* and *Phaffia rhodozyma* based on rDNA, IGS and ITS sequence analysis. J Ind Microbiol Biotechnol 21:677-681

Fell JW, Boekhout T, Fonseca A, Scorzetti G, Statzell-Tallman, A (2000) Biodiversity and systematics of basidiomycetous yeasts as determined by large-subunit rDNA D1/D2 domain sequence analysis. Int J Syst Evol Microbiol 50:1351-1371

Fischer G, James SA, Roberts IN, Oliver SG, Louis DJ (2000) Chromosomal evolution in *Saccharomyces*. Nature 405:451-454

Fuson GB, Price CW, Phaff HJ (1979) Deoxyribonucleic acid sequence relatedness among some members of the yeast genus *Hansenula*. Int J Syst Bacteriol 29:64-69

Gadanho M, Almeida JM, Sampaio JP (2003) Assessment of yeast diversity in a marine environment in the south of Portugal by microsatellite-primed PCR. Antonie van Leeuwenhoek 84:217-227

Geiser DM, Pitt JI, Taylor JW (1998) Cryptic speciation and recombination in the afla-toxin-producing fungus *Aspergillus flavus*. Proc Natl Acad Sci USA 95:388-393

Goffeau A, Barrell BG, Bussey H, Davis RW, Dujon B, Feldmann H, Galibert F, Hoheisel JD, Jacq C, Johnston M, Louis EJ, Mewes HW, Murakami Y, Philippsen P, Tettelin H, Oliver SG (1996) Life with 6000 genes. Science 274:563-567

Gojković Z, Knecht W, Zameitat E, Warneboldt J, Coutelis J-B, Pynyaha Y, Neuveglise C, Moller K, Loffler M, Piškur J (2004) Horizontal gene transfer promoted the evolution of the yeast ability to propagate under anaerobic conditions. Mol Gen Genomics 271:387-393

Graur D, Martin W (2004) Reading the entrails of chickens: molecular timescales of evolu-tion and the illusion of precision. Trends Genet 20:80-86

Groth C, Hansen J, Piškur J (1999) A natural chimeric yeast containing genetic material from three species. Int J Syst Bacteriol 49:1933-1938

Haber, JE (1998) Mating-type gene switching in *Saccharomyces cerevisiae*. Annu Rev Genet 32:561-599

Hausner G, Reid J, Klassen GR (1992) Do galeate-ascospore members of the Cephaloasca-ceae, Endomycetaceae and Ophiostomataceae share a common phylogeny? Mycologia 84:870-881

Hendriks L, Goris A, Van de Peer Y, Neefs J-M, Vancanneyt M, Kersters K, Berny J-F, Hennebert GL, De Wachter R (1992) Phylogenetic relationships among ascomycetes and ascomycete-like yeasts as deduced from small ribosomal subunit RNA sequences. Syst Appl Microbiol 15:98-104

James SA, Collins MD, Roberts IN (1996) Use of an rRNA internal transcribed spacer re-gion to distinguish phylogenetically closely related species of the genera *Zygosac-charomyces* and *Torulaspora*. Int J Syst Bacteriol 46:189-194

Johnston M (1999) Feasting, fasting and fermenting. Glucose sensing in yeast and other cells. Trends Genet 15:29-33

Keeling PJ, Roger AJ (1995) The selfish pursuit of sex. Nature 375:283

Kellis M, Birren BW, Lander, ES (2004) Proof and evolutionary analysis of ancient ge-nome duplication in the yeast *Saccharomyces cerevisiae*. Nature 428:617-624

Kurtzman CP (1984a) Synonymy of the yeast genera *Hansenula* and *Pichia* demonstrated through comparisons of deoxyribonucleic acid relatedness. Antonie van Leeuwenhoek 50:209-217

Kurtzman CP (1984b) Resolution of varietal relationships within the species *Hansenula anomala*, *Hansenula bimundalis*, and *Pichia nakazawae* through comparisons of DNA relatedness. Mycotaxon 19:271-279

Kurtzman CP (1987) Prediction of biological relatedness among yeasts from comparisons of nuclear DNA complementarity. In: de Hoog GS, Smith MTh, Weijman ACM (eds) The Expanding Realm of Yeast-like Fungi. Elsevier, Amsterdam, pp 459-468

Kurtzman CP (1990a) *Candida shehatae* – genetic diversity and phylogenetic relationships with other xylose-fermenting yeasts. Antonie van Leeuwenhoek 57:215-222

Kurtzman CP (1990b) DNA relatedness among species of *Sterigmatomyces* and *Fellomyces*. Int J Syst Bacteriol 40:56-59

Kurtzman CP (1993a) DNA-DNA hybridization approaches to species identification in small genome organisms. In: Zimmer EA, White TJ, Cann RL, Wilson AC (eds) Methods in Enzymology, Vol 224. Academic Press, New York, pp 335-348

Kurtzman CP (1993b) Systematics of the ascomycetous yeasts assessed from ribosomal RNA sequence divergence. Antonie van Leeuwenhoek 63:165-174

Kurtzman CP (2003) Phylogenetic circumscription of *Saccharomyces*, *Kluyveromyces* and other members of the Saccharomycetaceae, and the proposal of the new genera *Lachancea*, *Nakaseomyces*, *Naumovia*, *Vanderwaltozyma* and *Zygotorulaspora*. FEMS Yeast Res 4:233-245

Kurtzman CP, Fell JW (1998) The Yeasts, A Taxonomic Study, 4th edn, Elsevier Science BV, Amsterdam, pp 1055

Kurtzman CP, Robnett CJ (1994) Orders and families of ascosporgenous yeasts and yeast-like taxa compared from ribosomal RNA sequence similarities. In: Hawksworth DL (ed) Ascomycete systematics, problems and perspectives in the nineties. Plenum Press, New York, pp 249-258

Kurtzman CP, Robnett CJ (1995) Molecular relationships among hyphal ascomycetous yeasts and yeastlike taxa. Can J Bot 73 (Suppl 1):S824-S830

Kurtzman CP, Robnett CJ (1998) Identification and phylogeny of ascomycetous yeasts from analysis of nuclear large subunit (26S) ribosomal DNA partial sequences. Antonie van Leeuwenhoek 73:331-371

Kurtzman CP, Robnett CJ (2003) Phylogenetic relationships among yeasts of the 'Saccharomyces complex' determined from multigene sequence analyses. FEMS Yeast Res 3:417-432

Kurtzman CP, Sugiyama J (2001) Ascomycetous yeasts and yeastlike taxa. In: McLaughlin DJ, McLaughlin E, Lemke P (eds) The Mycota VII, Part A. Springer-Verlag, Berlin, pp 179-200

Kurtzman CP, Johnson CJ, Smiley MJ (1979) Determination of conspecificity of *Candida utilis* and *Hansenula jadinii* through DNA reassociation. Mycologia 71:844-847

Kurtzman CP, Smiley MJ, Johnson CJ, Wickerham LJ, Fuson GB (1980a) Two new and closely related heterothallic species, *Pichia amylophila* and *Pichia mississippiensis*: Characterization by hybridization and deoxyribonucleic acid reassociation. Int J Syst Bacteriol 30:208-216

Kurtzman CP, Smiley MJ, Johnson CJ (1980b) Emendation of the genus *Issatchenkia* Kudriavsev and comparison of species by deoxyribonucleic acid reassociation, mating reaction, and ascospore ultrastructure. Int J Syst Bacteriol 30:503-513

Kwast KE, Lai L-C, Menda N, James DT III, Aref S, Burke PV (2002) Genomic analyses of anaerobically induced genes in *Saccharomyces cerevisiae*: functional roles of Rox1 and other factors in mediating the anoxic response. J Bacteriol 184:250-265

Lachance MA, Daniel HM, Meyer W, Prasad GS, Gautam SP, Boundy-Mills K (2003) The D1/D2 domain of the large-subunit rDNA of the yeast species *Clavispora lusitaniae* is unusually polymorphic. FEMS Yeast Res 4:253-258

Landvik S (1996) *Neolecta,* a fruit-body producing genus of the basal ascomycetes, as shown by SSU and LSU rDNA sequences. Mycol Res 100:199-202

Liu YJ, Whelen S, Hall BD (1999) Phylogenetic relationships among ascomycetes: evidence from an RNA polymerase II subunit. Mol Biol Evol 16:1799-1808

Mannarelli BM, Kurtzman CP (1998) Rapid identification of *Candida albicans* and other human pathogenic yeasts by using short oligonucleotides in a PCR. J Clin Microbiol 36:1634-1641

Martini A, Phaff HJ (1973) The optical determination of DNA-DNA homologies in yeasts. Ann Microbiol 23:59-68

Meyer SA, Phaff HJ (1972) DNA base composition and DNA-DNA homology studies as tools in yeast systematics. In: Kochová-Kratochvílová A, Minarik E (eds) Yeasts as Models in Science and Technics. Publishing House of the Slovak Academy of Sciences, Bratislava, Czechoslovakia, pp 375-386

Møller K, Olsson L, Piškur J (2001) Ability for anaerobic growth is not sufficient for development of the petite phenotype in *Saccharomyces kluyveri.* J Bacteriol 183:2485-2489

Nagy M, Lacroute F, Thomas D (1992) Divergent evolution of pyrimidine biosynthesis between anaerobic and aerobic yeasts. Proc Natl Acad Sci USA 89:8966-8970

Naumov GI (2000) *Saccharomyces bayanus* var. *uvarum* comb. nov., a new variety established by genetic analysis. Mikrobiologiya 69:410-414

Naumov GI, James SA, Naumova ES, Louis EJ, Roberts IN (2000) Three new species in the *Saccharomyces sensu stricto* complex: *Saccharomyces cariocanus, Saccharomyces kudriavzevii* and *Saccharomyces mikatae.* Int J Syst Evol Microbiol 50:1931-1942

Nilsson-Tillgren T, Gjermansen C, Kielland-Brandt MC, Petersen JGL, Holmberg S (1981) Genetic differences between *Saccharomyces carlsbergensis* and *S. cerevisiae.* Analysis of chromosome III by single chromosome transfer. Carlsberg Res Commun 46:65-76

Nishida H, Sugiyama J (1993) Phylogenetic relationships among *Taphrina, Saitoella,* and other fungi. Mol Biol Evol 12:883-886

Nishida H, Sugiyama J (1994) Archiascomycetes: detection of a major new lineage within the Ascomycota. Mycoscience 35:361-366

O'Donnell K, Kistler HC, Tacke BK, Casper HH (2000) Gene genealogies reveal global phylogeographic structure and reproductive isolation among lineages of *Fusarium graminearum*, the fungus causing wheat scab. Proc Natl Acad Sci USA 97:7905-7910

Petersen RF, Marinoni G, Nielsen ML, Piškur J (2000) Molecular approaches for analyzing diversity and phylogeny among yeast species. In: Ernst JF, Schmidt A (eds) Dimorphism in Human Pathogenic and Apathogenic Yeasts. Karger, Basel, pp 15-35

Peterson SW, Kurtzman CP (1991) Ribosomal RNA sequence divergence among sibling species of yeasts. Syst Appl Microbiol 14:124-129

Piškur J, Langkjær RB (2004) Yeast genome sequencing: the power of comparative genomics. Mol Microbiol 53:381-389

Price CW, Fuson GB, Phaff HJ (1978) Genome comparison in yeast systematics: Delimitation of species within the genera *Schwanniomyces, Saccharomyces, Debaryomyces,* and *Pichia*. Microbiol Rev 42:161-193

Redhead SA, Malloch DW (1977) The Endomycetaceae: new concepts, new taxa. Can J Bot 55:1701-1711

Rokas A, Williams BL, King N, Carroll SB (2003) Genome-scale approaches to resolving incongruence in molecular phylogenies. Nature 425:798-804

Scorzetti G, Fell JW, Fonseca A, Statzell-Tallman A (2002) Systematics of basidiomycetous yeasts: a comparison of large subunit D1/D2 and internal transcribed spacer rDNA regions. FEMS Yeast Res 2:495-517

Sjamsuridzal W, Tajiri Y, Nishida H, Thuan TB, Kawasaki H, Hirata A, Yakota A, Sugiyama J (1997) Evolutionary relationships of members of the genera *Taphrina, Protomyces, Schizosaccharomyces*, and related taxa within the archiascomycetes: integrated analysis of genotypic and phenotypic characters. Mycoscience 38:267-280

Špírek M, Yang J, Groth C, Petersen RF, Langkjær RB, Naumova ES, Sulo P, Naumov GI, Piškur J (2003) *Saccharomyces* sensu lato chromosomes are highly dynamic. FEMS Yeast Res 3:363-373

Strathern JN, Herskowitz I (1979) Asymmetry and directionality in production of new cell types during clonal growth: the switching pattern of homothallic yeast. Cell 17:371-381

Sugiyama J (1998) Relatedness, phylogeny, and evolution of the fungi. Mycoscience 39:487-511

Swann EC, Taylor JW (1993) Higher taxa of basidiomycetes: An 18S rRNA gene perspective. Mycologia 85:923-936

Vaughan-Martini A, Kurtzman CP (1985) Deoxyribonucleic acid relatedness among species of the genus *Saccharomyces* sensu stricto. Int J Syst Bacteriol 35:508-511

Wilmotte A, Van de Peer Y, Goris A, Chapelle S, De Baere R, Nelissen B, Neefs J-M, Hennebert GL, De Wachter R (1993) Evolutionary relationships among higher fungi inferred from small ribosomal subunit RNA sequence analysis. Syst Appl Microbiol 16:436-444

Wolfe KH, Shields DC (1997) Molecular evidence for an ancient duplication of the entire yeast genome. Nature 387:708-713

Wolfe K (2004) Evolutionary genomics: Yeasts accelerate beyond BLAST. Current Biology 14:R392-R394

Kurtzman, Cletus P.
Microbial Genomics and Bioprocessing Research Unit, National Center for Agricultural Utilization Research, Agricultural Research Service, U.S. Department of Agriculture, Peoria, Illinois, USA
kurtzman@ncaur.usda.gov

Piškur, Jure
BioCentrum-DTU, Technical University of Denmark and Cell and Organism Biology, University of Lund, Sweden
jure.piskur@biocentrum.dtu.dk

Structural features of fungal genomes

Phatthanaphong Wanchanthuek, Peter F. Hallin, Rodrigo Gouveia-Oliveira, and David Ussery

Abstract

Eighteen fungal genomes have been sequenced to date from a variety of taxonomic groups, with fifteen *Ascomycota*, two *Basidomycota* and one *Microsporidia* species represented. The genomes vary in size more than tenfold, from approximately 2.5 Mbp to 38.8 Mbp. We have performed a computational analysis of DNA structural features of all 18 fungal genomes. The sequenced genomes can be visualised with Genome Atlases, which are graphical representations of the chromosomes, showing DNA structural properties (including the location of potentially highly expressed genes), DNA repeats, and DNA base-composition properties, such as AT-content and GC-skew. A comparison of DNA structural features in the various fungal genomes shows an over-representation of purine stretches of >10 bp in length; that is, there is a tendency for stretches containing only A's or G's on the same strand of the DNA helix. This strand bias is pronounced in all of the fungal chromosomes examined. The purine and pyrimidine/purine stretches are localized mainly within non-coding regions of the chromosomes. Another common structural feature for all of the fungal genomes is that the upstream promoter regions of genes are more AT rich than downstream coding regions. Codon usage patterns are different in the three phyla of fungi examined, as well as amino acid usage. Finally, a protein comparison of the predicted gene products gives an overview of the similarity based on protein homology and localization.

1 Introduction

This review will focus on comparison of structural features of the 18 fungal genomes that have been sequenced and are public as of the time of writing. The organisms are listed in Table 1, sorted alphabetically by genus name; a table of features can be found in our web pages[1].

In 1997, the sequence of the budding yeast *Saccharomyces cerevisiae* was published, making it the first eukaryote to have its complete genome sequenced and annotated (Goffeau et al. 1997). The *S. cerevisiae* genome has since been used as a model of eukaryote gene expression and function. Progress on sequencing other fungal genomes has been slow, even considering the size difference, compared to

---

[1] see website: http://www.cbs.dtu.dk/services/GenomeAtlas/suppl/Fung_rev

Topics in Current Genetics, Vol. 15
P. Sunnerhagen, J. Piškur (Eds.): Comparative Genomics
DOI 10.1007/4735_112 / Published online: 11 November 2005
© Springer-Verlag Berlin Heidelberg 2005

Table 1. Summary of the 18 fungal genomes discussed in this review

#	Phyla	Class	Organisms	Strain	Length (bp.)	% AT	tRNA	rRNA	ORF	Chr	Contig	Status	References
1	As	Sac	*Ashbya gossypii*	ATCC10895	8,741,431	48.2	194	4	4,718	7	-	completed	Dietrich et al. (2004)
2	As	Pez	*Aspergillus nidulans*	FGSCA4	30,095,918	49.6	187	12	9,342	8	248	incomplete	FGI (2003)[2]
3	As	Sac	*Candida albicans*	SC5314	14,851,000	66.5	166	36	6,419	8	1,213	incomplete	Odds et al. (2004)
4	Mi	-	*Encephalitozoon cuniculi*	GB-M1	2,497,519	52.7	44	25	1,996	11	-	completed	Katinka et al. (2001)
5	As	Pez	*Gibberella zeae*	PH-1	35,689,396	55.5	29	18	11,379	?	511	incomplete	FGI (2003)[2]
6	As	Sac	*Kluyveromyces waltii*	NCYC2644	10,613,225	55.6	240	2	5,230	8	457	incomplete	Kellis et al. (2004)
7	As	Pez	*Magnaporthe grisea*	70-15	38,760,322	47.7	168	38	10,856	7	2,273	incomplete	FGI (2003)[2]
8	As	Pez	*Neurospora crassa*	OR74A	38,044,345	50.0	413	74	9,897	7	821	incomplete	Katherine et al. (2004)
9	Ba	-	*Phanerochaete chrysosporium*	RP78	29,965,857	48.4	200	-	11,777	10	767	incomplete	Martinez et al. (2004) and JGI (2004)[1]
10	As	Sac	*Saccharomyces bayanus*	MCYC-623	11,931,212	59.7	275	6	4,970	16	678	incomplete	Kellis et al. (2003)
11	As	Sac	*Saccharomyces castellii*	NRRLY-126630	11,354,153	63.0	276	16	4,681	~9	721	incomplete	Cliften et al. (2003) and Petersen et al. 1999
12	As	Sac	*Saccharomyces cerevisiae*	S288C	12,457,860	61.6	274	25	5,844	16	-	completed	Goffeau et al. (1997)
13	As	Sac	*Saccharomyces kluyveri*	NRRLY-12651	11,028,823	58.3	263	6	2,975	8	2,446	incomplete	Cliften et al. (2003) and Petersen et al. 1999
14	As	Sac	*Saccharomyces kudriavzevii*	IFO1802	11,178,416	60.1	264	9	3,788	16	2,074	incomplete	Cliften et al. (2003)
15	As	Sac	*Saccharomyces mikatae*	IFO1818	10,825,581	61.8	251	8	4,525	16	2,902	incomplete	Kellis et al. (2003)
16	As	Schi	*Schizosaccharomyces pombe*	972	12,515,113	63.9	168	28	5,043	3	-	completed	Wood et al. (2002)
17	As	Pez	*Trichoderma ressei*	ATCC56765	34,485,773	46.1	210	-	8,620	-	1,094	incomplete	JGI (2004)[1]
18	Ba	-	*Ustilago maydis*	521	19,459,793	58.9	5	23	6,370	16	274	incomplete	FGI (2003)[2]

[1] FGI (Fungal Genome Initiative website: http://www.broad.mit.edu/annotation/)

[2] JGI (DOE Joint Genome Institute website: http://genome.jgi-psf.org/whiterot1/whiterot1.home.html)

Phyla: As = Ascomycota, Ba = Basidomycota, and Mi = Microsporidia

Class: Sac = *Saccharomycotina*, Schi = *Schizosaccharomycetes*, and Pez = *Pezizomycotina*

Table 1 footnote (overleaf). Summary of the 18 Fungal genomes publicly available at the time of writing (spring 2004). To find the number of tRNA and rRNA operons shown in the table, the tRNAs were determined using tRNAscan-SE in the "relaxed" mode algorithm (Lowe and Eddy 1997) whilst rRNAs were extracted from each of fungal genomes by BlastN comparison of the complete genome of *S. cerevisiae* with the 18s rRNA (accession number V01335). These blast results were converted to databases and pair wise comparisons were performed by using BlastN with the low complexity filter enabled. In addition, tRNAs and rRNA operons for *S. cerevisiae, Sc. pombe, A. gossypii, N. crassa*, and *E. cuniculi* were obtained from the literature (Kellis et al. 2003; Dietrich et al. 2004; Katherine et al. 2004; Wood et al. 2002 and Katinka et al. 2001). For two fungal genomes (*Phanerochaete chrysosporium* and *Trichoderma ressei*), no rRNAs were found which is likely due to their being excluded from assembly due to repeat families.

the bacterial genomes, which are now published almost weekly. The sequence of *Schizosaccharomyces pombe* was the second yeast genome sequenced (Wood et al. 2002) and the first filamentous fungus, *Neurospora crassa*, followed two years later (Katherine et al. 2004), although this sequence currently still contains many gaps. In addition, complete genomes have been sequenced of *Ashbya gossypii* (*Eremothecium gossypii*) (Dietrich et al. 2004) and *Kluyveromyces waltii* (Kellis et al. 2004), which are closely related to *S. cerevisiae*. Currently we count 18 fungal genomes publicly available, which include sequences from the Whitehead Institute sequencing efforts, the Fungal Genome Initiative, the sequence data released of *Candida albicans* from Stanford Genome Technology Center, and the *Sc. pombe* sequence from Sanger Center. In addition, there are 13 Hemiascomycetes genomes being sequenced as a project carried out at the Bordeaux Bioinformatics Center[2] (Souciet et al. 2000; Malpertuy et al. 2003). Dujon et al. (2004) have recently published the sequences and analysis of a diverse set of an additional four yeast genomes (*Candida glabrata, Kluyveromyces lactis, Debaryomyces hansenii*, and *Yarrowia lipolytica*). Since these genomes were not available at the time of writing, they are not included here, although more information about their genomes can be found in our supplemental web pages.

The substantial information content of any genome clearly requires the application of bioinformatic tools to handle the large amount of data. In this study, we systematically analyze 18 sequenced fungal genomes. First, DNA atlases are constructed to visualize DNA structural properties along the individual chromosomes. This is followed by an examination of other structural properties, such as the occurrence of purine or purine/pyrimidine stretches in the chromosomes. Structural profiles around promoter regions are also examined. Finally, these results are compared with the proteome similarities.

2 Overview of the fungal genomes

Fungi have been used in many important industrial processes, as well as in laboratories where they serve as models to study eukaryotic cells. For example, *Ascomy-*

---

[2] http://cbi.labri.ubordeaux.fr/Genolevures/-index.php

cota, such as *S. cerevisiae, Sc. pombe,* and *C. albicans* have been used in the first global studies of eukaryotic gene expression and gene function. In addition, comparative genomics of fungal genomes can shed light on the evolutionary history of the origins of many important biological processes found in more complex multicellular eukaryotes, and their experimental availability easily make fungi one of the most useful model systems in cell biology.

At the time of writing, 24 fungal genome projects are at various states of completion, and many of the completed sequences can be downloaded from the National Center for Biotechnology Information[3] (NCBI) or from web pages of the Genome Sequencing Center (see Table 1). Based on the taxonomy of fungi, the division of *Eumycota* are divided into three phyla: *Ascomycota, Basidomycota* (*Basidiomycetes*), and *Microsporidia.* The *Ascomycota* includes the majority of the sequenced genomes, with subclasses *Pezizomycotina, Saccharomycotina,* and *Schizosaccharomycetes.* Biological descriptions and a brief overview of these fungal genomes are briefly given below.

2.1 Ascomycota

2.1.1 Candida albicans

C. albicans belongs to the *Saccharomycotina* subphylum and it is the most common human fungal pathogen, although, it is generally a benign commensal that resides on the mucosal surfaces of most if not all animals. *C. albicans* is capable of causing superficial infections in their hosts, and severe systemic infections in immunocompromised hosts. *C. albicans* is diploid, and recent discoveries about *C. albicans* could lead to a deeper understanding of the sexual cycle and its role in virulence (Jones et al. 2004). The genome size of *C. albicans* is approximately 18 Mb, which is about a third larger that *S. cerevisiae,* and is distributed in eight pairs of chromosomes which are numbered from 1 (largest) to 7 (smallest), with an additional chromosome carrying the rDNA called R (Wickes et al. 1991). The preliminary sequence of the *C. albicans* genome is available[4] (Tzung et al. 2001) and is further described by Jones et al. (2004).

2.1.2 Saccharomyces sp.

The genome size of *Saccharomyces* yeast species is around 10 to 14 Mb, and it is organized in 8-16 chromosomes (Cliften et al. 2003; Kellis et al. 2003; Wolfe 2004). For example *S. cerevisiae, S. castellii* and *S. kluyveri* contain 16, 9, and 8 chromosomes, respectively. The genus *Saccharomyces* species consists of several species divided into the "*sensu stricto*" and the "*sensu lato*" groups. The *sensu stricto* group (strict sense) includes *S. cerevisiae, S. bayanus, S. kudriavzevii, S. mikatae,* and *S. paradoxus* and the *senso lato* group (broad sense) includes organ-

---

[3] http://www.ncbi.nlm.nih.gov/genomes/FUNGI/funtab.html
[4] http://www-sequence.stanford.edu/group/candida/

isms that are more divergent from S. cerevisiae such as *S. castellii*, *S. exiguous*, and *S. servazzii* (Piskur et al. 1998; Kurtzman and Fell 1998; Petersen et al. 1999), and the petite-negative group that is even more distant and includes *S. kluyveri* (Naumov 1996; Barnett 1992; Naumov et al. 2000; Fischer et al. 2000; Naumova et al. 2003). The genomes of these species differ in the number and organization of nuclear chromosomes and in the size and organization of mitochondrial DNA (mtDNA) (Marinoni et al. 1999).

The *senso stricto* species are so closely related that their genome organization is almost identical; only a few chromosome rearrangements have occurred in these species making their chromosome almost completely syntenic with their *S. cerevisiae* counterparts (Llorente et al. 2000). For decades, *S. cerevisiae* has been one of the best-characterized organisms in terms of genetics and physiology. The genome sequence project was started in 1992 by a consortium of yeast researchers. In 1996, it became the first completely sequenced eukaryote genome (Goffeau et al. 1996). The data are available through several public sequence databases; for example, through the Saccharomyces Genome Database (SGD) maintained at Stanford[5] (Kellis et al. 2003) and also the *Saccharomyces* Genome Sequencing at the Genome Sequencing Center maintained at Washington University Medical School[6] (Cliften et al. 2003).

2.1.3 Ashbya gossypii

The hemiascomycete *A. gossypii* was first described in 1929 as a cotton pathogen transmitted by sucking insects and has considerable commercial value as a major source for riboflavin and phytopathogenic filamentous fungi (Prillinger et al. 1997). The sequence of *A. gossypii* was completed by Dietrich et al. (2004) and its genome size is around 9 Mb, organized in seven chromosomes and these sequences can be downloaded from *Ashbya* Genome Database (AGD) maintained at the University of Basel[7]. *A. gossypii* shows filamentous growth with multinucleated and extensively branching hyphae. The fungus is a very promising experimental system because it has a small genome and haploid nuclei (Ayad-Durieux et al. 2000; Dietrich et al. 2004). Moreover, efficient gene targeting, propagation of plasmids, and growth on defined liquid and solid media are possible.

2.1.4 Kluyveromyces waltii

K. waltii has been proposed to be renamed to *Lachancea waltii* (Kurtzman et al. 2003). *K. waltii* is closely related to *Lachancea* (*Saccharomyces*) *kluyveri*. The assembled genome of *K. waltii* is approximately 10.7 Mb, organized in eight chromosomes. *K. waltii* is apparently more closely related to *S. cerevisiae* than *K. lac-*

---

[5] http://www.yeastgenome.org/
[6] http://genome.wustl.edu/projects/yeast/
[7] http://agd.unibas.ch/

tis (Kurtzman and Robnett 2003) but the two lineages diverged before the duplication event took place in the *S. cerevisiae* progenitor (Kellis et al. 2004).

2.1.5 Schizosaccharomyces pombe

In 1893, Lindner was the first author to describe fission yeast *Sc. pombe*. It is a single-celled free-living archiascomycete fungus sharing many features with cells of more complicated eukaryotes. It is distant in evolution from *Hemiascomycetes* (Sipiczki 1991; Vaughan 1991; Sunnerhagen 2002). This yeast diverged from budding yeast around 330-420 million years ago (Sipiczki 2000). The genome of *Sc. pombe* was initiated in the mid-1990s and the completed sequence has been available since 2001; it has a genome of approximately 13 Mb organized on three chromosomes, containing an estimated 4,940 ORFs (Wood et al. 2002). Its genome was the sixth eukaryotic genome to be sequenced[8]. It has attracted interest for cell biologists because its cell division is more typical of most eukaryotes and is distinct from that of the budding yeast (Herrero et al. 2003).

2.1.6 Aspergillus nidulans

Aspergillus nidulans (also known as *Emericella nidulans*) is an *Ascomycete* and is commonly used in genetics and cell biology. It is important because it is closely related to a large number of other *Aspergillus* species of industrial and medical significance, for example, *A. niger*, *A. oryzae*, *A. flavus*, and *A. fumigatus*. It grows rapidly as a filamentous fungus on solid or in liquid media under a variety of nutritional conditions. It is normally haploid, but can also be induced to grow as a heterokaryon or a vegetative diploid. It produces both asexual spores (conidia) and sexual spores (ascospores). It undergoes development to produce at least nine different cell types. *A. nidulans* has been used extensively to study the production and secretion of proteins (Sharma et al. 2001). The *A. nidulans* genome is approximately more than twice as large as the yeast genomes described above, with an estimated size of about 30 Mb. It has eight chromosomes containing an estimated 11,000–12,000 genes (Fungal Genome Initiative 2003).

2.1.7 Magnaporthe grisea

The genome of *M. grisea* is estimated to be approximately 40 Mb, with seven chromosomes (Fungal Genome Initiative 2003; Skinner et al. 1993). *M. grisea* is a haploid filamentous *Ascomycete* and causes serious disease on a wide variety of grasses including rice, wheat, and barley. *M. grisea*, like many foliar pathogens, is well adapted to attack and penetrate its host. All aerial parts of the plant are subject to invasion, but economic losses are most devastating when the panicle or node at the base of the panicle is infected and killed, resulting in loss of grain set (Bonman and MacKill 1988).

---

[8] http://www.sanger.ac.uk/Projects/S_pombe

2.1.8 Neurospora crassa

The *N. crassa* genome is approximately 38 Mb, organized in seven chromosomes, and from this around 1,100 genes (~10% of the expected total) have been annotated so far (Schulte et al. 2002; Fungal Genome Initiative 2003; Galagan et al. 2003). It is the first filamentous fungi to be sequenced (Schulte et al. 2002). One of the attractive features of *Neurospora* as a model organism is its complex yet genetically and biochemical tractable life cycle (Katherine et al. 2004). *N. crassa* was used as the model organism to test the idea that one gene equals one enzyme (Hudspeth 1992). The multicellular *Neurospora* possesses a large number of genes without homologues in *S. cerevisiae*, suggesting that *Neurospora* might be an alternative model for multi-cellular eukaryotes in many aspects of cell biology (Galagan et al. 2003). Among the unshared genomic equipment is an expanded group of sugar transporters, transcription factors, and environmental sensing pathways, plus diversified metabolic machinery. *Neurospora* also displays a number of gene-silencing mechanisms acting in the sexual or the vegetative phase of the life cycle (Selker 2002).

2.1.9 Trichoderma ressei

The *T. ressei* genome is approximately 34.5 Mb. It is a free-living fungus and is common in soil and root ecosystems. Recent discoveries show that *T. ressei* are opportunistic, avirulent plant symbionts, as well as being parasitic on other fungi. It produces a variety of compounds that can induce localized or systemic resistance responses, and this explains the lack of pathogenicity to plants (Harman et al. 2004). These root-microorganism associations cause substantial changes to the plant proteome and metabolism. Plants are protected from numerous classes of plant pathogens by responses that are similar to systemic acquired resistance and rhizobacteria-induced systemic resistance. Root colonization by *Trichoderma* sp. also frequently enhances root growth and development, crop productivity, resistance to abiotic stresses, and the uptake and use of nutrients.

2.1.10 Gibberella zeae (Fusarium graminearum)

The genome of *G. zeae* is approximately 40 Mb, and has 11,739 annotated genes (Fungal Genome Initiative 2003). *G. zeae* is a filamentous *Ascomycete*, which grows well on defined medium in pure culture as a haploid mycelial thallus. The organism can complete a sexual cycle in four-six weeks and abundant asexual sporulation occurs in less than one week. Although it is homothallic, out-crossing of strains is possible and can be promoted by manipulation of the mating type locus. A reproducible pathogenicity assay is adapted for both plants and animals, allowing the genetics of pathogenicity to be studied (O'Donnell et al. 2000).

2.2 Basidiomycota

2.2.1 Ustilago maydis

The genome of *U. maydis* is approximately 20 Mb, contained in 16 chromosomes and can exist as both haploid and diploid forms (Fungal Genome Initiative 2003). *U. maydis* is a basidiomycete fungal pathogen of maize and teosinte. The fungus induces tumours on host plants and forms masses of diploid teliospores. These spores germinate and form haploid meiotic products that can be propagated in culture as yeast-like cells. Haploid strains of opposite mating type fuse and form a filamentous, dikaryotic cell type that invades plant tissue to reinitiate infection (Kurtman and Robnett 2003).

2.2.2 Phanerochaete chrysosporium

The *P. chrysosporium* ("white rot" fungi) has been sequenced, and is about 30 Mbp, organized on ten chromosomes (Martinez et al. 2004). It is the first basidiomycete genome to be sequenced. The *P. chrysosporium* does not produce mushrooms for reproduction; instead, it forms effuse, very flat fruiting bodies that appear as a crust on wood. *P. chrysosporium* is the most intensively studied white-rot fungus. White-rot fungi degrade lignin, the polymer that surrounds and protects the cellulose microfibrils of plant cells. As part of their lignin-degrading enzyme system, these fungi produce unique peroxidases and oxidases that are also capable of degrading compounds related to lignin that are present in toxic wastes and pesticides (Martinez et al. 2004). These enzymes have great potential for environmental and biotechnological applications. Unlike some white-rot fungi, *P. chrysosporium* leaves the white cellulose of wood nearly untouched (Blanchette 1991; Eriksson et al. 1990). It also has a very high optimal growth temperature, which allows it to grow on composting wood. These characteristics suggest several roles for *P. chrysosporium* in biotechnology, which has attracted considerable interest for application in bioprocesses such as organopollutant degradation and fibre bleaching (Kirk et al. 1987; Kersten et al. 1987).

2.3 Microsporidia

2.3.1 Encephalitozoon cuniculi

E. cuniculi is an obligate intracellular parasite infesting a wide range of eukaryotes ranging from protozoans to humans. Lacking mitochondria and peroxisomes, these unicellular eukaryotes were first considered a deeply branching protist lineage (Vossbrinck et al. 1987) that diverged before the endosymbiotic event that led to mitochondria. The discovery of a gene for a mitochondrial type chaperone

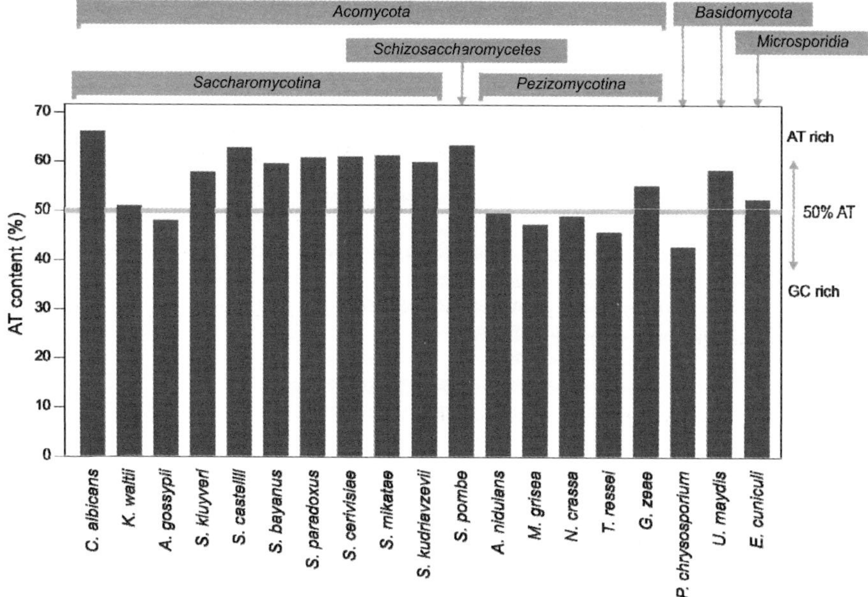

Fig. 1. AT-content for 18 different fungal genomes. Note that the AT content shown in the figure is the average for the genomic DNA sequences available, and are not reflective of the mtDNA. The genomes are grouped into three phyla and further subgrouped according to taxonomy.

(Germot et al. 1997; Hirt et al. 1997) combined with molecular phylogenetic analysis (Baldauf et al. 2000; Keeling et al. 2000) later implied that microsporidia are atypical fungi that lost mitochondria during evolution. Instead, they recruit the mitochondria of their host for energy supply. The *E. cuniculi* genome is the smallest fungal genome sequenced to date, with a size of about 2.5 Mb, organized on 11 tiny chromosomes with sizes ranging from 217-315 kb (Katinka et al. 2001).

3 A global view of fungal genomes

Simple average values for each genome, such as AT-content and genome size can be used as a first comparison of the sequenced fungal genomes. Typical representatives of the three phyla are given in Figure 1. The 48-65% AT-content in *Sachharomycotina* is higher than for *Pezizomycotina.* The two *Basidomycota* genomes have a large variation with 48 and 59% AT-content, respectively. Table 1 summarizes further details the summary of genome properties for each of the different 18 fungal genomes. A web version of these fungal genomes is also available on the supplemental web page and the fungal sequence can also be downloaded[9].

---

[9] http://www.cbs.dtu.dk/services/GenomeAtlas/suppl/Fung_rev

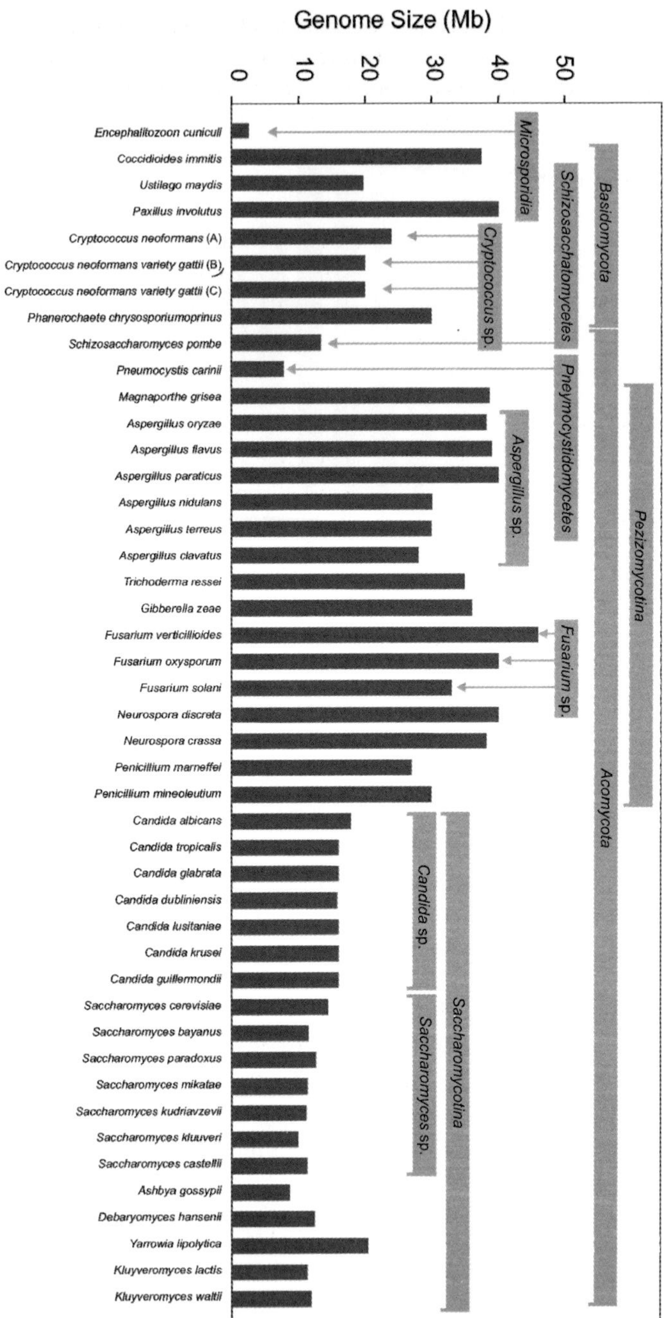

Fig. 2. Fungal genome size variation, including phyla of Ascomycota, Basidomycota, and Microsporidia.

We have compared the genome length for 46 fungal genomes (including incompletely sequenced species) as shown in Figure 2. Again, the three fungal phyla, *Microsporidia, Basidiomycota*, and *Ascomycota*, are represented. First, *Microsporidia* are presently only represented by one genome (*E. cuniculi*) and it is also the smallest fungal genome, with a size around 2.5 Mb, and containing 1,996 genes. The genome of *P. involutus* is currently the largest of *Basidiomycota* with a size approximately 40 Mb, contained 11,777 ORFs genes whereas *U. maydis* has the smallest genome size. *Saccharomycotina* have the smallest genomes of the *Acomycota*, with *Yarrowia lipolytica* having an exceptionally large genome within this group (approximately 20 Mb) whilst *A. gossypii* has the smallest genome size (8.8 Mb), containing 4,718 genes. *A. nidulans, M. grisea, N. crassa, T. ressei*, and *G. zeae* belong to *Pezizomycotina*, which all have relatively large genomes, around 25-45 Mb. *Fusarium verticilloides* is currently the largest of *Pezizomycotina*, with a size approximately 48 Mb, containing at least 10,856 genes.

The large variation in fungal genome sizes, which can be seen in Figure 2, may result from either gene loss, when a species specializes to live in a highly specific ecological niche, or genes are duplicated or acquired via horizontal gene transfer (Kellis et al. 2004; Piskur and Langkjaer 2004). Gene duplication and the presence of repeat elements are undoubtedly responsible for an increase in genome size (Wolfe 2004; Dujon et al. 2004).

4 Genome Atlases for visualization of DNA structural properties in chromosomes

Genome Atlas plots for individual fungal chromosomes were constructed using Genewiz software. The underlying data were obtained by in-house developed software. Genewiz can be applied to visualize any complete microbial chromosome, and is available upon request (Pedersen et al. 2000). To generate Genome Atlas plots, a number of parameters are calculated for the DNA double helix based on the nucleotide sequence. The parameters can be divided into three categories: structural parameters, DNA repeats, and base composition properties. The values for these parameters are visualized along the chromosome by the intensity of the colour (Jensen et al. 1999). Note that the Genome Atlas for all publicly available sequenced chromosomes can be downloaded from our web page[10] (Hallin and Ussery 2004). A Genome Atlas for *S.cerevisiae* chromosome XII is shown in Figure 3. Briefly, several properties of the DNA are displayed. First, Lane A shows the curvature of the double helix based on a 21 bp window; blue regions are significantly more curved than average for the chromosome and yellow regions are significantly less curved. Lane B describes how easily the DNA melts, based on a dinucleotide model of the stacking energy; red regions are predicted to melt more easily than the chromosomal average, and green regions require more energy than

---

[10] http://www.cbs.dtu.dk/services/GenomeAtlas/

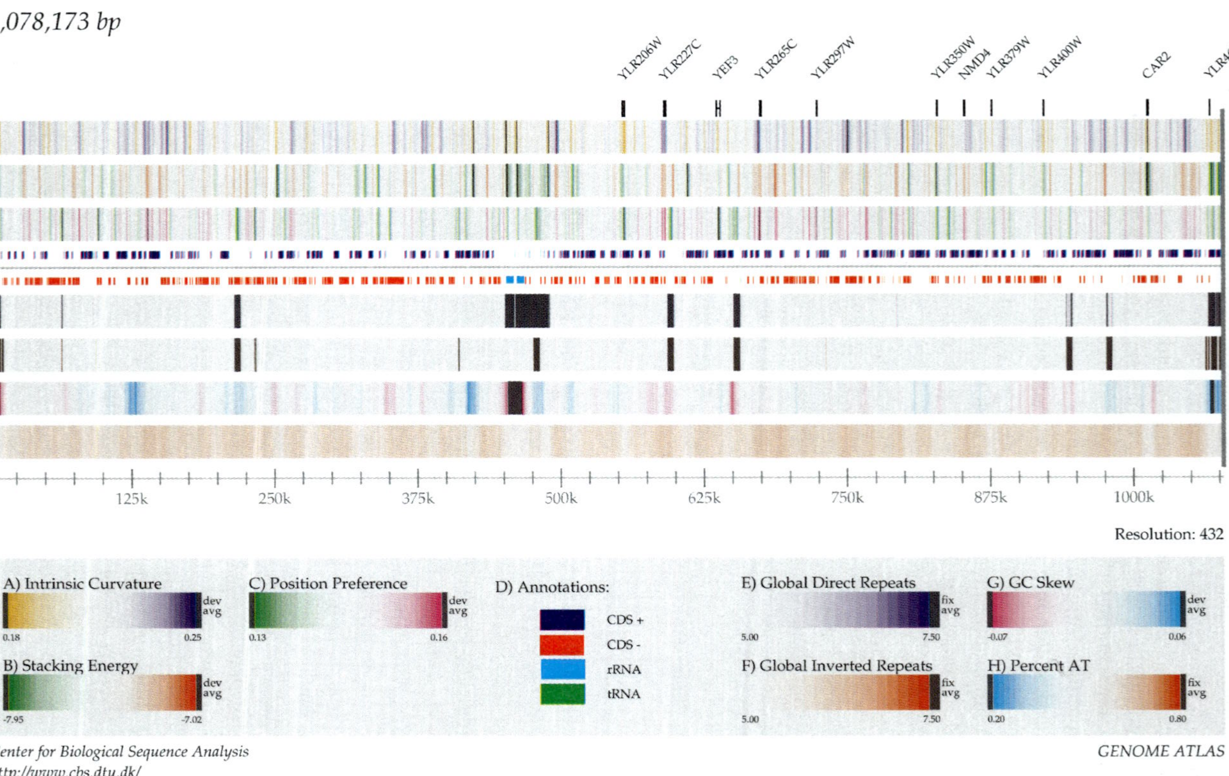

Fig. 3. The Genome Atlas for *S. cerevisiae* chromosome XII.

average to melt. Lane C represents "position preference", which is a measure of chromatin accessibility, with low values (green) corresponding to regions which would tend to exclude nucleosomes and hence have a more open chromatin structure, and magenta applying to the opposite situation. The green regions often correspond with more highly expressed genes (Pedersen et al. 2000). Lane D graphically represents ORFs, with blue and red blocks representing genes encoded on the positive and negative strand, respectively. Light blue and green are reserved for rRNA and tRNA genes. Lanes E and F display the location of repeated areas within this chromosome; direct repeats are present on the same strand and inverted repeats on opposite strands (see Section 5). GC-skew represents the bias of G's towards one strand of the DNA helix. Finally, a running average of the AT-content is shown as the bottom line on the plot; red regions are more AT rich, and turquoise regions are more GC rich (Skovgaard et al. 2002). Note that the overall pinkish colour in the bottom lane indicates that this chromosome is slightly AT rich with a fairly even distribution.

The Genome Atlas of *S. cerevisiae* chromosome XII is treated in more detail below. In addition, Genome Atlases of *Sc. pombe* chromosome I, *A. gossypii* chromosome VII and *E. cuniculi* chromosome I were produced as these chromosomes contain rRNAs operons, which contain 10, 14, and 7 genes, respectively, and these data can be obtained form our Genome Atlas database[11].

4.1 The *S. cerevisiae* chromosome XII Genome Atlas

The Genome Atlas for *S. cerevisiae* chromosome XII shown in Figure 3 contains 543 open reading frames (ORFs), with a 61.5 % AT-content. The total length is 1,078,173 bp and it contains 22 tRNAs of which seven are predicted to contain introns (Hani and Feldmann 1998; Johnston et al. 1997). One common feature is that tRNA genes are often grouped together and these clusters tend to be dispersed throughout the genome and in some cases are involved in processing transcripts of flanking genes. For example, the tRNA genes in *Sc. pombe* are organized in clusters (Hani and Feldmann 1998). Many tRNAs are near yeast retrotransposons elements (Ty elements), probably as a result of preferred integration sites by these elements. Four of the six retrotransposons on this chromosome are of the Ty1 type and the other two are Ty2 elements. There are about 25 rRNA operons, clustered in one location (around 455-490 kb) near to the middle of the chromosome (Lane D of Fig. 3). The boundaries of the rDNA repeats are found in non-transcribed regions downstream of the 35s rDNA (the left, or centromere-proximal, boundary) and 5s rDNA (the right, centromere-distal, boundary) (Johnston et al. 1997).

In Figure 3, the location of the rRNA operons stands out by having strong stacking interaction (green area in Lane B) and low intrinsic curvature (yellow area in Lane A) whereas position preference is low (Lane G). It has been hypothesized that genes with a low position preference have a potential to be highly ex-

---

[11] http://www.cbs.dtu.dk/services/Genome-Atlas

pressed (Pedersen et al. 2000). This region also contains *ASP3-1* and *YLR154W*-F. A highly negative stacking energy is indicative of more thermodynamically stable (usually GC-rich) region and likely has a lower mutation rate (Jensen et al. 1999). The expected high expression potential and low mutation frequency are congruent with our predictions based on DNA structural properties.

The chromosome contains seven relatively large global direct and inverted repeats (Lanes E and F) located around the centromere, Ty element, telomere, and rRNA genes clusters. There is one region in the chromosome with a high GC-skew (Lane G), at approximately 473 kb, around the rDNA repeats. There does not seem to be any regularity in the variation in nucleotide composition, as may be the case for other yeast chromosomes (Johnston et al. 1997).

The regions with potentially highly expressed genes are indicated by low position preference (*e.g.* green in Lane C), and include *ENT2, ENT4, YPS3, ASP3-1, CAR2, ORM2, YRF1-4, YLR154W*-F, and *ICT1*. *ENT2* and *ENT4* respond to clathrin binding, whereas *YPS3, ASP3-1* and *CAR2* belong to genes involved in enzymatic functions.

5 Comparison of significance of DNA structure in fungal genomes

5.1 DNA Repeats

Repeats are multiple copies of the same sequence at a different location on DNA sequence. The frequent occurrence of local repeats within a chromosome may be indicative of an elevated level of recombination. The fraction of repeat sequences is a simple and important measure of the genome's properties (Warren 1996 and Primrose 1998). The three main types of repeats we calculate are simple oligonucleotide repeats, local repeats (within a 100 bp window), and global repeats (100 bp searched against the entire length of the chromosome). Local and global repeats can be further divided into direct (same strand) and inverted (opposite strands) repeats (Jensen et al. 1999). Certain repeats can promote mutation as they have a tendency to be amplified during replication in the region where they occur. Inverted repeats are lying on opposite strands of the DNA and run in opposite directions. These repeats can form cruciform structures.

Global direct and global inverted repeats are calculated with the same basic algorithm as described by Jensen et al. (1999). The fraction of repeat sequences for four fungal genomes is shown in Figure 4. The average level of global repeats is around 2%. Global repeats (Fig. 4A) are less frequent in *A. gossypii* and *E. cuniculi* chromosomes compared to the other fungal chromosomes. Variation in repeat fraction between chromosomes of one species can be substantial, as with *S. cerevisiae*. In contrast to the low level of repeats at the global level, local repeats (4B) are found in larger numbers in *E. cuniculi*, with approximately 6% of the genome, containing local repeats. These relatively high fractions of repeats in fungal genomes suggest that they are important for genome plasticity, since they can be

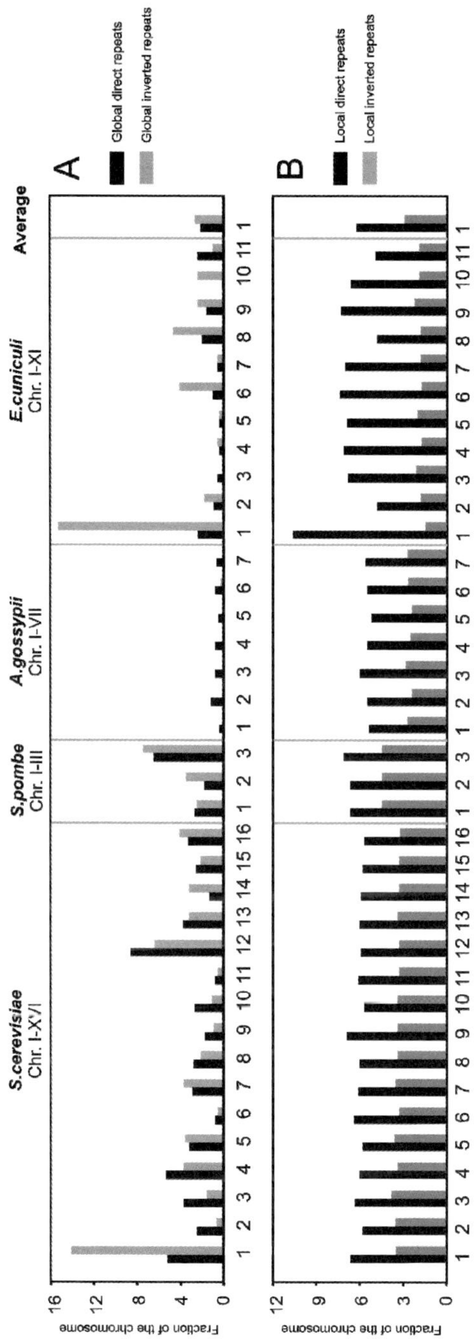

Fig. 4. Repeat sequences in fungal genomes. Individual chromosomes are represented. (A): Global repeats (match >80%), (B): Local repeats (match > 80%).

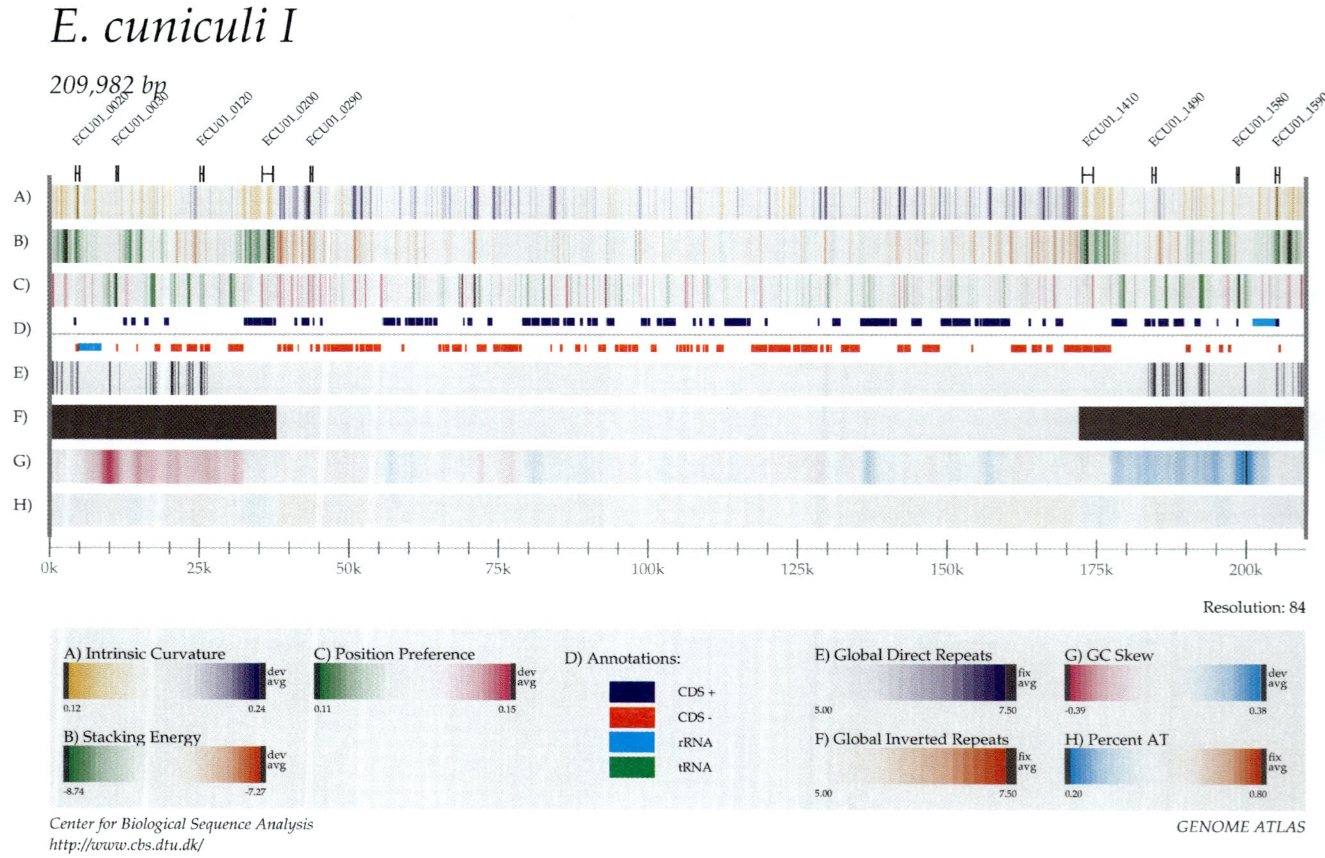

Fig. 5. Genome Atlas for *E. cuniculi* chromosome I. Note the high level for global inverted repeats.

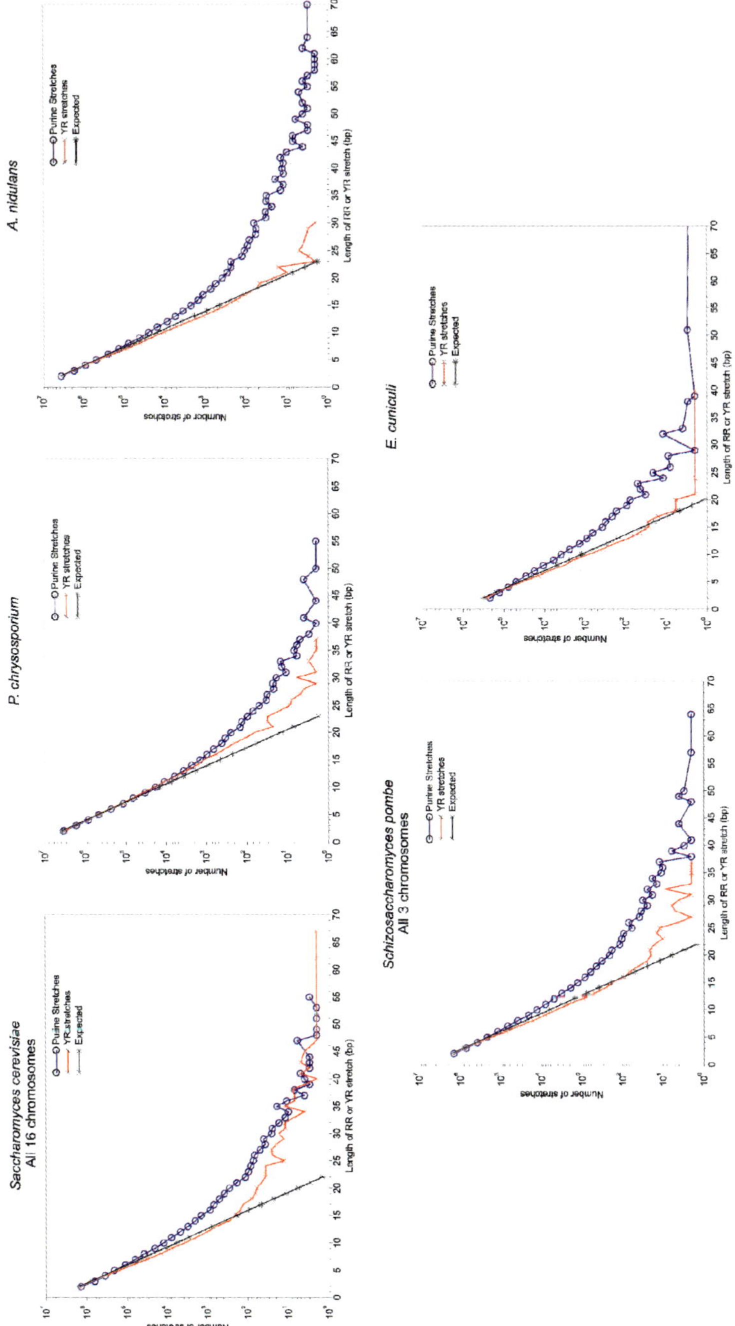

Fig. 6. Observed versus expected frequencies of purine and purine/pyrimidine stretches (YR) in fungal genomes.

involved in several processes, such as recombination, inversion, deletion, translocation, and transposition. A Genome Atlas plot for *E. cuniculi* chromosome 1 is shown in Figure 5. Note the large inverted repeat (dark red in Lane E) that covers the first 40 kbp of the beginning and end of the chromosome. Because this chromosome is small, it represents a large fraction of the total length of the chromosome.

5.2 Bias in purine stretches

Bias of long runs of purines on the same strand and stretches of alternating purines/pyrimidines have been observed in eukaryotic genomes. To investigate this in fungal genomes, we made plots to visualize the strand bias purine stretches and purine/pyrimidine (YR) regions throughout the fungal chromosome (Fig. 6). The expected purine and alternating YR stretches are calculated as described previously (Ussery et al. 2002). The observed and expected frequencies of purine and purine/pyrimidine stretches of five fungal genomes are shown in Figure 6. In all cases, the observed purine stretches deviate markedly from the expectation, whereas for YR stretches, the deviation between expected and observed is generally minor, and greatest for *S. cerevisi*ae (Fig. 5). Many purine stretches prefer an A-DNA helix, which is also the conformation found in DNA/RNA helices; purine stretches are often found at the long terminal repeats (LTRs) of insertion sequences. Some of YR sequences can be less stable, and associated with regions of the chromosome which will melt more readily (*e.g.* TATA) or regions likely to be involved in chromosomal breakage, such as $(CG)_n$ runs.

6 Analyses of promoter area and coding regions

6.1 Identifying putative promoter areas

Promoter areas of fungal chromosomes were predicted based on location and DNA structural properties. All the genes in each fungal genome were aligned at the translation start site, and a window of +/- 400 base pairs was chosen for which the average AT-content and plotted as a Z-score, as shown in Figure 7. In addition, within the chosen window, plots of several DNA structural parameters are made (intrinsic curvature, stacking energy, DnaseI sensitivity, and position preference).

We have chosen four fungal genomes, which represent different AT-contents: *S. cerevisiae* (61.6 %), *Sc. pombe* (63.9 %), *A. gossypii* (48.2 %), and *E. cuniculi* (52.7 %). As can be seen in Figure 7, in *E. cuniculi* intrinsic curvature and stacking energy increase in a narrow window from -100 to -20 relative to translation start. Within this window AT-content decreases as the position preference increases. Predicted DnaseI sensitivity is conversely related to these features. Combining these findings identifies the most likely position for promoters. In contrast, the other fungal chromosomes have a higher AT-content, resulting in a decrease of

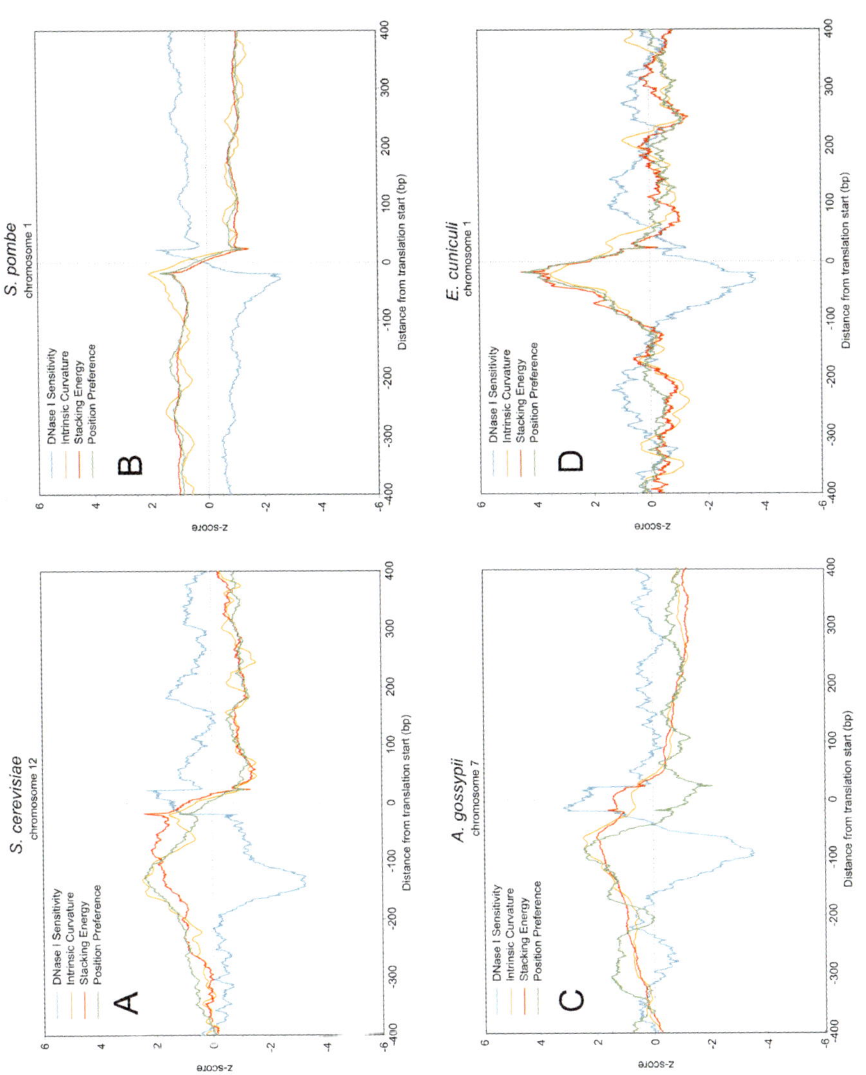

Fig. 7. Structural properties profile in putative promoter regions for individual fungal chromosomes. The profiles were calculated from upstream sequences aligned at the transcription start. The plots have been smoothed using a running average with window size 31 bp and normalized based on the genomic average and standard deviation.

the position preference, intrinsic curvature and stacking energy. These chromosomes all display a significant drop in DNase sensitivity (roughly 2-3 Z-scores), upstream of translation in a 100bp window. The stacking energy is generally higher upstream of the translation start but the window of promoter presence is less well defined than it is for *E. cuniculi*. For *Sc. pombe*, the increase in stacking energy seems to extend beyond the 400 bp window examined. These differences may be dictated by the coding fraction of the chromosomes examined, with tightly packed *E. cuniculi*, allowing close proximity of promoters to translation starts only.

6.2 Codon usage comparisons

Since there are 64 possible codons and only 20 amino acids, the genetic code contains degeneracies, and the same amino acid can be encoded by multiple codons. Based on the wobble hypothesis (Bonitz et al. 1980), a given tRNA gene can utilize certain codons, which differ only in the third position. Additionally, the codon ATG (codes for Methionine) also serves as a special translation initiation signal, and three codons (TGA, TAG, TAA) are dedicated translation termination (stop) signals. The stop codon UGA is translated as tryptophan in many *Ascomycetes*, whereas some other fungi translate the stop codon UAG as leucine (Paquin et al. 1997). Analysis of codon usage has been used to identify highly expressed genes (Cancilla et al. 1995b; Freirepicos et al. 1994; Gharbia et al. 1995). Atypical codon usage has also been used to infer that genes have been acquired by horizontal transfer (Delorme et al. 1994; Groisman et al. 1992; Medigue et al. 1991).

Codon usage star plots for five different fungal genomes (from five classes in table 1) are shown in Figure 8A. A striking overrepresentation of AAA (Lysine), UUU (Phenylalanine) and GAA (Glutamic acid), is apparent in both *S. cerevisiae* and *Sc. pombe*. GAG (Glutamic acid) and CUC (Leucine) are frequency used in *A. nidulans*, *P. chrysosporium* and *E. cuniculi*. The frequency of amino acid usage is plotted in Figure 8B (note that this ignores which codon is used). Here it can be seen that leucine (L) and serine (S) are frequently used in all fungal genomes examined.

Figure 8C details the bias of third position of the codon. A preference for A and T (U) is commonly observed among *S. cerevisiae* and *Sc. pombe* genes whilst a bias of C and G is found in *A. nidulans*, *P. chrysosporium* and *E. cuniculi* genes. *Microsporidia* is different in the third position codon bias, compared to the other fungal genomes. This might be a reflection of the adaptation to a symbiotic/parasitic lifestyle (Wittner and Weiss 1999; Keeling et al. 2000). Note that plots of codon usage, position bias, and amino acid usage can be found in our web page[12].

The codon usage of *Sc. pombe* is similar to both *S. cerevisiae* and *C. albicans* (the latter not shown*)*, and the same optimal codons appear to be selected in highly expressed genes in all species. This contrasts with intergenic DNA, which in *Sc. pombe* and *S. cerevisiae* has similar G+C-contents, but in *C. albicans* is more A+T-rich (67 %). Thus, in all essential features, codon usage in *Sc. pombe* is very similar to that in *S. cerevisiae*, even though intergenic regions in these two species have undergone extensive mutation. We conclude that the factors influencing overall codon usage, namely mutational biases and the abundances of particular

---

[12] http://www.cbs.dtu.dk/services/GenomeAtlas/suppl/Fung_rev

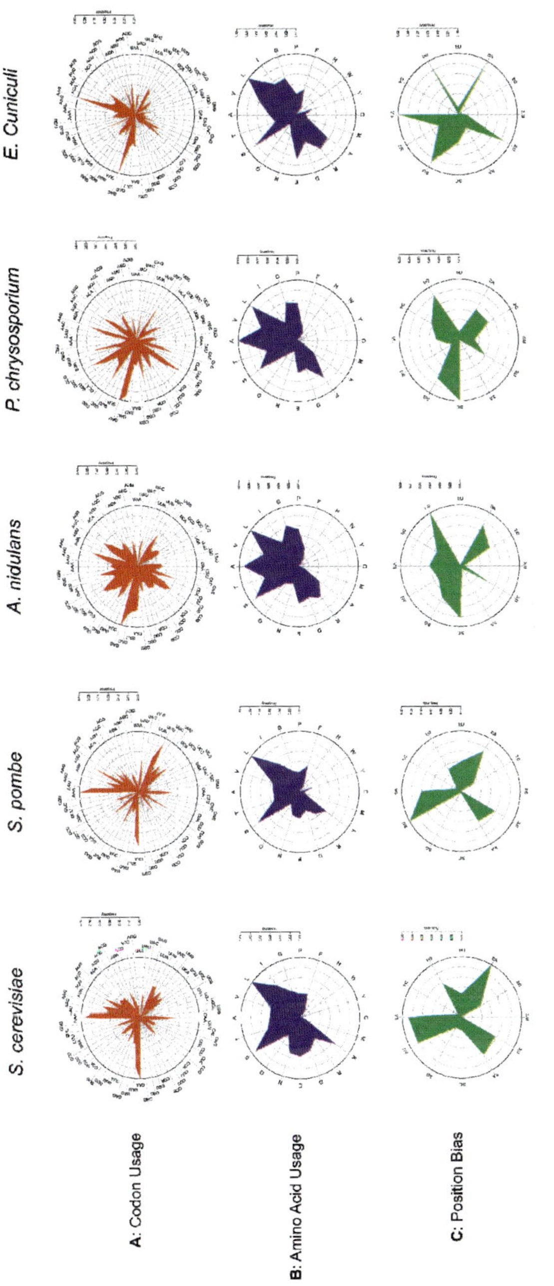

Fig. 8. Codon usage "star-plots" for *S. cerevisiae, Sc. pombe, A. nidulans, P. chrysosporium* and *E. cuniculi* genomes. (A) Codon usage for all open reading frames within a given genome. Note that several of the codons are rarely used, whilst others are quite common (B): Amino acid usage and (C): The bias of positions within all codons.

tRNAs, have not diverged between the two species (Lloyd and Sharp 1993). In addition, the GC-content of the *A. nidulans* and *P. chrysosporium* genome are close to 50%, indicating little overall mutational bias, and so the codon usage of lowly expressed genes is as expected in the absence of selection pressure. Most of the optimal codons are C- or G-ending, making highly expressed genes more G+C-rich at third base positions.

7 Proteome comparisons

Using BLAST, an all-against-all comparison of 18 publicly available fungal genomes has been carried out. This resulted in a matrix of blast reports from which alignments were counted. Homology was then expressed relative to the species listed in top of the resulting matrix (shown in Fig. 9). For homologies within genomes (internal homology), the highest homology is shown in red whereas grey represents <10% homology. Homology between genomes (external homology) is presented in green for homology of at least 80% and in grey for <40% (Fig. 9).

As apparent from Figure 9, *C. albicans* has the highest internal homology with deep red representing around 88% similarity. This is probably due to a duplication in the *C. albicans* genome (Jones et al. 2004).

The most obvious homology is found within the *Sachharomycotina* (the dark green block in the matrix, Fig. 9), with a surprising homology between *A. gossypii*, *K. waltii* and the *senso stricto Sachharomyces*. The homology within the *senso stricto Sachharomyces* is not higher than within the *senso lato* group or between the two groups, all having inter-species homologies between 50-93% of the proteome. Homology within the *Pezizomycotin*, between *G. zeae*, *T. ressei*, *N. crassa*, and *M. grisea* is less apparent.

C. albicans and *S. cerevisiae* are proposed to share a common ancestor ca. 200 million years ago, and the speciation of *Saccharomyces* and *Kluyveromyces* yeast diverged form the *Ashbya*-like yeast lineages, starting ca. 100-150 million years ago (Berbee and Taylor 2001). The recently complete sequenced of the *A. gossypii* revealed an unexpected high degree of gene homology and order conservation with *S. cerevisiae*. Both genomes differ substantially in their GC-content that 38.3% GC-content was found in *S. cerevisiae* whilst 51.9 % GC-content was found in *A. gossypii*. However, 95 % of the 4,700 genes *A. gossypii* proteins coding genes were found to have a homology in *S. cerevisiae*. Thus, it became important to reinvestigate the *S. cerevisiae* genome syntenic regions leading to an improved annotation (Brachat et al. 2003). *A. gossypii* and *K. waltii* have therefore currently been proposed to study genome duplication in a common descendant led to the creation of baker's yeast (Dietrich et al. 2004; Kellis et al. 2004). Furthermore, the fission yeast *Sc. pombe* and *S. cerevisiae* are as different from each other as either is from *Saccharomycotina*: their ancestors separated about 420 to 330 million years ago (Sipiczki 2000).

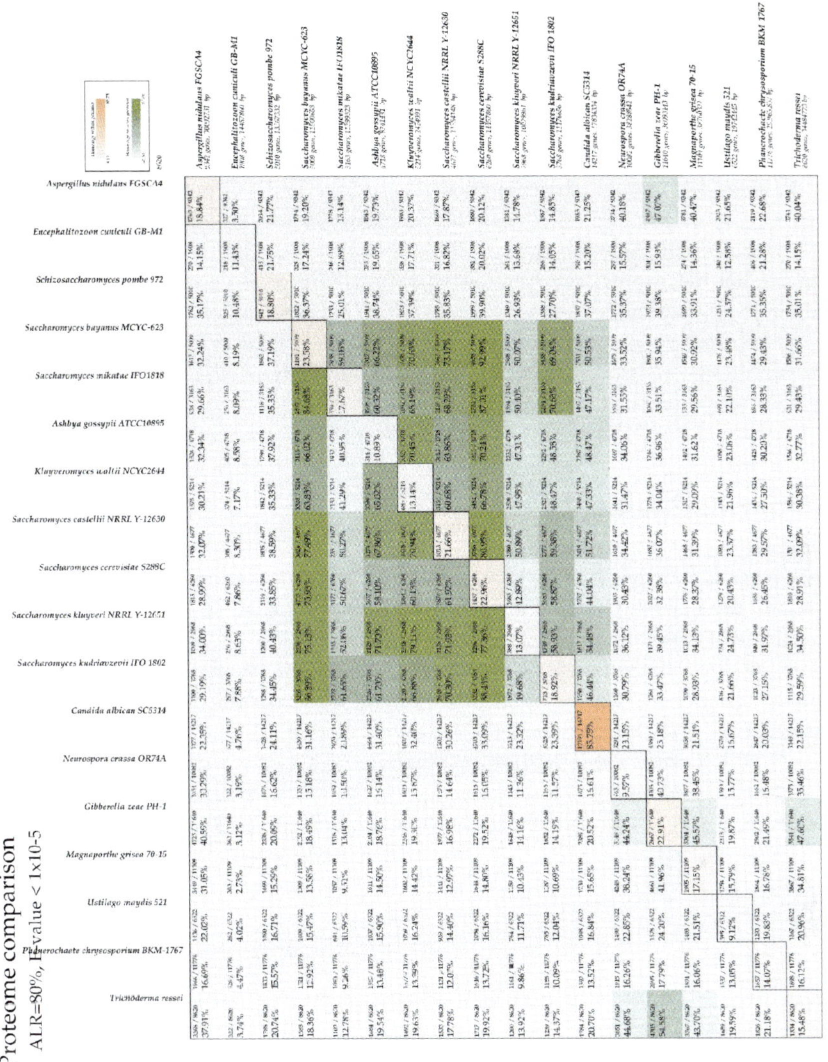

Fig. 9. Proteome comparison of fungal genomes. Each square in the matrix represents the blast results of the genome on the top row vs. the genome on the side. Note that this is not symmetrical because the genomes are of different sizes.

8 Conclusions

We have compared 18 different fungal species, using several different approaches. The Genome Atlases are used to plot various DNA structural properties of complete chromosomes. This way flexible DNA regions can be related to highly expressed gene (*e.g.* rRNA operons), and presence of location of DNA repeats can be indicated. DNA repeats in fungal genome seem to contribute to genome plasticity, since they are involved in several processes, such as recombination, inversion, deletion, translocation, and transposition. Our analysis showed that there is over-representation of purine stretches of >10 bp in length and this is pronounced in all of the analyzed fungal chromosomes. The purine and pyr/pur stretches are localized mainly in non-coding regions of the chromosome. Comparison of closely related species will improve future gene annotation (Brachat et al. 2003) so that we can predict more precisely how many genes there really are within the genome. Genomic comparison (Proteome comparison) helps to understand evolutionary relationships between species, and has pointed out crucial molecular events in fungal evolutionary history, such as large-scale duplications and other events that led to the diversity in fungal genomes.

Acknowledgements

This work was supported by grants form the Danish National Research Foundation and the Danish Center for Scientific Computing (DCSC). The authors would like to thank the people at CBS for their help and PW thanks the department of Biotechnology, Faculty of Agro-Industry, Chiangmai University, Thailand. Preliminary sequence data for the fungal genome were obtained from the Sanger institute, GSC (Genome Sequencing Center, the Washington University in St. Louis) FGI (Fungal Genome Initiative, MIT), NCBI (the National Center for Biotechnology Information), SGD (Saccharomyces Genome Database, Stanford University) and JGI (DOE Joint Genome Institute, the University of California).

References

ACT (Artemis Comparison Tool) (2003): http://www.sanger.ac.uk/Software/ACT/

Ayad-Durieux Y, Knechtle P, Goff S, Dietrich F, Philippsen P (2000) A PAK-like protein kinase is required for maturation of young hyphae and septation in the filamentous ascomycete *Ashbya gossypii*. J Cell Sci 24:4563-4575

Baldauf SL, Roger AI, Wenk-Siefert I, Doolittle WF (2000) A kingdom-level phylogeny-ofeukaryotes based on combined protein data. Science 290:972-977

Barnett JA (1992) The taxonomy of the genus *Saccharomyces* Meyen ex Reess: a short review for non-taxonomists. Yeast 8:1-23

Berbee ML, Taylor JW (2001) Systematics and evolution. In: McLaughlin DJ, McLaughlin EG, Lemke PA (eds) The Mycota VIIB. Berlin: Springer-Verlag pp 229-245

Blanchette RA (1991) Delignification by wood-decay fungi. Annu Rev Phytopathol 29:381–398

Bonitz SG, Berlani R, Coruzzi G, Li M, Macino G, Nobrega FG, Nobrega MP, Thalenfeld BE, Tzagoloff A (1980) Codon recognition rules in yeast mitochondria. Proc Nat Acad Sci USA 77:3167-3170

Bonman JM, MacKill DJ (1988) Durable resistance to rice blast disease. Oryza 25:103-110

Brachat S, Dietrich FS, Voegeli S, Zhang Z, Stuart L, Lerch A, Gates K, Gaffney T, Philippsen P (2003) Reinvestigation of the *Saccharomyces cerevisiae* genome annotation by comparison to the genome of a related fungus: *Ashbya gossypii*. Genome Biol 4(R):45

Cancilla MR, Hillier AJ, Davidson BE (1995) *Lactococcus lactis* glyceraldehyde-3-phosphate dehydrogenase gene, gap - further evidence for strongly biased codon usage in glycolytic pathway genes. Microbiology 141:1027-1036

Candida albicans pilot project (2004) http://alces.med.umn.edu/Candida.html

Cliften P, Sudarsanam P, Desikan A, Fulton L, Fulton B, Majors J, Waterston R, Cohen BA, Johnston M (2003) Finding functional features in *Saccharomyces* genomes by phylogenetic footprinting. Science 301:71-76

Delorme C, Godon JJ, Ehrlich SD, Renault P (1994) Mosaic structure of large regions of the *Lactococcus lactis* subsp cremoris chromosome. Microbiology 140:3053-3060

Dietrich FS, Voegeli S, Brachat S, Lerch A, Gates K, Steiner S, Mohr C, Pohlmann R, Luedi P, Choi S, Wing RA, Flavier A, Gaffney TD, Philippsen P (2004) The *Ashbya gossypii* genome as a tool for mapping the ancient *Saccharomyces cerevisiae* genome. Science 304:304-307

Dlakic M, Ussery D, Brunak S (2004) DNA bendability and nucleosome positioning in transcriptional regulation In: Ohyama T (ed) DNA Conformation in Transcription. Georgetown: Landes Bioscience

Dujon B, Sherman D, Fischer G, Durrens P, Casaregola S, Lafontaine I, de Montigny J, Marck C, Neuveglise C, Talla E, Goffard N, Frangeul L, Aigle M, Anthouard V, Babour A, Barbe V, Barnay S, Blanchin S, Beckerich JM, Beyne E, Bleykasten C, Boisrame A, Boyer J, Cattolico L, Confanioleri F, de Daruvar A, Despons L, Fabre E, Fairhead C, Ferry-Dumazet H, Groppi A, Hantraye F, Hennequin C, Jauniaux N, Joyet P, Kachouri R, Kerrest A, Koszul R, Lemaire M, Lesur I, Ma L, Muller H, Nicaud JM, Nikolski M, Oztas S, Ozier-Kalogeropoulos O, Pellenz S, Potier S, Richard GF, Straub ML, Suleau A, Swennen D, Tekaia F, Wesolowski-Louvel M, Westhof E, Wirth B, Zeniou-Meyer M, Zivanovic I, Bolotin-Fukuhara M, Thierry A, Bouchier C, Caudron B, Scarpelli C, Gaillardin C, Weissenbach J, Wincker P, Souciet JL (2004) Genome evolution in yeasts. Nature 430:35-44

Eriksson K–E, Blanchette RA, Ander P (1990) Microbial and enzymatic degradation of wood and wood components. New York: Springer-Verlag

Freire-Picos MA, Gonzalez-Siso MI, Rodriguez-Belmonte E, Rodriguez-Torres AM, Ramil E, Cerdan ME (1994) Codon usage in *Kluyveromyces lactis* and in yeast cytochrome c-encoding genes. Gene 139:43-49

Fischer G, James SA, Roberts IN, Oliver SG, Louis EJ (2000) Chromosomal evolution in *Saccharomyces*. Nature 405:451-454

Fungal Genome Initiative: A white paper for fungal comparative genomics, June 10 2003. http://www.broad.mit.edu/cgi-bin/annotation/fungi/fgi/

Galagan JE, Sarah EC, Katherine AB, Eric US, Nick DR, David J, William F, Li-Jun M, Serge S, Seth P, Bushra R, Timothy E, Reinhard E, Shunguang W, Cydney BN, Jonathan B, Matthew E, Dayong Q, Peter I, Deborah BP, Mary AN, Margaret WW, Claude PS, John AK, Edward LB, Alex Z, Ulrich S, Gregory OK, Gregory J, Werner M,

Chuck S, Edward M, David G, Alice R, Karen F, Jerome N, Nicole ST, Robert B, Sante G, Michael K, Manolis K, Evan M, Cord B, Stephen R, Dmitrij F, Svetlana K, Carolyn R, Robert LM, David DP, Scott K, Carlo C, Giuseppe M, David C, Weixi L, Robert JP, Stephen AO, Colin CD, Louise G, Marc JO, J AB, Rodger V, Oded Y, Michael P, Stephan S, Jay D, Alan R, Rodolfo A, Donald ON, Lisa AA, Gertrud M, Daniel JE, Michael F, Ian P, Matthew SS, Eric SL, Chad N, Bruce B (2002) The genome sequence of the filamentous fungus *Neurospora crassa*. Nature 422:859-868

Gaskell J, Dieperink E, Cullen D (1991) Genomic organization of lignin peroxidase genes of *Phanerochaete chrysosporium*. Nucleic Acids Res 19:599–603

Germot A, Philippe H, Le Guyader H (1997) Evidence for loss oftnitochondria in microsporidia from a mitochondrial-type HSP70 in *Nosema locustae*. Mol Biochem Parasitol 87:159-168

Gharbia SE, Williams JC, Andrews DMA, Shah HN (1995) Genomic clusters and codon usage in relation to gene-expression in oral gram-negative anaerobes. Anaerobe 1:239-262

Goffeau A, Barrell BG, Bussey H, Davis RW, Dujon B, Feldmann H, Galibert F, Hoheisel JD, Jacq C, Johnston M, Louis EJ, Mewes HW, Murakami Y, Philippsen P, Tettelin H, Oliver SG (1996) Life with 6000 genes. Science 274:563-567

Goffeau A et al. (1997) The yeast genome directory. Nature 387:5-6

Groisman EA, Saier MHJ, Ouchman H (1992) Horizontal transfer of a phosphatase gene as evidence for mosaic structure of the *Salmonella*. EMBO J 11:1309-1316

Hallin PF, Ussery D (2004) CBS Genome Atlas Database: a dynamic storage for bioinformatics results and sequence data. Bioinformatics 20:3682-3686

Hani J, Feldmann H (1998) tRNA genes and retroelements in the yeast genome. Nucleic Acids Research 26:689-696

Harman GE, Howell CR, Viterbo A, Chet I, Lorito M (2004) *Trichoderma* species-opportunistic, avirulent plant symbionts. Nat Rev Microbiol 2:43-56

Herrero E, de la Torre MA, Valentin E (2003) Comparative genomics of yeast species: new insights into their biology. Int Microbiol 6:183-190

Hirt RR, Healy B, Vossbrink CR, Canning EU, Embley TM (1997) A mitochondrial Hsp70 orthologue in *Vairimorpha necatrir*. Molecular evidence that microsporidia once contained mitochondria. Curr Biol 7:995-998

Hudspeth MES (1992) The fungal mitochondrial genome -- a broader perspective. In: Arora DK, Elander RP, Mukerji KG (eds) Handbook of Applied Mycology. Vol. 4: Fungal Biotechnology. New York: Marcel Dekker pp 213-241

Jain R, Rivera MC, Moore JE, Lake JA (2002) Horizontal gene transfer in microbial genome evolution. Theor Popul Biol 61:489-495

Jensen LJ, Friis C, Ussery D (1999) Three views of microbial genomes. Res Microbiol 150:773-777

Johnston M, Hillier L, Riles L, other members of the Genome Sequencing Center, Albermann K, André B, Ansorge W, Benes V, Brückner M, Delius H, Dubois E, Düsterhöft A, Entian KD, Floeth M, Goffeau A, Hebling U, Heumann K, Heuss-Neitzel D, Hilbert H, Hilger F, Kleine K, Kötter P, Louis EJ, Messenguy F, Mewes HW, Miosga T, Möst D, Müller-Auer S, Nentwich U, Obermaier B, Piravandi E, Pohl TM, Portetelle D, Purnelle B, Rechmann S, Rieger M, Rinke M, Rose M, Scharfe M, Scherens B, Scholler P, Schwager C, Schwarz S, Underwood AP, Urrestarazu LA, Vandenbol M, Verhasselt P, Vierendeels F, Voet M, Volckaert G, Voss H, Wambutt R, Wedler E,

Wedler H, Zimmermann FK, Zollner A, Hani J, Hoheisel JD (1997) The nucleotide sequence of *Saccharomyces cerevisiae* chromosome XII. Nature 387:87-90

Jones T, Federspiel NA, Chibana H, Dungan J, Kalman S, Magee BB, Newport G, Thorstenson YR, Agabian N, Magee PT, Davis RW, Scherer S (2004) The diploid genome sequence of *Candida albicans*. Proc Natl Acad Sci USA 101:7329-7334

José RH, Alfredo ME (1998) The fungus *Ustilago maydis*, from the aztec cuisine to the research laboratory. Int Microbiol1:149–158

Katherine AB, Lisa AA, Oded Y, Michael F, Gloria ET, Nick DR, Stephan S, Deborah B, John P, Nora P, Michael P, Marta G, Ulrich S, Gertrud M, Frank EN, Alan R, Claude S, James EG, Jay CD, Jennifer JL, David C, Hirokazu I, Rodolfo A, Michael P, Eric US, Matthew SS, George AM, Ian P, Rowland D, Daniel JE, Alex Z, Eric RK, Rebecca O'R, Frederick B, Jane Y, Chizu I, Keiichiro S, Wataru S, Robert P (2004) Lessons from the genome sequence of *Neurospora crassa*: tracing the path from genomic blueprint to multicellular organism. Microbiol Mol Biol Rev 68:1–108

Katinka MD, Duprat S, Cornillot E, Metenier G, Thomarat F, Prensier G, Barbe V, Peyretaillade E, Brottier P, Wincker P, Delbac F, El Alaoui S, Peyret P, Saurin W, Gouy M, Weissenbach J, Vivares CP (2001) Genome sequence and gene compaction of the eukaryote parasite *Encephalitozoon cuniculi*. Nature 414:450-453

Keeling PJ, Luker MA, Palmer JD (2000) Evidence from bate-tubulin phylogeny that microsporidia evolved from within the fungi. Mol Biol Evol 17:23-31

Kellis M, Birren BW, Lander ES (2004) Proof and evolutionary analysis of ancient genome duplication in the yeast *Saccharomyces cerevisiae*. Nature 428:617-624

Keogh RS, Seoighe C, Wolfe KH (1998) Evolution of gene order and chromosome number in *Saccharomyces, Kluyveromyces*, and related fungi. Yeast 14:443-457

Kersten PJ, Kirk TK (1987) Involvement of a new enzyme, glyoxal oxidase, in extracellular H_2O_2 production by *Phanerochaete chrysosporium*. J Bacteriol 169:2195-2201

Kellis M, Patterson N, Endrizzi M, Birren B, Lander ES (2003) Sequencing and comparison of yeast species to identify genes and regulatory element. Nature 423:241-254

Kirk TK, Farrell R (1987) Enzymatic 'combustion': Themicrobial degradation of lignin. Annu Rev Microbiol 41:465–505

Kurtzman CP, Fell JW (1998) Definition, classification and nomenclature of the yeasts. In: Kurtzman CP, Fell JW (eds) The Yeasts, A Taxonomic Study. 4th edn. Amsterdam: Elsevier Science BV pp 3-5

Kurtzman CP, Robnett CJ (2003) Phylogenetic relationships among yeast of the 'Saccharomyces complex' determined for multigen sequence analyses. FEMS Yeast Res 3:417-432

Kurtzman CP, Lynch M, Force A (2003) Phylogenetic circumscription of *Saccharomyces, Kluyveromyces* and other members of the *Saccharomycetaceae*, and the proposal of the new genera *Lachancea, Nakaseomyces, Naumovia, Vanderwaltozyma*, and *Zygotorulaspora*. FEMS Yeast Res 4:233–245

Langkjaer RB, Nielsen ML, Daugaard PR, Liu W, Piskur J (2000) Yeast chromosomes have been significantly reshaped during their evolutionary history. J Mol Biol 304:271-288

Lindner P (1893) *Schizosaccharomyces pombe* n. sp. neuer Gärungserreger. *Wochenschr f Brauerei* 10:1298-1300

Llorente B, Malpertuy A, Neuveglise C, de Monigny J, Aigle M, Artiguenave F, Blandin G, Bolotin-Fukuhara M, Bon E, Brottier P, Casaregola S, Durrens P, Gaillardin C, Lepingle A, Ozier-Kalogeropoulos O, Potier S, Saurin W, Tekaia F, Toffano-Nioche

C, Wesolowski-Louvel M, Wincker P, Weissenbach P, Souciet J-L, Dujon B (2000) Genomic exploration of the hemiascomycetous yeasts: 18: Comparative analysis of chromosome maps and synteny with *S. cerevisiae*. FEBS Lett 487:101-112

Lloyd AT, Sharp PM (1993) Synonymous codon usage in *Kluyveromyces lactis*. Yeast 9:1219-1228

Lowe TM, Eddy SR (1997) tRNAscan-SE: a program for improved detection of transfer RNA genes in genomic sequence. Nucleic Acids Research 25:955-964

Marinoni G, Manuel M, Petersen RF, Hvidtfeldt J, Sulo P, Piskur J (1999) Horizontal transfer of genetic material among *Saccharomyces* yeasts. J Bacteriol 181:6488-6496

Malpertuy A, Dujon B, Richard GF (2003) Analysis of microsatellites in 13 hemiascomycetous yeast species: mechanisms involved in genome dynamic. J Mol Evol 56:730-741

Martinez D, Larrondo LF, Putnam N, Gelpke MD, Huang K, Chapman J, Helfenbein KG, Ramaiya P, Detter JC, Larimer F, Coutinho PM, Henrissat B, Berka R, Cullen D, Rokhsar D (2004) Genome sequence of the lignocellulose degrading fungus *Phanerochaete chrysosporium* strain RP78. Nat Biotechnol 22:695-700

McGuire AM, Hughes JD, Church GM (2000) Conservation of DNA regulatory motifs and discovery of new motifs in microbial genomes. Genome Res 10:744-757

Medigue C, Rouxel T, Vigier P, Henaut A, Danchin A (1991) Evidence for horizontal gene transfer in *Escherichia coli* speciation. J Mol Biol 222:851-856

Mewes HW, Albermann K, Bähr M, Frishman D, Glrissner A, Hani J, Heumann K, Kleine K, Maierl A, Oliver SG, Pfeiffer F, Zollner A (1997) Overview of the yeast genome. Nature 387:7-65

Monschau N, Sahm H, Stahmann K (1998) Threonine aldolase overexpression plus threonine supplementation enhanced riboflavin production in *Ashbya gossypii*. Appl Environ Microbiol 64:4283-4290

Naumov GI (1996) Genetic identification of biological species in the *Saccharomyces* senso stricto complex. J Ind Microbiol 17:295-320

Naumov GI, James SA, Naumova ES, Louis EJ, Roberts LN (2000) Three new species in the *Sacchromyces* senso stricto complex: *Sacchromyces cariocanus, Sacchromyces kudriavzevii*, and *Sacchromyces mikatae*. Int J Syst Evol Microbiol 50:1931-1942

Naumova ES, Bulat SA, Mironenko NV, Naumov GI (2003) Differentiation of six sibling species in the *Saccharomyces* senso stricto complex by multilocus enzyme electrophoresis and UP-PCR analysis. Antonie van Leeuwenhoek 83:155-163

O'Donnell K, Kistler HC, Tacke BK, Casper HH (2000) Gene genealogies reveal global phylogeographic structure and reproductive isolation among lineages of *Fusarium graminearum,* the fungus causing wheat scab. Proc Natl Acad Sci USA 97:7905-7910

Odds FC, Brown JP, Gow AR (2004) *Candida albicans* genome sequence: a platform for genomics in the absence of genetics. Genome Biol 5:230-238

Ohno S (1970) Evolution by gene duplication. London: George Allen and Unwin

Paquin B, Laforest MJ, Forget L, Roewer I, Wang Z, Longcore J, Lang BF (1997) The fungal mitochondrial genome project: evolution of fungal mitochondrial genomes and their gene expression. Curr Genet 31:380-395

Pedersen AG, Jensen LJ, Staerfeldt HH, Brunak S, Ussery D (2000) A DNA atlas for *Escherichia coli*. J Mol Biol 299:907-930

Pennacchio LA, Rubin EM (2001) Genomic strategies to identify mammalian regulatory sequences. Nat Rev Genet 2:100-109

Petersen RF, Nilsson-Tillgren T, Piskur J (1999) Karyotypes of *Saccharomyces* senso lato species Int J Syst Bacteriol 49:1925-1931

Petersen L, On SLW, Ussery D (2002) Visualization and Significance of DNA Structural Motifs in the *Campylobacter jejuni* genome. Genome Lett 1:16-25

Phanerochaete chrysosporium and *Trichoderma ressei* sequencing genome project (2004) ftp://ftp.jgi-psf.org/pub/JGI_data/

Piskur J, Smole S, Groth C, Petersen RF, Pedersen MB (1998) Structure and genetic stability of mitochondrial genomes vary among yeasts of the genus *Saccharomyces*. Int J Syst Bacteriol 48:1015-1024

Piskur J, Langkjaer RB (2004) Yeast genome sequencing: the power of comparative genomics. Mol Microbiol 53:381-389

Prillinger H, Schweigkofler W, Breltenbach M, Briza P, Staudacher E, Lopandic K, Molnaur O, Weigang F, Ibl M, Ellinger A (1997) Phytopathogenic filamentous (*Ashbya, Eremothecium*) and dimorphic fungi (Holleya, Nematospora) with needle-shaped *Ascospores* as new members within the *Saccharomycetaceae*. Yeast 13:945–960

Primrose SB (1998) The organization and structure of genomes. In: Principles of Genome Analysis. Massachusetts: Blackwell Science Ltd pp 17-44

Raeder U, Thompson W, Broda P (1989) RFLP-based genetic map of *Phanerochaete chrysosporium* ME446: lignin peroxidase genes occur in clusters. Mol Microbiol 3:911-918

Rokas A, Barry LW, Nicole K, Sean BC (2003) Genome-scale approaches to resolving incongruence in molecular phylogenies. Nature 425:798-804

Saccharomyces genome database (2004) http://www.sequence.stanford.edu/Saccharomyces

Saccharomyces genome sequencing (2004) http://genome.wustl.edu/projects/yeast/

Schulte U, Becker I, Mewes HW, Mannhaupt G (2002) Large scale analysis of sequences from *Neurospora crassa*. J Biotechnol 94:3-13

Selker EU (2002) Repeat-induced gene silencing in fungi. Adv Genet 46:439–450

Sharma R, Chisti Y, Banerjee UC (2001) Production, purification, characterization, and applications of lipase. Biotechnol Adv 19:627-662

Sipiczki M (1995) Phylogenesis of fission yeasts. Contradictions surrounding the origin of century old genus. Antonie van Leeuwenhoek 68:119-149

Sipiczki M (2000) Where does fission yeast sit on the tree of life? Genome Biol 1(2):1-4

Skinner DZ, Budde AD, Farman ML, Smith JR, Leung H (1993) Genome organization of *Magnaporthe grisea*: genetic map, electrophoretic karyotype, and occurrence of repeated DNAs. Theor of Appl Genet 87:545-557

Skovgaard M, Jensen LJ, Friis C, Stærfeldt HH, Worning P, Brunak S, Ussery D (2002) The atlas visualization of genomewide information. In: Wern B, Dorrell N (eds) Functional Microbial Genomics (Method in Microbiology) London: Academic Press pp 49-63

Souciet JL, Aigle M, Artiguenave F, Blandin G, Bolotin-Fukuhara M, Bon E, Brottier P, Casaregola S, de Montigny J, Dujon J, Durrens P, Gaillardin C, Lépingle A, Llorente B, Malpertuy A, Neuvéglise C, Ozier-Kalogéropoulos O, Potier S, Saurin W, Tekaia F, Toffano-Nioche C, Wésolowski-Louvel M, Wincker P, Weissenbach J (2000) Genomic exploration of the hemiascomycetous yeasts: 1. A set of yeast species for molecular evolution studies. FEBS Letter 487:3-12

Spirek M, Yang J, Groth C, Petersen RF, Langkjaer RB, Naumova ES, Sulo P, Naumov GI, Piskur J (2003) High-rate evolution of *Saccharomyces* senso lato chromosomes. FEMS Yeast Res 3:363-373

Steiner S, Wendland J, Wright MC, Philippsen P (1995) Homologous recombination as the main mechanism for DNA integration and cause of rearrangements in the filamentous ascomycete *Ashbya gossypii*. Genetics 140:973-987

Sunnerhagen P (2002) Prospects for functional genomic in *Schizosaccharomyces pombe*. Curr Genet 42:73-84

Talbot NJ, Salch YP, Ma M, Hamer JE (1993) Karyotypic variation with clonal lineages of the rice blast fungus (*Magnaporthe grisea*). Appl Environ Microbiol 59:585-593

The yeast comparative genomics (2004) http://www.broad.mit.edu/annotation/fungi/comp_yeasts/downloads.html

Tzung KW, Williams RM, Scherer S, Federspiel N, Jones T, Hansen N, Bivolarevic V, Huizar L, Komp C, Surzycki R, Tamse R, Davis RW, Agabian N (2001) Genomics evidence for a complete sexual cycle in *Candida albicans*. Proc Natl Acad Sci USA 98:3249-3253

Ussery D, Soumpasis DM, Brunak S, Storfeldt HH, Worning P, Krogh A (2002) Bias of purine stretches in sequenced chromosomes. Comput Chem 26:531-541

Ussery D, Hallin PF (2004) Genome update: AT content in sequenced prokaryotic genome. Microbiology 150:749-752

Vaughan MA (1991) Evaluation of phylogenetic relationships among fission yeast by nDNA/nDNA reassociation and conventional taxonomic criteria. Yeast 7:73-78

Vossbrinck CR, Maddox JV, Friedman S, Debrunner-Vossbrinck BA, Woese CR (1987) Ribosomal RNA sequence suggests microsporidia are extremely ancient eukaryotes. Nature 326:411-414

Warren ST (1996) The expanding world of trinucleotide repeats. Science 271:1374-1375

Wendland J, Pohlmann R, Dietrich F, Steiner S, Mohr C, Philippsen P (1999) Compact organization of rRNA genes in the filamentous fungus *Ashbya gossypii*. Curr Genet 35:618-25

Wickes B, Staudinger J, Magee BB, Kwon-Chung KJ, Magee PT, Scherer S (1991) Physical and genetic mapping of *Candida albicans*: several genes previously assigned to chromosome 1 map to chromosome R, the rDNA-containing linkage. Infect Immun 59:2480–2484

Wittner M, Weiss LM (1999) The *Microsporidia* and *Microsporidiosis*. American Society of Microbiology, Washington DC

Wolfe K (2004) Evolutionary Genomics: Yeasts accelerate beyond BLAST dispatch. Curr Biol 14:R392–R394

Wolfe K, Shields DC (1997) Molecular evidence for an ancient duplication of the entire yeast genome. Nature 387:708-713

Wood V, Gwilliam R, Rajandream MA, Lyne M, Lyne R, Stewart A, Sgouros J, Peat N, Hayles J, Baker S, Basham D, Bowman S, Brooks K, Brown D, Brown S, Chillingworth T, Churcher C, Collins M, Connor R, Cronin A, Davis P, Feltwell T, Fraser A, Gentles S, Goble A, Hamlin N, Harris D, Hidalgo J, Hodgson G, Holroyd S, Hornsby T, Howarth S, Huckle EJ, Hunt S, Jagels K, James K, Jones L, Jones M, Leather S, McDonald S, McLean J, Mooney P, Moule S, Mungall K, Murphy L, Niblett D, Odell C, Oliver K, O'Neil S, Pearson D, Quail MA, Rabbinowitsch E, Rutherford K, Rutter S, Saunders D, Seeger K, Sharp S, Skelton J, Simmonds M, Squares R, Squares S, Stevens K, Taylor K, Taylor RG, Tivey A, Walsh S, Warren T, Whitehead S, Wood-

ward J, Volckaert G, Aert R, Robben J, Grymonprez B, Weltjens I, Vanstreels E, Rieger M, Schafer M, Muller-Auer S, Gabel C, Fuchs M, Dusterhoft A, Fritzc C, Holzer E, Moestl D, Hilbert H, Borzym K, Langer I, Beck A, Lehrach H, Reinhardt R, Pohl TM, Eger P, Zimmermann W, Wedler H, Wambutt R, Purnelle B, Goffeau A, Cadieu E, Dreano S, Gloux S, Lelaure V, Mottier S, Galibert F, Aves SJ, Xiang Z, Hunt C, Moore K, Hurst SM, Lucas M, Rochet M, Gaillardin C, Tallada VA, Garzon A, Thode G, Daga RR, Cruzado L, Jimenez J, Sanchez M, del Rey F, Benito J, Dominguez A, Revuelta JL, Moreno S, Armstrong J, Forsburg SL, Cerutti L, Lowe T, McCombie WR, Paulsen I, Potashkin J, Shpakovski GV, Ussery D, Barrell BG, Nurse P, Cerrutti L (2002) The genome sequence of *Schizosaccharomyces pombe*. Nature 415:871-880
Yanagida M (2003) The model unicellular eukaryote, *Schizosaccharomyces pombe*. Genome Biol 3:comment2003.1–2003.4

Gouveia-Oliveira, Rodrigo
Center for Biological Sequence Analysis, BioCentrum-DTU, Building 208, Technical University of Denmark, DK-2800 Kgs. Lyngby, Denmark

Hallin, Peter F.
Center for Biological Sequence Analysis, BioCentrum-DTU, Building 208, Technical University of Denmark, DK-2800 Kgs. Lyngby, Denmark

Ussery, David
Center for Biological Sequence Analysis, BioCentrum-DTU, Building 208, Technical University of Denmark, DK-2800 Kgs. Lyngby, Denmark
dave@cbs.dtu.dk

Wanchanthuek, Phatthanaphong
Center for Biological Sequence Analysis, BioCentrum-DTU, Building 208, Technical University of Denmark, DK-2800 Kgs. Lyngby, Denmark

Duplication of genes and genomes in yeasts

Simon Wong and Kenneth H. Wolfe

Abstract

The molecular evolution of the group of yeasts closely related to *Saccharomyces cerevisiae* has been profoundly affected by an ancient polyploidy event that resulted in duplication of the whole genome. This event occurred in the common ancestor of the *Saccharomyces sensu stricto* and *sensu lato* species, including *Candida glabrata*. Recent progress in genome sequencing has allowed the molecular sorting-out process after genome duplication to be investigated in detail. The loci where both copies of the gene were retained, as opposed to deletion of one copy, appear to be those that have either been subject to selection for high dosage of the gene product, or where functional divergence between the two copies was achieved rapidly.

1 The 'true' yeasts

The kingdom Fungi consists of a vast range of eukaryotic organisms found in diverse environments. Most fungi are composed of hyphae – filamentous, threadlike structures often congregating into systems called mycelia. However, some fungi assume unicellular forms known as yeasts. In addition, some species are dimorphic, incorporating both structural forms in their life cycles depending on environmental conditions. Although many fungal species are unicellular, those in the phylum Ascomycota are often referred to as the 'true' yeasts due to their predominantly unicellular life cycles. This phylum comprises three classes: Archiascomycetes (e.g. *Schizosaccharomyces pombe*), Euascomycetes (e.g. *Neurospora crassa*), and Hemiascomycetes (e.g. *Saccharomyces cerevisiae*). This review focuses on the class Hemiascomycetes, which includes some of the most important yeasts for basic, applied and medical research, features that have made Hemiascomycetes the focus of extensive genomics research.

Saccharomyces cerevisiae (bakers' yeast) is the most renowned and best studied yeast. Its natural capability to produce ethanol by fermentation and carbon dioxide by respiration has been exploited for millennia in the brewing and baking industries. Its economic importance has provided much impetus for basic research into this yeast. Beginning with the work of Winge and Lindegren in the 1930s (reviewed in Mortimer 1993a; 1993b), the ability to perform crosses with *S. cerevisiae* and its tractability in the laboratory made it an attractive research tool

Topics in Current Genetics, Vol. 15
P. Sunnerhagen, J. Piškur (Eds.): Comparative Genomics
DOI 10.1007/b105770 / Published online: 7 January 2005

in classical genetics. Combined with modern molecular techniques, it has become one of the best characterized eukaryotic model organisms.

Apart from *S. cerevisiae*, many related yeasts are also widely employed for the production of different wine, beer, and bread (Demain et al. 1998). However, some species can produce other important compounds such as vitamins, citric acid, and lipids. *Candida utilis* is used for the production of animal feed as well as the flavoring substances ethyl acetate and acetaldehyde. Many species of the genera *Candida*, *Debaryomyces*, *Pichia*, and *Yarrowia* can utilize hydrocarbons as sole carbon sources and could potentially be used to clean up oil spills. With recent advances in recombinant DNA technology, a number of yeasts have been developed as host organisms for the production of heterologous protein such as human hormones and enzymes of commercial interest (Gellissen and Hollenberg 1997).

Within the Hemiascomycetes, many species of the genus *Candida* are opportunistic pathogens of humans. They cause a range of diseases and are often associated with immunocompromised patients (Hazen 1995; Calderone 2002). The principal yeast pathogen for human is *Candida albicans*, which is the most common species isolated from bloodstream infections. However, other species such as *C. tropicalis*, *C. dubliniensis*, and *C. glabrata* are emerging concerns as they are less susceptible to some antifungal drugs and their incidence has increased relative to that of *C. albicans*. The medical importance of the *Candida* species has stimulated much research interest and the development of accurate strain detection systems.

Fig. 1 (overleaf). Phylogeny of hemiascomycetes in the '*Saccharomyces* complex', redrawn from Kurtzman (2003). Species whose genomes have been extensively sequenced (> 3x coverage) are highlighted (Goffeau et al. 1996; Cliften et al. 2003; Kellis et al. 2003; Dietrich et al. 2004; Dujon et al. 2004; Kellis et al. 2004). *Eremothecium gossypii* is synonymous with *Ashbya gossypii*. The tree is based on parsimony analysis of six genes. Numbers on internal branches are bootstrap percentages, and branches where no number is shown recurred in <50% of bootstrap replicates. Names on the right are new genus names proposed by Kurtzman (2003). The probable phylogenetic position of the whole-genome duplication (WGD) event is shown by an arrow. The positioning of the WGD after the divergence between *Zygosaccharomyces* and the upper clades is based on the presence of 7 chromosomes in the type strain of *Z. rouxii* (Torok et al. 1993; Sychrova et al. 2000) and the extensive colinearity of gene order between *Z. rouxii* and outgroup species such as *K. waltii* or *K. lactis*, contrasting with its 1:2 relationship to *S. cerevisiae* (Wong et al. 2002; J. Gordon and K. H. Wolfe, unpublished results). The positioning of the WGD before the divergence of *Tetrapisispora* and *Vanderwaltozyma* from the upper clades is less certain and is inferred from the presence of about 20 chromosomes in *K. yarrowii* and about 13 in *K. polysporus* (Belloch et al. 1998).

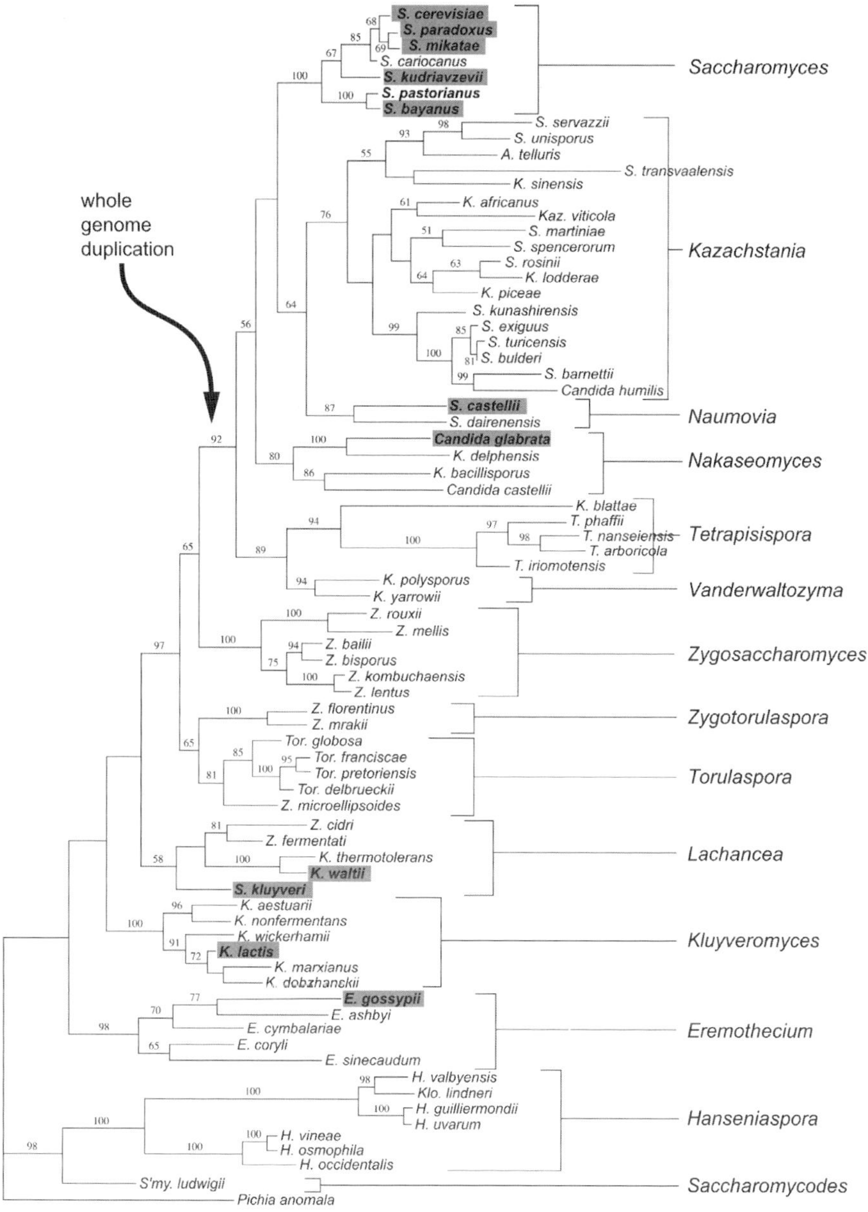

2 Taxonomy and phylogeny

A consistent and well established phylogenetic relationship is fundamental to infer evolutionary events within a group of species. Traditionally, yeast taxonomy has been based on phenotypic and metabolic features often prone to ambiguity. This is especially true in closely related species such as the hemiascomycete yeasts. To address this issue, various sequence-based reconstructions of the phylogeny of the ascomycetes have been carried out. Some of the more comprehensive studies involved the use of 18S rRNA sequences (Cai et al. 1996; James et al. 1997; Keogh et al. 1998), partial 26S rRNA sequences (Kurtzman and Robnett 1998), the complete rDNA repeating unit (Wong et al. 2003), and the mitochondrial cytochrome-c oxidase II (*cox2*) gene (Belloch et al. 2000). Although the phylogenetic trees obtained from these studies are reasonably congruent, significant progress was achieved by a recent phylogenetic analysis that combined sequence data from multiple loci and included almost all the known species in the group of Hemiascomycetes called the 'Saccharomyces complex' (Kurtzman and Robnett 2003). Their analysis placed 75 species associated with the genera *Saccharomyces* and *Kluyveromyces* into 14 clades. Kurtzman (2003) subsequently used this phylogeny as the basis for proposing a reorganization of the taxonomy of this group of species (Fig. 1).

Species in the current genus *Saccharomyces* can be divided into three groups (Fig. 1). The *Saccharomyces sensu stricto* species, including *S. cerevisiae*, are in a homogeneous group. The phylogenetic relationships within the *sensu stricto* were recently re-examined using a 106-gene data set from whole genome sequences, which produced a tree slightly different from that in Figure 1 (*S. paradoxus* clustered with *S. cerevisiae* instead of with *S. mikatae*; Rokas et al. (2003); see also Phillips et al. (2004) and Holland et al. (2004)).

The *sensu stricto* group is phylogenetically distinct from the *Saccharomyces sensu lato* species, which form a heterogeneous group that is not monophyletic with respect to other species in the genera *Kluyveromyces* and *Candida*. Many of the *Saccharomyces sensu lato* species have been placed into a new genus, *Kazachstania*, by Kurtzman (2003) (Fig. 1). The third *Saccharomyces* group consists of just one species, *S. kluyveri*, which is phylogenetically distant from the *sensu stricto* and *sensu lato* groups. Physiologically, *S. kluyveri* is quite dissimilar to the other *Saccharomyces* yeasts. Most notably, it is unable to form true petite mutants (Moller et al. 2001). It can utilize pyrimidines and purines as sole sources of nitrogen (Gojkovic et al. 1998). Cytogenetic analysis have revealed that its karyotype reveals a lack of small chromosomes and it contains roughly half the number of chromosomes found in other *Saccharomyces* species (Petersen et al. 1999). The latter, along with subsequent studies, have established that it diverged from other yeasts before the whole genome duplication event leading to other *Saccharomyces* lineages (Wolfe and Shields 1997; Wong et al. 2002). Taken together, the placement of *S. kluyveri* in the genus *Saccharomyces* seems questionable and Kurtzman (2003) placed it in a new genus, *Lachancea*.

Kurtzman and Robnett (2003) highlighted the disparity between the well-defined phylogenetic clades in their analysis and the way in which species were grouped into genera under current systematic treatments (summarized in Kurtzman and Fell 1998). Many well-known genera, including *Saccharomyces*, *Kluyveromyces*, *Zygosaccharomyces*, and *Candida* were polyphyletic. This led Kurtzman (2003) to propose sweeping revisions to the taxonomy of this group of species, whereby the 14 well-supported clades in the multi-gene phylogenetic tree (Kurtzman and Robnett 2003) became 14 genera that are probably monophyletic. Rapid progress in genomics has resulted in almost-complete genome sequences becoming available from representatives of six of these 14 clades (Fig. 1). However, even though each of the 14 clades seems reasonably robust, there is still some doubt about the branching order of these clades relative to one other. Many of the internal branches along the 'spine' of the tree in Figure 1 have low bootstrap confidence, and analyses based on complete genome sequences tend to arrange *S. kluyveri*, *K. waltii*, *K. lactis*, and *E. gossypii* into one or two monophyletic groups as opposed to the three separate lineages represented by *Lachancea*, *Kluyveromyces*, and *Eremothecium* in Figure 1 (Hittinger et al. 2004; J. Gordon, D. Scannell, K. Byrne, S. Wong and K. H. Wolfe, unpublished results).

3 Yeast genome sequencing projects

In 1995, *Haemophilus influenzae* became the first free-living organism to have its genome completely sequenced (Fleischmann et al. 1995). While the genome of this bacterium is only around 1.8 Mb in size, it heralded the genomics era where the full complement of genes of an organism can be systematically identified and analyzed. At the same time, researchers worldwide were busy sequencing the appreciably larger genomes of various eukaryotes. In 1996, *S. cerevisiae* became the first eukaryotic genome to be completely sequenced (Goffeau et al. 1996). The sequence was determined by a large consortium of laboratories over several years, beginning with chromosome III in 1992 (Oliver et al. 1992). It consists of 16 chromosomes that add up to approximately 14 Mb, much smaller than those of other model eukaryotes such as *Arabidopsis thaliana* (125 Mb) or *Drosophila melanogaster* (137 Mb). Yet, it still poses a substantial challenge for researchers trying to decipher its contents. The number of protein-coding genes was originally estimated to be in the region of 6,200 but has since been modified to a more conservative 5,500 to 5,700 (Wood et al. 2001; Kellis et al. 2003). Unlike multicellular organisms, the genome of *S. cerevisiae* is very compact with around 70% of the total sequence coding for genes. In addition, only around 4% of genes contain introns, greatly assisting the annotation process. Repetitive elements in the genome are represented by the yeast retrotransposons, the Ty elements, which occur in about 50 copies often associated with tRNA genes (Hani and Feldmann 1998). But these make up a relatively small proportion of the genome compared with multicellular eukaryotes such as human, where over 50% of the genome can be classified as repetitive DNA (Baltimore 2001). Hence, the compact nature of the *S. cerevisiae* genome permits useful comparative genomics studies to be carried out using relatively small amounts of sequence data from other similar yeasts.

Subsequently, a number of fungal genomes have been completely sequenced. They include the genomes of *Schizosaccharomyces pombe* (in class Archiascomycetes; Wood et al. 2002) and *Neurospora crassa* (in class Euascomycetes; Galagan et al. 2003). These species are so distantly related to *S. cerevisiae* that, although some interesting comparisons can be made in terms of their proteome contents and organism-specific biology, there is almost no conservation of gene order along chromosomes between these species and *S. cerevisiae* and many genes do not fall into simple one-to-one orthology relationships between these genomes.

In 2003, extensive genome sequence data became available for several other *Saccharomyces sensu stricto* species. *S. mikatae* and *S. bayanus* were independently sequenced to 7x coverage by Kellis et al. (2003) and to 3x coverage by Cliften et al. (2003); Kellis et al. also sequenced *S. paradoxus* (7x), and Cliften et al. also sequenced *S. kudriavzevii* (3x). The close relationship between these yeasts and *S. cerevisiae* means that their genomes are almost identical in organization, with few chromosomal rearrangements disrupting syntenic regions. However, it proved to be extremely useful in the identification of rapidly evolving regulatory elements.

Comprehensive sequence information has also become available in the past year from more distantly related hemiascomycetes, as summarized in Figure 1 (Cliften et al. 2003; Dietrich et al. 2004; Dujon et al. 2004; Kellis et al. 2004). The result is that we now have sequences from two species in (or close to) the *sensu lato* group (*S. castellii* and *Candida glabrata*), and four species that are somewhat more distantly related (*K. waltii*, *S. kluyveri*, *K. lactis*, and *E. gossypii*). In addition to the species highlighted in Figure 1, which covers only the 'Saccharomyces complex' (Kurtzman and Robnett 2003), the genome sequences of several other more distantly related hemiascomycetes are known: the genome sequences of *Candida albicans*, *Debaryomyces hansenii*, and *Yarrowia lipolytica* sequences are public (Dujon et al. 2004; Jones et al. 2004), and the *Pichia angusta (Hansenula polymorpha)* sequence is available under restricted terms (Ramezani-Rad et al. 2003). Lastly, more limited amounts of random sequence information from the genomes of several other yeasts were produced by the Génolevures project (Souciet et al. 2000).

4 The origin of new genes

There are four possible ways for a new gene to emerge during evolution: (i) duplication of an existing gene, (ii) combination of parts of different genes to create a mosaic gene, (iii) *de novo* generation of a gene from non-coding DNA, and (iv) horizontal transfer of a gene from another species. While examples of all four routes have been documented (reviewed in Wolfe and Li 2003), by far the most common way to create new genes is by gene duplication.

In his classic book *Evolution by Gene Duplication*, Ohno (1970) proposed that biodiversity evolved in big leaps by the creation – through duplication – of novel, redundant genetic raw material. Some three decades later, this mechanism of genome evolution is universally accepted. In Ohno's view, the original copy of a gene retained the original function (a sort of backup mechanism) while the extra copy was free to vary in sequence. Under this hypothesis, a newly formed copy of a gene faces one of two possible alternative outcomes: either it is lost from the genome due to the accumulation of deleterious mutations (nonfunctionalization), or else it is preserved in the genome by virtue of acquiring a novel role that is selectively advantageous (neofunctionalization). Since deleterious mutations occur more frequently than beneficial ones, it was expected that most new gene duplicates would quickly pick up an inactivating mutation that would turn them into pseudogenes, eventually becoming deleted from the genome, but occasionally the extra copy of a gene would survive because it acquired a sequence change that conferred a beneficial new function. Ohno's model predicts that, in cases where a duplicated gene has survived, the rate of sequence change in the new copy of the gene will have been faster than in the original copy. The problem, of course, is that it is usually not possible to know which member of a pair of paralogous sequences is the 'original' gene and which is the 'copy'. In fact, the distinction is meaningless for some types of duplication (e.g. polyploidy) and only makes sense in some very specific circumstances where it is possible to tell which copy is derived from which (e.g. in the case of retrotransposed mammalian genes that have lost introns).

In the decades following Ohno's work, it has become apparent that all genomes contain many large gene families, which indicates that gene duplication has been a major force in organismal evolution. However, the ubiquity of gene duplication has led to a problem: there have been so many duplications that it is hard to see how they can all have involved the gain of novel gene functions. A solution to this problem was proposed by Lynch and Force, who suggested that subfunctionalization could provide a mechanism of gene preservation in the immediate aftermath of gene duplication (Force et al. 1999; Lynch and Force 2000). Subfunctionalization is a process whereby a gene with multiple functions (e.g. a gene whose expression is induced under several distinct conditions) becomes duplicated, and random inactivation of some of the functions in each of the daughter copies results in selection against loss of either of the daughters from the genome. Subfunctionalization is, thus, a mechanism whereby mutations that are not adaptive (i.e. most mutations) can lead to the preservation of both copies of a duplicated gene because the daughters both perform subsets of the parent's suite of functions. Later on, it is possible (but not essential) that further mutations could result in the gain of a new function (neofunctionalization) by one of the daughters.

For yeast, however, subfunctionalization is not expected to be an effective mechanism of duplicate gene preservation (Lynch and Force 2000). This is because the population sizes of yeast species are very large. For subfunctionalization to happen, loss-of-subfunction alleles must become fixed by genetic drift at the daughter loci. But when the effective population size exceeds $10^6 - 10^7$ individuals, which is almost certainly true for yeast species, the very long time required for a neutral loss-of-subfunction allele to drift to fixation in the population means that the allele is very likely to acquire a second, inactivating, mutation in transit before it can become fixed (Lynch and Force 2000). This means that loss-of-subfunction alleles will not drift to fixation, so subfunctionalization will not occur in yeast.

So, have most of the duplicated genes in the yeast genome therefore been retained because they have novel functions? Before tackling this question in Section

8, we will review one of the sources of duplicated genes in the *Saccharomyces* complex of species.

5 Whole genome duplication

One of the most dramatic ways to increase the gene repertoire of an organism involves the duplication of the entire genome (polyploidization). Genomic data has provided evidence of ancient polyploidization events in many species that are now genetically diploid – a situation referred to as paleopolyploidy. Paleopolyploid species include plants such as *Arabidopsis* and the cereals (Blanc et al. 2003; Paterson et al. 2004), ray-finned fishes such as the zebrafish and *Fugu* (Taylor et al. 2001; Vandepoele et al. 2004), tetrapods such as frogs of the genus *Xenopus* (Hughes and Hughes 1993), and a large clade of yeasts in the *Saccharomyces* complex (Fig. 1).

Fig. 2 (overleaf). (a). Illustration of our model of gene order evolution following whole-genome duplication (WGD). The box at the top shows a hypothetical region of chromosome containing ten genes numbered 1–10. After WGD, the whole region is briefly present in two copies. However, many genes subsequently return to single-copy state because there is no evolutionary advantage to maintaining both copies. In this example, only genes 1, 6 and 10 remain duplicated. However, the arrangement of these three homolog pairs in the post-WGD species (bottom) would be sufficient to allow the sister regions to be detected using that genome sequence alone. Also, the order of genes in sister regions in post-WGD species have well-defined relationships to the gene order that existed in the pre-WGD genome (top), which will also be similar to the gene order seen in any species that diverged from the WGD lineage before the WGD occurred. Based on Keogh et al. (1998). Figure 2 (b). An example of gene order relationships between parts of two sister regions in *S. cerevisiae* (from chromosomes X and XI), and the homologous single chromosome regions from *A. gossypii*, *K. lactis* and *K. waltii*. In this representation, each rectangle represents a gene and homologs are arranged as vertical columns. Arrows below the rectangles show transcriptional orientation. Gray lines connect adjacent genes but do not indicate the actual gene spacing on the chromosome. In this example, the *S. cerevisiae* genes *TOR1* and *TOR2* are the only pair of ohnologs in the region and there is a single *TOR1/TOR2* ortholog in the other species. Nine other genes have all returned to a single-copy state following WGD. Apart from the post-WGD gene losses in *S. cerevisiae* and the presence of a gene (9923) in *K. waltii* that has no ortholog in the other species, there have been no other rearrangements of the region in any species. This image is a screenshot from a Yeast Gene Order Browser (YGOB) currently under development in our laboratory (K. Byrne and K. H. Wolfe, unpublished).

(a)

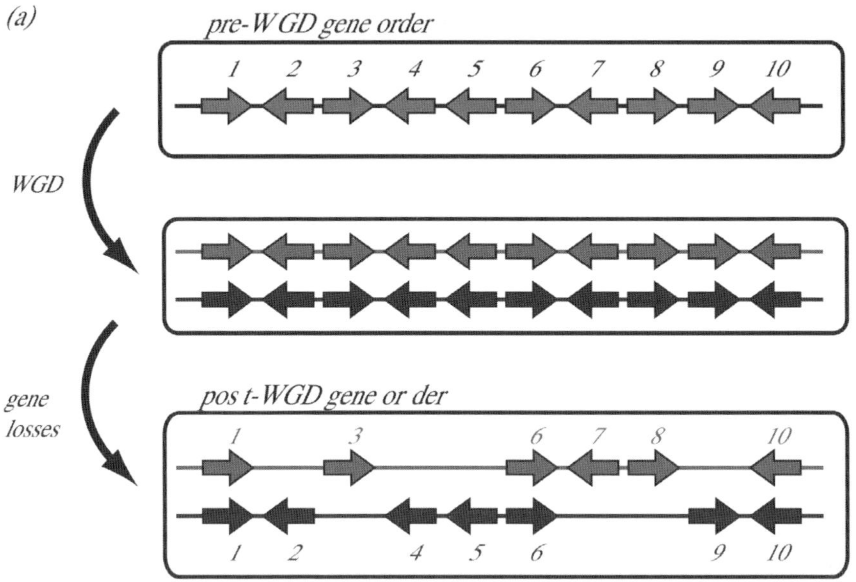

(b)

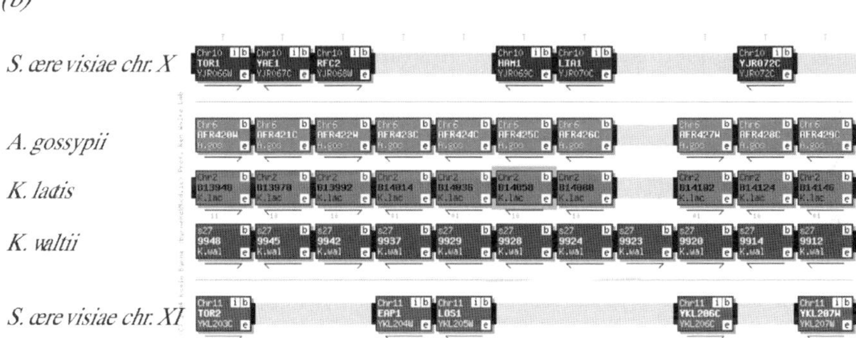

Ohno (1970) envisaged that whole genome duplication (WGD) provides a simple mechanism to generate vast numbers of duplicated genes. His name is so often associated with this process that gene duplicates produced by polyploidization are sometimes referred to as 'ohnologs' (Wolfe 2001). There are compelling advantages for polyploidy in evolution. In a polyploid species every gene is copied, including all the necessary regulatory elements. The relative stoichiometric proportions of all the gene products are also preserved in this process, minimizing potential damaging dosage effects caused by gene copy number imbalance (Papp et al. 2003). Polyploidy also generates fully redundant biochemical pathways, allowing freedom for radical biochemical innovation that can lead to major evolutionary transitions. Duplicating all the components of a pathway would be impossible by other means unless all the genes are physically close. Therefore, polyploidization is able to provide great genetic flexibility without some of the problems associated with smaller scale duplication events.

In 1997, we proposed that *S. cerevisiae* is a paleopolyploid species derived from an ancestor whose genome duplicated in a single event roughly 10^8 years ago (Wolfe and Shields 1997). The hypothesis was that subsequent chromosomal translocation and gene loss events have shaped the *S. cerevisiae* genome into its current form. By assessing the locations of duplicated genes in the *S. cerevisiae* genome, several pieces of supporting evidence for this model were found, including (i) that approximately half of the genome could be paired into sister regions where a series of genes on one chromosome had a series of homologs on another chromosome; (ii) that the large sister regions did not overlap with one another; and (iii) that the overall orientation of duplicated regions, with respect to centromeres and telomeres, had largely remained the same. That a whole genome duplication (WGD) has occurred during the evolution of *S. cerevisiae* was confirmed in 2004 through the sequencing of the genomes of several species that separated from the *S. cerevisiae* lineage prior to the WGD (Dietrich et al. 2004; Dujon et al. 2004; Kellis et al. 2004; Fig. 2).

6 Intraspecific detection of genome duplication

Genomic data from a particular species provides two ways to uncover a past polyploidization event, using information from that species alone. Both methods are dependent on the presence of ohnologs. The map-based approach involves the matching up of chromosomes, or parts of chromosomes, that can be linked by homologs located in each sister region. In the example cartooned in Figure 2a, genes 1, 6, and 10 occur as duplicated pairs in the current genome and would allow identification of the whole region between genes 1 and 10 as a duplicated 'block'. Wolfe and Shields (1997) identified 55 such duplicated blocks in *S. cerevisiae*, and subsequent work verified 52 of these with a further 32 possible paired regions (Seoighe and Wolfe 1999a). Such fragmented blocks are proposed to have arisen by WGD with subsequent reciprocal translocation between chromosomes. This view is supported by the fact that nearly all pairs of sister regions are interchromosomal, as confirmed by independent analyses (Friedman and Hughes 2001; Cavalcanti et al. 2003). Physically, the identified duplicated blocks covered about 50% of the genome (compared to a theoretical expectation of 100% for a complete genome duplication), highlighting the limitation of intraspecific block detection methods.

Extensive loss of gene duplicates has occurred in *S. cerevisiae*. It was estimated that 16% of the total gene set are ohnologs, meaning that only 8% of duplicates were preserved from the pre-polyploid ancestor. Therefore, the low number of ohnologs retained in yeast is likely to cause the sizes of sister regions found by map-based approaches to be underestimated, and may even cause some regions to lie undetected in the case of small sister regions where every pair of duplicated genes has been reduced to single-copy (Seoighe and Wolfe 1998; Kellis et al. 2004). An additional factor in yeast is the estimated 10^8 years of sequence divergence, which means that some ohnologs may not be identified using standard sequence similarity searches, further diminishing the coverage of sister regions.

Despite these shortcomings, there is convincing evidence from the *S. cerevisiae* genome itself that this species is a paleopolyploid. Llorente et al. (2000b) proposed a contradictory model where the duplicated blocks were produced by independent segmental duplications occurring at different times. The key to the resolution of the two hypotheses lies on the organization of sister regions. Under the WGD model, blocks produced by a single genome duplication event cannot overlap with each other, because the blocks are the surviving fragments of structures that were originally whole duplicated chromosomes (Wolfe and Shields 1997; Keogh et al. 1998). Under the alternative segmental duplication model, parts of chromosomes can be duplicated multiple times, creating significant overlaps between sister regions. The data clearly reveal that the majority of blocks do not overlap. Despite this, recent experimental evidence has shown that segmental duplications of large sections of chromosome can be formed in *S. cerevisiae* during artificial evolution experiments (Koszul et al. 2003), but for unknown reasons segmental duplications do not seem to have contributed significantly to the actual evolutionary history of the *S. cerevisiae* genome.

An alternative to map-based approaches to study paleopolyploidy is to use phylogenetic tree-based approaches to re-trace the origin of duplicated genes. The expectation is that ohnologs produced by a single round of genome duplication should be the same age. Friedman and Hughes (2001) tested this by estimating when, in evolutionary time, pairs of *S. cerevisiae* ohnologs diverged (i.e. the time they were duplicated). They identified 28 blocks containing genes that seemed to have duplicated simultaneously and 11 blocks containing varying amounts of relatively young duplicated genes. However, most of these recent duplicates are located within subtelomeric regions (regions near the telomeres of chromosomes) which can be subject to gene homogenizing effects (Wolfe 2001).

It is important to note that although a single genome duplication event has undoubtedly taken place in the lineage leading to *S. cerevisiae*, there are many other groups of paralogous genes in *S. cerevisiae* that were not formed by this event. The ancestral organism that underwent WGD was itself a complex eukaryote with numerous gene families – many of which expanded further via the WGD. With the availability of complete genome sequences, there is now good evidence that genes are often duplicated as tandem repeats, creating locally clustered multigene families. This is exemplified by the *SUC*, *MAL*, and *MEL* gene families in *S. cerevisiae* (Carlson et al. 1985; Michels et al. 1992; Turakainen et al. 1994). Interestingly, these tandem duplicates tend to be located in subtelomeric regions. It is known that recombinational exchanges, a process that can generate tandem repeats, are relatively frequent near chromosome ends compared to the rest of the genome (Pryde and Louis 1997). There is extensive population variation in the repertoire of some subtelomeric genes even between different yeast strains, and these genes often play adaptive roles, such as the utilization of different carbon sources or resistance to stresses (e.g. Maciaszczyk et al. 2004; Nomura and Takagi 2004). As well as tandem repeats, some other gene families have copies with highly similar sequences at dispersed locations around the genome, but these were not formed by the WGD, for example the pyruvate decarboxylase (*PDC1/ PDC5/ PDC6/ THI3*) gene family (Moller et al. 2004).

7 Interspecific detection of genome duplication

Evidence of genome duplication can be obscured by events such as extensive gene loss, chromosomal rearrangements and independent gene duplications. As described above, the ohnolog approach to infer polyploidization is useful but it is limited in its ability to detect small sister regions in a genome. This can be greatly supplemented with genomics data from related species. Sister regions in *S. cerevisiae* are interspersed with 'singletons' – genes that were duplicated *en bloc* but have subsequently returned to single-copy (in Fig. 2a, genes 2, 3, 4, 5, 7, 8, and 9 are singletons). They have little informative value in intraspecific comparative mapping because only ohnologs can be considered. However, singletons can be brought into play using genomics data from an outgroup species that diverged before polyploidization. Immediately after genome duplication, every ancestral chromosomal region corresponds to a pair of duplicated blocks in the polyploid genome. In terms of gene order, it follows that every pair of neighboring genes is also duplicated. Due to the stochastic nature of gene loss after diploidization, a pair of previously adjacent genes may end up as singletons residing on different chromosomes, although still within the same duplicated block (*e.g.* genes 2 and 3 in Fig. 2a). Without nearby ohnologs to act as anchors, the pairing of the region would have been impossible to detect intraspecifically. Yet, the gene adjacency relationship is readily preserved in the genome of a species that diverged from the *S. cerevisiae* lineage before the WGD occurred (*e.g.* a species with the 'pre-WGD' gene order shown at the top of Fig. 2a). Therefore, ancestral gene order information can be invaluable in providing the missing links between sister regions.

Several early studies used fragmentary gene order information from other species to study the origins of sister regions in *S. cerevisiae* (Keogh et al. 1998; Ozier-Kalogeropoulos et al. 1998; Seoighe and Wolfe 1999a; Ladrière et al. 2000; Langkjaer et al. 2000; Llorente et al. 2000b; Wong et al. 2002). These studies suggested that species such as *K. lactis* had 'pre-WGD' gene orders similar to what is inferred to have existed in an ancestor of *S. cerevisiae* before the genome duplication happened. These findings have now been comprehensively confirmed through the sequencing of the complete genomes of three pre-WGD species: *E. gossypii* (Dietrich et al. 2004), *K. waltii* (Kellis et al. 2004), and *K. lactis* (Dujon et al. 2004), each of which shows a 1:2 gene order relationship to sister regions in the *S. cerevisiae* genome (Fig. 2b).

In contrast, 'post-WGD' species (e.g. *S. bayanus, S. mikatae, S. paradoxus*) showed extensive gene order conservation with *S. cerevisiae* (Keogh et al. 1998; Fischer et al. 2001). Most of the disruptions in synteny in these species have been attributed to genome rearrangements such as translocations after the WGD event.

However, some genomic regions in post-WGD species can also display an apparent pre-WGD organization (Langkjaer et al. 2000; Llorente et al. 2000b; Fischer et al. 2001). This can be explained by species divergence after WGD but before the process of gene loss is complete, resulting in differential gene loss between sister regions (Seoighe and Wolfe 1999b).

Phylogenetic analysis of genes in related species provides another way of detecting polyploidization. Ohnologs retained by a paleopolyploid genome are predicted to be present as singletons in species that diverged before genome duplication. Unless the complete genome sequence of an outgroup species is available, straightforward gene counting to investigate this 2:1 relationship, as attempted by Llorente et al. (2000a), may not be reliable due to the confounding influence of multigene families. In order to obtain a clearer picture, phylogenetic methods are required. A pair of ohnologs in one species is expected to be more closely related to one another than to their ortholog in a species that diverged before the duplication event. This is represented by an A(BC) topology in a phylogenetic tree, where A corresponds to the gene in the outgroup (pre-WGD) species and the ohnologs in the post-WGD species are denoted by B and C. Other possible topologies, C(AB) and B(AC), can reflect shared older gene duplication events followed by gene loss (i.e. misidentification of putative ohnologs), or rapid sequence divergence of one gene, causing aberrant phylogenetic tree reconstruction. Under perfect circumstances, trees drawn only from *S. cerevisiae* ohnologs and their orthologs in a pre-WGD species should all assume the A(BC) topology. Furthermore, the timing of the duplication event (the coalescence date) should be uniform among different ohnolog pairs.

Due to the lack of appropriate outgroup sequences, Wolfe and Shields (1997) obtained coalescent dates for only 12 pairs of duplicated genes and concluded that the genome duplication event had occurred on the order of magnitude of 10^8 years ago. This date is consistent with the results from a later analysis using a larger ohnolog data set with *C. albicans* genes acting as outgroups (Pal et al. 2001). However, some relatively young ohnolog pairs were found in both studies. There are several possible reasons why two pairs of genes that in fact duplicated simultaneously might appear to be different ages. The age can be underestimated if gene conversion has acted to homogenize the sequences at any time after their initial duplication. Aberrantly old date estimates can result if the pre-WGD genome contained a pair of tandemly duplicated genes that were already different in sequence, and each sister region in the post-WGD species retained one of these paralogs (Smith et al. 1999). In the context of a genome doubling process, a set of ohnologs may have apparent variable duplication dates due to the asynchronous nature of the diploidization process, as may have happened in maize (Gaut and Doebley 1997; Wolfe 2001).

Apart from irregular coalescent dates, there is also another difficulty in using of tree-based methods to place the WGD event on the phylogeny of hemiascomycetes. Trees constructed from *S. cerevisiae* ohnologs and their putative pre-polyploidization orthologs do not always conform to the A(BC) topology. For example, the citrate synthase genes of *S. cerevisiae* (*ScCIT1* and *ScCIT2*) are ohnologs based on their genomic locations, but phylogenetic analysis grouped

ScCIT1 with *SkCIT1* from *S. kluyveri* to the exclusion of *ScCIT2*, even though extensive gene order information indicates that *S. kluyveri* is a pre-WGD species (Langkjaer et al. 2000). More recently, Langkjaer et al. (2003) analyzed the phylogenetic relationship of 38 *S. cerevisiae* ohnolog pairs and their orthologs in five other yeasts. Surprisingly, significant proportions of orthologs from *S. kluyveri* and *K. lactis* (58% and 28% respectively) grouped with one member of their corresponding ohnolog pair. The authors arrived at the conclusion that the WGD event pre-dated the speciation of *S. kluyveri* and *K. lactis* from *S. cerevisiae*, and that different ohnolog pairs diverged in sequence independently at different times. The complete genome sequence data now available from the *S. kluyveri* and *K. lactis* genomes makes this hypothesis untenable, however, because these species are clearly pre-WGD (they only have one locus orthologous to each ohnolog pair in *S. cerevisiae*). We suspect that the phylogenetic trees reported by Langkjaer et al., including the *CIT* tree, have been affected by long branch attraction, an artifact of phylogenetic methodology that causes erroneous tree topologies and can arise if sequences have very unequal evolutionary rates (M. A. Fares and K. H. Wolfe, unpublished results).

8 Genes lost, genes kept

The occurrence of WGD during the evolution of *S. cerevisiae* enables us to study the evolutionary fates of a large sample of genes (i.e. every gene in the genome) that were all duplicated simultaneously. Some of these genes survived in two copies, whereas many others went back to being single copies. Studying the functions of these sets of genes provides some answers to the question of how duplicated genes can survive in a species where subfunctionalization is impossible. There seem to have been two major mechanisms by which duplicates formed by WGD survived: selection for increased dosage, and neofunctionalization.

For some types of gene, the presence of additional copies in the genome can confer a selective advantage even without any divergence in the function of the loci. This can occur through 'dosage' effects if a cell gains a competitive advantage simply by merit of having higher quantities of the protein or RNA encoded by the gene. This concept is familiar from examples such as the tandem amplification of metallothionein genes in response to high concentrations of copper (Fogel and Welch 1982), or the correlation between the numbers of copies of tandem repeats of the rDNA array and cell division rate (Rustchenko et al. 1993). Selection for increased dosage is the likely reason why almost every gene for cytosolic ribosomal proteins has been retained in duplicate following the WGD in an ancestor of *S. cerevisiae*; for most of these highly expressed gene pairs there has been little or no divergence in the sequence of the two copies and they are probably being homogenized by gene conversion. Similarly, selection for increased dosage probably underlay the retention, after WGD, of duplicated genes for chaperones such as *SSB1/SSB2* and *HSP82/HSC82*.

For other pairs of genes, neofunctionalization is the probable reason why both copies have been retained in the genome. One of the clearest examples of apparent *en masse* neofunctionalization of genes duplicated by WGD is in the establishment of a set of gene isoforms specialized for growth under highly anaerobic conditions. Well studied examples of aerobic/anaerobic gene pairs include *CYC1*/*CYC7*, and *COX5A*/*COX5B*. Microarray experiments identified ten ohnolog pairs that display alternate expression profiles under aerobic or hypoxic conditions and suggested that these are only the tip of the iceberg: one-quarter of *S. cerevisiae* ohnolog pairs have at least one member that shows differential expression depending on oxygen levels (Kwast et al. 2002). The group of species that are descended from the WGD (Fig. 1) also show other evidence of adaptation towards specialization for rapid anaerobic growth: all are likely petite-positive (meaning that they can dispense with their mitochondrial if grown on a fermentable carbon source; Piskur 2001), and their genomes are depleted of genes coding for oxygen-requiring peroxisomal oxidase enzymes (S. Wong and K. H. Wolfe, unpublished results). Another example of neofunctionalization after WGD is the formation of a specialized myosin heavy chain (Myo4) that is involved specifically in setting up the asymmetry between mother and daughter cells, while its ohnolog (Myo2) is not involved in this process and continues to carry out the more usual functions of myosin (Bohl et al. 2000). A further example of neofunctionalization is the evolution of Gal3 into an inducer of galactose catabolism, whereas its ohnolog Gal1 retains enzymatic activity as a galactokinase (Platt et al. 2000).

It is still unclear what fraction of the duplicated genes retained in *S. cerevisiae* after WGD were retained for dosage reasons, and what fraction underwent neofunctionalization. In fact, the two processes are not mutually exclusive, and some gene pairs that were originally retained for dosage reasons may subsequently have undergone functional divergence. Kellis et al. (2004) searched for examples of ohnolog pairs where one copy shows evidence of significantly accelerated evolution, as expected under Ohno's model, and found evidence of acceleration in 76 out of 457 ohnolog pairs (17%). Although there are many possible causes of such an acceleration, this result suggests that neofunctionalization may have occurred in many of the retained pairs. Kellis et al. pointed to several examples where the faster-evolving member of the pair also seemed to be the one with the more 'derived' function. Their result contrasts with an earlier study of ohnologs in tetraploid *Xenopus* (an organism with a much lower population size, making it much more likely that subfunctionalization will be a major factor in the retention of duplicate frog genes): Hughes and Hughes (1993) did not find any evidence of sequence acceleration in either copy of the *Xenopus* gene pairs.

Can any generalizations be made about which genes are retained after a WGD and which become single-copy again? Genome duplication provides a unique opportunity to compare the fates of duplicated genes in different functional categories because, unlike the case for individual gene duplications in a genome, all the ohnologs are the same age so those that have survived in duplicate have survived for the same length of time. One of the most striking early results about the WGD in *S. cerevisiae* was that almost all the genes for cytosolic ribosomal proteins were retained in duplicate, and genes coding for protein kinases and other signal transduction components were also significantly over-represented among the ohnologs (Seoighe and Wolfe 1999b). Was this an accident, or was this outcome somehow inevitable? An indication that some types of genes might have higher probabilities of survival after a polyploidy has recently come from analyses of the genome of *Arabidopsis thaliana*. This plant underwent several successive polyploidizations during its evolution, the most recent of which was about 24-40 million years ago and so is considerably younger than the yeast WGD. For the most recent WGD in *Arabidopsis*, signal transduction (*i.e.* protein kinases and protein phosphatases) is among the categories of gene function that are over-represented among the retained genes, just like in yeast (Blanc and Wolfe 2004; Seoighe and Gehring 2004). Ribosomal proteins are also over-represented. Furthermore, the same types of gene tend to have been retained in duplicate after each round of WGD in *Arabidopsis* – that is, genes that were retained in duplicate after the earlier rounds of duplication are more likely also to have been retained in duplicate in the recent WGD (Seoighe and Gehring 2004). This suggests that there is a degree of inevitability to the sorting-out process after a WGD. Diversifying a signal transduction pathway by retaining duplicate genes for many of its components would be a powerful way to increase the regulatory complexity of an organism following a WGD. Interestingly, though, the end players in signal transduction cascades – transcription factors – are over-represented among the ohnologs in *Arabidopsis* but not in *Saccharomyces* (Seoighe and Wolfe 1999b; Blanc and Wolfe 2004; Seoighe and Gehring 2004).

However, yeasts and plants are so distantly related that perhaps one should not read too much into these apparently convergent results. It would be preferable to make comparisons about the outcomes of WGDs in groups of more closely-related species, as is now becoming possible in yeast species. It is particularly notable that *C. glabrata*, which is a descendant of the same WGD event as in *S. cerevisiae* (Fig. 1), does not retain two copies of most cytosolic ribosomal protein genes (Dujon et al. 2004). This suggests that species-specific factors can also strongly affect the outcome of a WGD. Further investigation of these types of questions should lead to a better understanding of the (r)evolutionary effects that the WGD had on yeast biology.

References

Baltimore D (2001) Our genome unveiled. Nature 409:814-816

Belloch C, Barrio E, Garcia MD, Querol A (1998) Inter- and intraspecific chromosome pattern variation in the yeast genus *Kluyveromyces*. Yeast 14:1341-1354

Belloch C, Querol A, Garcia MD, Barrio E (2000) Phylogeny of the genus *Kluyveromyces* inferred from the mitochondrial cytochrome-c oxidase II gene. Int J Syst Evol Microbiol 50:405-416

Blanc G, Hokamp K, Wolfe KH (2003) A recent polyploidy superimposed on older large-scale duplications in the *Arabidopsis* genome. Genome Res 13:137-144

Blanc G, Wolfe KH (2004) Functional divergence of duplicated genes formed by polyploidy during *Arabidopsis* evolution. Plant Cell 16:1679-1691

Bohl F, Kruse C, Frank A, Ferring D, Jansen RP (2000) She2p, a novel RNA-binding protein tethers *ASH1* mRNA to the Myo4p myosin motor via She3p. EMBO J 19:5514-5524

Cai J, Roberts IN, Collins MD (1996) Phylogenetic relationships among members of the ascomycetous yeast genera *Brettanomyces*, *Debaryomyces*, *Dekkera*, and *Kluyveromyces* deduced by small-subunit rRNA gene sequences. Int J Syst Bacteriol 46:542-549

Calderone RA (2002) *Candida* and Candidiasis. ASM Press, Washington D.C.

Carlson M, Celenza JL, Eng FJ (1985) Evolution of the dispersed *SUC* gene family of *Saccharomyces* by rearrangements of chromosome telomeres. Mol Cell Biol 5:2894-2902

Cavalcanti AR, Ferreira R, Gu Z, Li WH (2003) Patterns of gene duplication in *Saccharomyces cerevisiae* and *Caenorhabditis elegans*. J Mol Evol 56:28-37

Cliften P, Sudarsanam P, Desikan A, Fulton L, Fulton B, Majors J, Waterston R, Cohen BA, Johnston M (2003) Finding functional features in *Saccharomyces* genomes by phylogenetic footprinting. Science 301:71-76

Demain AL, Phaff HJ, Kurtzman CP (1998) The industrial and agricultural significance of yeasts. In: Fell JW (ed) The Yeasts, A Taxonomic Study. Elsevier, Amsterdam

Dietrich FS, Voegeli S, Brachat S, Lerch A, Gates K, Steiner S, Mohr C, Pohlmann R, Luedi P, Choi S, Wing RA, Flavier A, Gaffney TD, Philippsen P (2004) The *Ashbya gossypii* genome as a tool for mapping the ancient *Saccharomyces cerevisiae* genome. Science 304:304-307

Dujon B, Sherman D, Fischer G, Durrens P, Casaregola S, Lafontaine I, de Montigny J, Marck C, Neuvéglise C, Talla E, Goffard N, Frangeul L, Aigle M, Anthouard V, Babour A, Barbe V, Barnay S, Blanchin S, Beckerich J-M, Beyne E, et al. (2004) Genome evolution in yeasts. Nature 430:35-44

Fischer G, Neuvéglise C, Durrens P, Gaillardin C, Dujon B (2001) Evolution of gene order in the genomes of two related yeast species. Genome Res 11:2009-2019

Fleischmann RD, Adams MD, White O, Clayton RA, Kirkness EF, Kerlavage AR, Bult CJ, Tomb JF, Dougherty BA, Merrick JM, et al. (1995) Whole-genome random sequencing and assembly of *Haemophilus influenzae Rd*. Science 269:496-512

Fogel S, Welch JW (1982) Tandem gene amplification mediates copper resistance in yeast. Proc Natl Acad Sci USA 79:5342-5346

Force A, Lynch M, Pickett FB, Amores A, Yan YL, Postlethwait J (1999) Preservation of duplicate genes by complementary, degenerative mutations. Genetics 151:1531-1545

Friedman R, Hughes AL (2001) Gene duplication and the structure of eukaryotic genomes. Genome Res 11:373-381

Galagan JE, Calvo SE, Borkovich KA, Selker EU, Read ND, Jaffe D, FitzHugh W, Ma LJ, Smirnov S, Purcell S, Rehman B, Elkins T, Engels R, Wang S, Nielsen CB, Butler J, Endrizzi M, Qui D, Ianakiev P, Bell-Pedersen D, et al. (2003) The genome sequence of the filamentous fungus *Neurospora crassa*. Nature 422:859-868

Gaut BS, Doebley JF (1997) DNA sequence evidence for the segmental allotetraploid origin of maize. Proc Natl Acad Sci USA 94:6809-6814

Gellissen G, Hollenberg CP (1997) Application of yeasts in gene expression studies: a comparison of *Saccharomyces cerevisiae*, *Hansenula polymorpha* and *Kluyveromyces lactis* - a review. Gene 190:87-97

Goffeau A, Barrell BG, Bussey H, Davis RW, Dujon B, Feldmann H, Galibert F, Hoheisel JD, Jacq C, Johnston M, Louis EJ, Mewes HW, Murakami Y, Philippsen P, Tettelin H, Oliver SG (1996) Life with 6000 genes. Science 274:546, 563-567

Gojkovic Z, Paracchini S, Piskur J (1998) A new model organism for studying the catabolism of pyrimidines and purines. Adv Exp Med Biol 431:475-479

Hani J, Feldmann H (1998) tRNA genes and retroelements in the yeast genome. Nucleic Acids Res 26:689-696

Hazen KC (1995) New and emerging yeast pathogens. Clin Microbiol Rev 8:462-478

Hittinger CT, Rokas A, Carroll SB (2004) Parallel inactivation of multiple GAL pathway genes and ecological diversification in yeasts. Proc Natl Acad Sci USA 101:14144-14149

Holland BR, Huber KT, Moulton V, Lockhart PJ (2004) Using consensus networks to visualize contradictory evidence for species phylogeny. Mol Biol Evol 21:1459-1461

Hughes MK, Hughes AL (1993) Evolution of duplicate genes in a tetraploid animal, *Xenopus laevis*. Mol Biol Evol 10:1360-1369

James SA, Cai J, Roberts IN, Collins MD (1997) A phylogenetic analysis of the genus *Saccharomyces* based on 18S rRNA gene sequences: description of *Saccharomyces kunashirensis* sp. nov. and *Saccharomyces martiniae* sp. nov. Int J Syst Bacteriol 47:453-460

Jones T, Federspiel NA, Chibana H, Dungan J, Kalman S, Magee BB, Newport G, Thorstenson YR, Agabian N, Magee PT, Davis RW, Scherer S (2004) The diploid genome sequence of *Candida albicans*. Proc Natl Acad Sci USA 101:7329-7334

Kellis M, Patterson N, Endrizzi M, Birren B, Lander ES (2003) Sequencing and comparison of yeast species to identify genes and regulatory elements. Nature 423:241-254

Kellis M, Birren BW, Lander ES (2004) Proof and evolutionary analysis of ancient genome duplication in the yeast *Saccharomyces cerevisiae*. Nature 428:617-624

Keogh RS, Seoighe C, Wolfe KH (1998) Evolution of gene order and chromosome number in *Saccharomyces*, *Kluyveromyces* and related fungi. Yeast 14:443-457

Koszul R, Caburet S, Dujon B, Fischer G (2003) Eucaryotic genome evolution through the spontaneous duplication of large chromosomal segments. EMBO J 23:234-243

Kurtzman CP, Fell JW (1998) The Yeasts, a taxonomic study. Elsevier, Amsterdam

Kurtzman CP, Robnett CJ (1998) Identification and phylogeny of ascomycetous yeasts from analysis of nuclear large subunit (26S) ribosomal DNA partial sequences. Antonie van Leeuwenhoek 73:331-371

Kurtzman CP (2003) Phylogenetic circumscription of *Saccharomyces*, *Kluyveromyces* and other members of the *Saccharomycetaceae*, and the proposal of the new genera *Lachancea*, *Nakaseomyces*, *Naumovia*, *Vanderwaltozyma* and *Zygotorulaspora*. FEMS Yeast Res 4:233-245

Kurtzman CP, Robnett CJ (2003) Phylogenetic relationships among yeasts of the 'Saccharomyces complex' determined from multigene sequence analyses. FEMS Yeast Res 3:417-432

Kwast KE, Lai LC, Menda N, James DT 3rd, Aref S, Burke PV (2002) Genomic analyses of anaerobically induced genes in *Saccharomyces cerevisiae*: functional roles of Rox1 and other factors in mediating the anoxic response. J Bacteriol 184:250-265

Ladrière JM, Georis I, Guerineau M, Vandenhaute J (2000) *Kluyveromyces marxianus* exhibits an ancestral *Saccharomyces cerevisiae* genome organization downstream of *ADH2*. Gene 255:83-91

Langkjaer RB, Nielsen ML, Daugaard PR, Liu W, Piskur J (2000) Yeast chromosomes have been significantly reshaped during their evolutionary history. J Mol Biol 304:271-288

Langkjaer RB, Cliften PF, Johnston M, Piskur J (2003) Yeast genome duplication was followed by asynchronous differentiation of duplicated genes. Nature 421:848-852

Llorente B, Durrens P, Malpertuy A, Aigle M, Artiguenave F, Blandin G, Bolotin-Fukuhara M, Bon E, Brottier P, Casaregola S, Dujon B, de Montigny J, Lepingle A, Neuveglise C, Ozier-Kalogeropoulos O, Potier S, Saurin W, Tekaia F, Toffano-Nioche C, Wesolowski-Louvel M, et al. (2000a) Genomic Exploration of the Hemiascomycetous Yeasts: 20. Evolution of gene redundancy compared to *Saccharomyces cerevisiae*. FEBS Lett 487:122-133

Llorente B, Malpertuy A, Neuveglise C, de Montigny J, Aigle M, Artiguenave F, Blandin G, Bolotin-Fukuhara M, Bon E, Brottier P, Casaregola S, Durrens P, Gaillardin C, Lepingle A, Ozier-Kalogeropoulos O, Potier S, Saurin W, Tekaia F, Toffano-Nioche C, Wesolowski-Louvel M, et al. (2000b) Genomic Exploration of the Hemiascomycetous Yeasts: 18. Comparative analysis of chromosome maps and synteny with *Saccharomyces cerevisiae*. FEBS Lett 487:101-112

Lynch M, Force A (2000) The probability of duplicate gene preservation by subfunctionalization. Genetics 154:459-473

Maciaszczyk E, Wysocki R, Golik P, Lazowska J, Ulaszewski S (2004) Arsenical resistance genes in *Saccharomyces douglasii* and other yeast species undergo rapid evolution involving genomic rearrangements and duplications. FEMS Yeast Res 4:821-832

Michels CA, Read E, Nat K, Charron MJ (1992) The telomere-associated *MAL3* locus of *Saccharomyces* is a tandem array of repeated genes. Yeast 8:655-665

Moller K, Olsson L, Piskur J (2001) Ability for anaerobic growth is not sufficient for development of the petite phenotype in *Saccharomyces kluyveri*. J Bacteriol 183:2485-2489

Moller K, Langkjaer RB, Nielsen J, Piskur J, Olsson L (2004) Pyruvate decarboxylases from the petite-negative yeast *Saccharomyces kluyveri*. Mol Genet Genomics 270:558-568

Mortimer RK (1993a) Øjvind Winge: Founder of yeast genetics. In: Hall MN, Linder P (eds) The Early Days of Yeast Genetics. Cold Spring Harbor Laboratory Press, New York, p 3-16

Mortimer RK (1993b) Carl C. Lindegren: Iconoclastic father of *Neurospora* and yeast genetics. In: Hall MN, Linder P (eds) The Early Days of Yeast Genetics. Cold Spring Harbor Laboratory Press, New York, p 17-38

Nomura M, Takagi H (2004) Role of the yeast acetyltransferase Mpr1 in oxidative stress: Regulation of oxygen reactive species caused by a toxic proline catabolism intermediate. Proc Natl Acad Sci USA 101:12616-12621

Ohno S (1970) Evolution by Gene Duplication. George Allen and Unwin, London

Oliver SG, van der Aart QJ, Agostoni-Carbone ML, Aigle M, Alberghina L, Alexandraki D, Antoine G, Anwar R, Ballesta JP, Benit P, et al. (1992) The complete DNA sequence of yeast chromosome III. Nature 357:38-46

Ozier-Kalogeropoulos O, Malpertuy A, Boyer J, Tekaia F, Dujon B (1998) Random exploration of the *Kluyveromyces lactis* genome and comparison with that of *Saccharomyces cerevisiae*. Nucleic Acids Res 26:5511-5524

Pal C, Papp B, Hurst LD (2001) Highly expressed genes in yeast evolve slowly. Genetics 158:927-931

Papp B, Pal C, Hurst LD (2003) Dosage sensitivity and the evolution of gene families in yeast. Nature 424:194-197

Paterson AH, Bowers JE, Chapman BA (2004) Ancient polyploidization predating divergence of the cereals, and its consequences for comparative genomics. Proc Natl Acad Sci USA 101:9903-9908

Petersen RF, Nilsson-Tillgren T, Piskur J (1999) Karyotypes of *Saccharomyces sensu lato* species. Int J Syst Bacteriol 49:1925-1931

Phillips MJ, Delsuc F, Penny D (2004) Genome-scale phylogeny and the detection of systematic biases. Mol Biol Evol 21:1455-1458

Piskur J (2001) Origin of the duplicated regions in the yeast genomes. Trends Genet 17:302-303

Platt A, Ross HC, Hankin S, Reece RJ (2000) The insertion of two amino acids into a transcriptional inducer converts it into a galactokinase. Proc Natl Acad Sci USA 97:3154-3159

Pryde FE, Louis EJ (1997) *Saccharomyces cerevisiae* telomeres. A review. Biochemistry (Mosc) 62:1232-1241

Ramezani-Rad M, Hollenberg CP, Lauber J, Wedler H, Griess E, Wagner C, Albermann K, Hani J, Piontek M, Dahlems U, Gellissen G (2003) The *Hansenula polymorpha* (strain CBS4732) genome sequencing and analysis. FEMS Yeast Res 4:207-215

Rokas A, Williams BL, King N, Carroll SB (2003) Genome-scale approaches to resolving incongruence in molecular phylogenies. Nature 425:798-804

Rustchenko EP, Curran TM, Sherman F (1993) Variations in the number of ribosomal DNA units in morphological mutants and normal strains of Candida albicans and in normal strains of *Saccharomyces cerevisiae*. J Bacteriol 175:7189-7199

Seoighe C, Wolfe KH (1998) Extent of genomic rearrangement after genome duplication in yeast. Proc Natl Acad Sci USA 95:4447-4452

Seoighe C, Wolfe KH (1999a) Updated map of duplicated regions in the yeast genome. Gene 238:253-261

Seoighe C, Wolfe KH (1999b) Yeast genome evolution in the post-genome era. Curr Opin Microbiol 2:548-554

Seoighe C, Gehring C (2004) Genome duplication led to highly selective expansion of the Arabidopsis thaliana proteome. Trends Genet 20:461-464

Smith NGC, Knight R, Hurst LD (1999) Vertebrate genome evolution: a slow shuffle or a big bang? Bioessays 21:697-703

Souciet J, Aigle M, Artiguenave F, Blandin G, Bolotin-Fukuhara M, Bon E, Brottier P, Casaregola S, de Montigny J, Dujon B, Durrens P, Gaillardin C, Lepingle A, Llorente B, Malpertuy A, Neuveglise C, Ozier-Kalogeropoulos O, Potier S, Saurin W, Tekaia F, et al. (2000) Genomic exploration of the hemiascomycetous Yeasts: 1. A set of yeast species for molecular evolution studies. FEBS Lett 487:3-12

Sychrova H, Braun V, Potier S, Souciet JL (2000) Organization of specific genomic regions of *Zygosaccharomyces rouxii* and *Pichia sorbitophila*: comparison with *Saccharomyces cerevisiae*. Yeast 16:1377-1385

Taylor JS, Van de Peer Y, Braasch I, Meyer A (2001) Comparative genomics provides evidence for an ancient genome duplication event in fish. Philos Trans R Soc Lond B Biol Sci 356:1661-1679

Torok T, Rockhold D, King AD, Jr. (1993) Use of electrophoretic karyotyping and DNA-DNA hybridization in yeast identification. Int J Food Microbiol 19:63-80

Turakainen H, Kristo P, Korhola M (1994) Consideration of the evolution of the *Saccharomyces cerevisiae MEL* gene family on the basis of the nucleotide sequences of the genes and their flanking regions. Yeast 10:1559-1568

Vandepoele K, De Vos W, Taylor JS, Meyer A, Van De Peer Y (2004) Major events in the genome evolution of vertebrates: Paranome age and size differ considerably between ray-finned fishes and land vertebrates. Proc Natl Acad Sci USA 101:1638-1643

Wolfe KH, Shields DC (1997) Molecular evidence for an ancient duplication of the entire yeast genome. Nature 387:708-713

Wolfe KH (2001) Yesterday's polyploids and the mystery of diploidization. Nat Rev Genet 2:333-341

Wolfe KH, Li W-H (2003) Molecular evolution meets the genomics revolution. Nature Genet 33 Suppl:255-265

Wong S, Butler G, Wolfe KH (2002) Gene order evolution and paleopolyploidy in hemiascomycete yeasts. Proc Natl Acad Sci USA 99:9272-9277

Wong S, Fares MA, Zimmermann W, Butler G, Wolfe KH (2003) Evidence from comparative genomics for a complete sexual cycle in the "asexual" pathogenic yeast *Candida glabrata*. Genome Biol 4:R10

Wood V, Rutherford KM, Ivens A, Rajandream M-A, Barrell B (2001) A re-annotation of the *Saccharomyces cerevisiae* genome. Compar Funct Genomics 2:143-154

Wood V, Gwilliam R, Rajandream MA, Lyne M, Lyne R, Stewart A, Sgouros J, Peat N, Hayles J, Baker S, Basham D, Bowman S, Brooks K, Brown D, Brown S, Chillingworth T, Churcher C, Collins M, Connor R, Cronin A, et al. (2002) The genome sequence of *Schizosaccharomyces pombe*. Nature 415:871-880

Wolfe, Kenneth H.
Department of Genetics, Smurfit Institute, University of Dublin, Trinity College, Dublin 2, Ireland
khwolfe@tcd.ie

Wong, Simon
Department of Genetics, Smurfit Institute, University of Dublin, Trinity College, Dublin 2, Ireland

Telomeres in fungi

Marita Cohn, Gianni Liti, David BH Barton

Abstract

Telomeres are the functional elements concluding and defining each linear chromosome in eukaryotes. They play an essential role in protecting genetic material and preventing genome loss during cell division. At the same time, and in stark contrast, they are remarkably dynamic regions: initial analyses of yeast genomes have shown, through comparative genomics, that regions close to telomeres are prone to rearrangements and duplication and thus are particularly variable between strains and species. This propensity for variation leads to the birth of new and alternative gene functions and helps to accelerate genome evolution and divergence. However, this special property, while making telomeric regions of even greater scientific interest, complicates investigation. Firstly, repetitive DNA is problematic to clone and sequence properly. Secondly, the reoccurring rearrangements and associated lack of synteny between the telomeric regions of even very closely related species creates daunting challenges for the comparative approach. This drives the development of special cloning and bioinformatic strategies. Such efforts should be fruitful, since a comparative approach of telomeres and subtelomeres promises many insights of significance to the research of ageing and cancer, chromosome dynamics in cell division, and the processes of evolution and speciation.

1 Introduction

Like the protective aglets on the end of shoelaces, telomeres are the caps on the ends of linear chromosomes. Special nucleotide sequences and protein complexes at the telomeres distinguish these purposeful ends from the damaged ends left by an accidental double-strand DNA break. Thus, chromosome ends with normal telomeres avoid the fate of a naked DNA terminus: they escape from degradation by nucleases, avoid the attentions of DNA repair mechanisms, and fail to set off alarm signals at DNA damage checkpoints. However, since DNA polymerases are unable to faithfully reproduce linear DNA all the way to the lagging 3' end (the well-known "end replication problem"; Olovnikov 1971, 1973; Watson 1972), multiple rounds of replication can lead to the erosion of telomeres down cell generations, until eventually DNA checkpoints, genome loss, or gross chromosomal rearrangements trigger cell arrest or cell death (Fig. 1). Even though this erosion probably contributes to the aging process in higher eukaryotes, these organisms

Topics in Current Genetics, Vol. 15
P. Sunnerhagen, J. Piškur (Eds.): Comparative Genomics
DOI 10.1007/4735_106 / Published online: 11 October 2005
© Springer-Verlag Berlin Heidelberg 2005

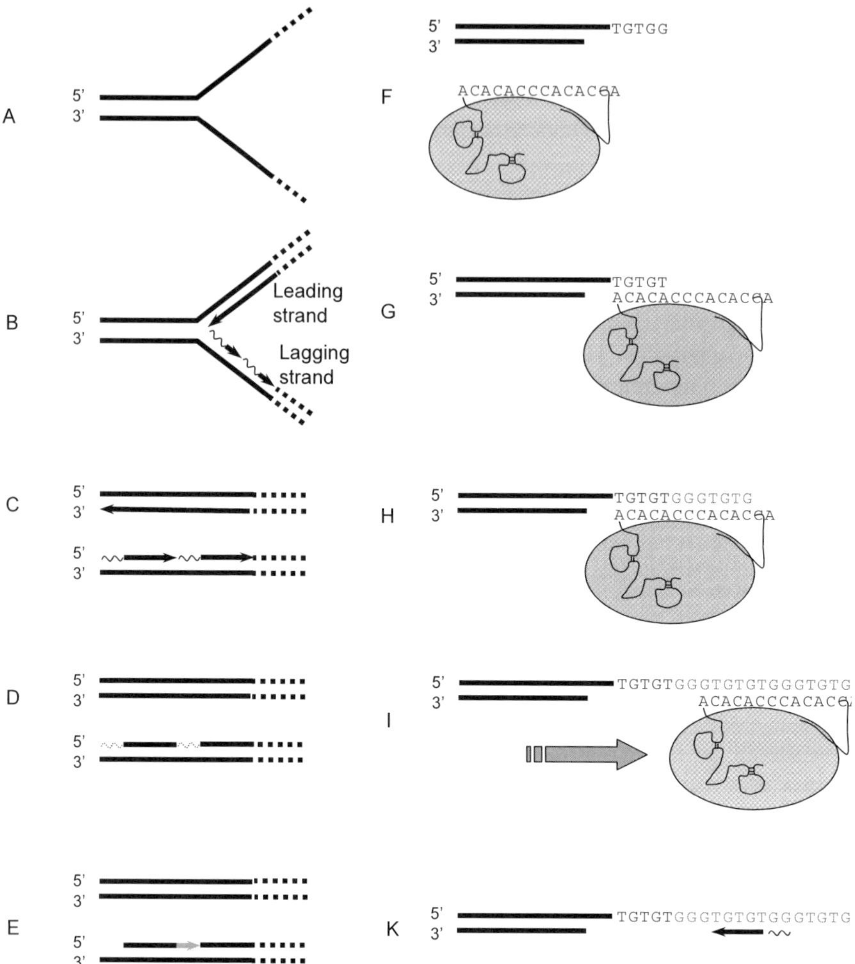

Fig. 1. End-replication problem (A-E) and telomere extension by telomerase (F-L). DNA is unzipped (A) and replication is initiated by the leading and the lagging strands (B). Only the replication fork moving to the left is shown. Dotted lines indicate the DNA continuing towards the centromere. The leading strand proceeds continuously whereas the lagging strand discontinuously extends Okazaki fragments (C) from multiple RNA primers (~~). RNA primers are replaced (D) with DNA and ligated. The terminal RNA primer cannot be replaced generating a shorter DNA molecule (E) each replication round. In order to overcome telomere attrition telomerase binds to DNA (F-G) and synthesises new DNA by reverse transcription from an RNA template (H). Telomerase transpositions extend the synthesis (I) and other enzymes ensure synthesis of the complementary end (K).

normally exploit the inherent shortcomings of telomeres as a built-in defence against unlimited (cancerous) division of somatic cells. Where telomere loss must be countered (in germ cells, e.g., or in single-celled organisms), there are fascinating specialised mechanisms for rebuilding telomeric sequences. In general these mechanisms (and the repeat sequences on which they operate) are well conserved even between different kingdoms. Within ascomycetous yeasts however, there is quite unexpected diversity, as we will describe.

Aside from the telomeres themselves, it has been known for many years that the genomic regions close to telomeres are uncommonly variable and prone to re-arrangements and duplication. This subtelomeric plasticity is yet another telomere-related phenomenon that can be exploited by organisms for their benefit: many parasitic eukaryotes, for example, seem to use it to adapt quickly to host defences.

For their potential significance in ageing and cancer (e.g. Zakian 1985; Kruk et al. 1995; Neidle and Parkinson 2002), in parasitic immunology, in normal chromosome dynamics as well as in the processes of evolution and speciation, telomeres and their neighbouring subtelomeric domains are inviting objects for study, and a comparative approach promises many insights. One problem confounds us however: telomeres and subtelomeres happen to be particularly difficult to sequence.

2 Telomeres in sequencing projects

If *S. cerevisiae* is the first eukaryote to be completely sequenced, it is in large part because it is the first eukaryote for which there was a concerted effort to sequence the 'difficult bits'. The Whole Genome Shotgun methodology used so successfully in major genome sequencing projects is thwarted by lengthy and highly repetitive or duplicated regions of genome. Telomeres and subtelomeres fall into this category in every eukaryote known, as do centromeres in higher eukaryotes. Firstly, telomeric or other repetitive reads may be under-represented because their very nature interferes with the cloning or sequencing processes. Secondly, it is very difficult to reliably assemble sequence reads that lie within large areas of repeated sequence; it is akin to correctly assembling a jigsaw puzzle where all the pieces are blue sky, and square! Finally, these regions cannot necessarily be assembled on the basis of synteny with a reference organism, because (in budding yeast at least) rearrangements in subtelomeric regions seem to break this synteny even between very closely related species.

Special cloning strategies have been developed to overcome these hurdles and specifically sequence telomeric and subtelomeric regions, methods that are now being applied to organisms other than yeast. These often use *S. cerevisiae* as a host, rather than *E. coli* where repetitive sequences can be unstable.

Half-YAC vectors have been used to fill the subtelomeric gaps of the human genome project (Riethman et al. 1989, 2001). This method consists in cloning partially digested high molecular weight human DNA into a YAC vector that contains *Tetrahymena* telomeric repeats in only one arm. This will enrich for inserts that provide a telomeric repeat in the other end because this then creates a stable linear mini-chromosome. The vector also contains a selectable yeast marker

(*URA3*), and an origin of replication (ARS1) next to a centromeric sequence (CEN4).

Telomere-associated recombination (TAR) cloning (Larionov et al. 1996; Noskov et al. 2003; Becker et al. 2004) is a strategy for targeting a specific chromosome end using an anchor sequence (500bp) unique to that end. The anchor sequence is cloned into a half-YAC vector between a negative selectable marker (*CYH2*) and the rest (*URA3*, CEN4, ARS1, and telomeric repeat array). The linearised vector is co-transformed with the whole genomic DNA from the organism of interest into an *S. cerevisiae* strain where the ligase IV gene (*DNL4*) has been deleted. This deletion prevents non-homologous end joining (NHEJ) and favours homologous recombination (HR) as a mode of repair. Clones that lost *CYH2* can be selected and characterised. This method has been used successfully to clone the six chromosome ends of *Schizosaccharomyces pombe* (Sharp and Louis, personal communication), as well as the telomeres of *Pneumocystis carinii* (Underwood et al. 1994, 1996), *Trypanosoma brucei* (Becker et al. 2004) and others.

The 32 telomeres of *S. cerevisiae*, however, were cloned using an early and more time-consuming method, in which an *URA3*-based vector was integrated into the repeat array of each telomere (Louis and Borts 1995). In each case it was necessary to determine which of the telomeres had been tagged: both intact and digested preparations of chromosomal DNA were separated in a pulse field gel electrophoresis (PFGE) and probed with unique vector sequences. These blots were compared to the physical map (Link and Olson 1991) to uniquely determine each marked telomere. Using the appropriate restriction enzymes the *URA3* integrated vector could be rescued, ligated, and transformed into *E. coli* for sequencing.

3 The structure and maintenance of telomeric repeats

The telomeres of *S. cerevisiae* remain to this day the most fully characterised of any organism. Their terminal sequence is a series of imperfect ($TG_{1-3}/C_{1-3}A$) repeats typically about 300 bp long (Shampay and Blackburn 1988; Wright et al. 1992), with the TG-rich strand forming a single-stranded 3' overhang. The length of this overhang is typically 12-14 nt for much of the cell cycle (Larrivee et al. 2004) but peaks at >25 nt during S phase (Wellinger et al. 1993). It serves as the substrate for telomere extension so it must be maintained as a single strand: either DNA polymerases are prevented from replicating it, or else any complementary strand that is synthesised on it is then degraded by nuclease activity.

As in most other eukaryotes (*Drosophila* is a notable exception), telomeres in *S. cerevisiae* are normally maintained through the action of a dedicated enzyme called telomerase. This enzyme contains an RNA moiety, part of which serves as the template for complementary binding to and extension of the single-strand overhang. It is, in other words, a specialised reverse transcriptase (Fig. 1). The template component often complements about 1.5 repeat units (e.g. REPEATREP). In each round of polymerization, a single repeat is synthesised on the end of the 3' overhang, but telomerase may perform several rounds of polymerization before dissociating from the end. After extension of the overhang, all or some of the complementary strand is presumably synthesised by conventional DNA polymerases.

Given that telomeric repeats are copied from a fixed RNA template, why then are the *S. cerevisiae* repeats so irregular? Until the identification of *S. cerevisiae* telomerase activity (Cohn and Blackburn 1995), researchers had supposed that this irregularity indicated a telomere maintenance mechanism other than telomerase in yeasts. The *S. cerevisiae* RNA template, as well as the putative RNA template in its near relatives, is complementary to the sequence TGTGTGGGTGTGGT (Dandjinou et al. 2004). The telomerase probably synthesises the irregular repeats by using alternative start and stop positions on this low complexity template (Forstemann and Lingner 2001).

Saccharomyces castellii and *Saccharomyces dairenensis* contain multiple types of variant repeat forms that can be grouped according to length (Cohn et al. 1998). Constituting, respectively, 26% and 36% of the total repeats, the variants are distributed randomly along the telomeric repeat array. These variants could also be accounted for by the use of alternative segments of a single telomerase template (Chappell and Lundblad 2004; authors' interpretation). In *Candida tropicalis* the telomeric DNA repeats are of two distinct types, differing in a single nucleotide position. It may be, in this case, that the two variant forms are instead produced by two different telomerase RNA genes (McEachern and Blackburn 1994). *Schizosaccharomyces pombe* also has irregular repeats, which do not conform to any simple repeated pattern (Matsumoto et al. 1987; Lue and Peng 1997), but to our knowledge the RNA template still remains unidentified. Variant telomeric repeats are, in fact, the exception rather than the rule, but they are a distinct advantage when analyzing the lengths and maintenance dynamics of telomeres since they can be used to distinguish individual telomeres on the basis of their different combinations along the sequence array of each individual telomere (Wahlin et al. 2003).

Most fungi, and indeed most eukaryotes, have quite regular telomeric repeats. Indeed, from most fungi to vertebrates, the sequence of those repeats is remarkably conserved: TTAGGG in mammals, some plants and most of the fungal kingdom (*Neurospora crassa, Podospora anserina,* and *Pneumocystis carinii;* Schechtman 1990; Javerzat et al. 1993; Underwood et al. 1996). The related repeats TTAGG and TTTAGGG are prevalent in insects (Frydrychova et al. 2004) and in plants (Fajkus and Zentgraf 2002), respectively. Among the ascomycetous yeasts, however, even those with regular repeats still show significant diversity in the length (8-26 bp) and composition of the repeat unit (see Table 1).

4 Telomerase structure

In both humans and yeast, the telomerase enzyme needs only two components for catalytic activity in vitro (Beattie et al. 1998): the template-containing RNA subunit (*TER*) and a catalytic protein subunit (*TERT*: telomerase reverse transcriptase). In *S. cerevisiae*, these are encoded by the genes *TLC1* and *EST2*, respectively. In *TLC1*, RNA stem-loop structures around the template are essential for the binding of protein subunits (Chappell and Lundblad 2004; Dandjinou et al. 2004; Zappulla and Cech 2004), and through such loops Est2p binds *TLC1* directly.

There are two other regulatory protein components of telomerase in *S. cerevisiae*, encoded by *EST1* and *EST3*. Est1p clearly helps to recruit the holoenzyme to its target: it also binds the RNA subunit *TLC1* (Lin and Zakian 1995; Steiner et al. 1996; Zhou et al. 2000), as well as the single-stranded telomere overhang itself (Virta-Pearlman et al. 1996; Zhou et al. 2000), and also the overhang-binding protein Cdc13p (Qi and Zakian 2000).

Genes for *TERT* (the catalytic protein subunit) were first identified by comparing the putative sequences of two yeast species, *S. cerevisiae* and *Schiz. pombe*, with the ciliate *Euplotes aediculatus* and with human sequence; an excellent example of the power of comparative genomics (Lingner et al. 1997). The biochemical evidence from some of those species was combined with the genetic evidence from others to determine which genes, showing high amino acid conservation across the species, coded for the catalytic protein of telomerase. The alignment showed conserved amino acid sequence motifs from retroviral and retrotransposon reverse transcriptases, and mutations of specific amino acids in the putative *TERT* proteins were shown to eliminate telomerase activity both in vivo and in vitro (Lingner et al. 1997; Nakamura et al. 1997). This is a good example of the way that comparative biochemical analyses and comparative genomics approaches can work in parallel to deliver important insights. Another notable example is the very first isolation of yeast telomerase activity, which was done in *S. castellii*, and which led to the subsequent isolation in *S. cerevisiae* (Cohn and Blackburn 1995). Thus, using comparative genomics, comparative biochemistry, as well as other methods, newly sequenced yeast species will become interesting models that will be helpful to reveal yet unidentified proteins and pathways in *S. cerevisiae*.

A simple BLASTP search can readily identify *EST2* homologues amongst the recently released genome sequences of other *Saccharomyces* species (*S. paradoxus*, *S. mikatae*, *S. bayanus*, and *S. castellii*) although none of these have yet been genetically or biochemically characterised (Cliften et al. 2003; Kellis et al. 2003). Though total amino acid identities diverge substantially (86% and 74% for *S. paradoxus* and *S. bayanus*, respectively), all show conservation of the residues previously determined to be of importance to reverse transcriptase function (Lingner et al. 1997; authors analysis). In *C. albicans* two different *TERT* genes were identified bioinformatically; conserved *TERT* motifs were used to search the *C. albicans* sequencing project database for regions showing high sequence similarity (Metz et al. 2001). These orthologues, *CaTERT1* and *CaTERT2*, are either different genes or different alleles of the same gene. They differ from each other at five amino acid positions, and both have ~34% identity to *S. cerevisiae* Est2p.

Table 1. Telomeric repeats in fungi

Species	Sequence	Reference
Neurospora crassa	TTAGGG	Schechtman 1990
Fusarium oxysporum	TTAGGG	Powell and Kistler 1990
Histoplasma capsulatum	TTAGGG	Woods and Goldman 1992
Cladosporium fulvum	TTAGGG	Coleman et al. 1993
Podospora anserina	TTAGGG	Javerzat et al. 1993
Ustilago maydis	TTAGGG	Guzman and Sanchez 1994
Magnaporthe grisea	TTAGGG	Farman and Leong 1995
Pneumocystis carinii	TTAGGG	Underwood et al. 1996
Aspergillus nidulans	TTAGGG	Bhattacharyya and Blackburn 1997
Aspergillus oryzae	TTAGGGTCAACA	Kusumoto et al. 2003
Cryptococcus neoformans	$TTAG_{3-5}$	Edman 1992
Schizosaccharomyces pombe	$T_{1-2} ACA_{0-1} C_{0-1} G_{1-6}$	Matsumoto et al. 1987
Saccharomyces cerevisiae	$TG_{2-3}(TG)_{1-6}$	Shampay et al. 1984
Saccharomyces exiguus	$TG_{2-3}(TG)_{1-6}$	Cohn et al. 1998
Saccharomyces castellii,	TCTGGGTG	Cohn et al. 1998
Saccharomyces dairenensis	TCTGGGTG	Cohn et al. 1998
Saccharomyces kluyveri	GACATGCGTACTGTGAGGTCTGGGTG	Cohn et al. 1998
Candida albicans	TCTAACTTCTTGGTGTACGGATG	McEachern and Hicks 1993
Candida guilliermondii	TACTGGTG	McEachern and Blackburn 1994
Candida glabrata	CTGTGGGGTCTGGGTG	McEachern and Blackburn 1994
Candida maltosa	CAGACTCGCTTGGTGTACGGATG	McEachern and Blackburn 1994
Candida tropicalis	TCACGATCATTGGTGTA(A/C)GGATG	McEachern and Blackburn 1994
Candida pseudotropicalis	TGATTAGTTATGTGGTGTACGGATT	McEachern and Blackburn 1994
Kluyveromyces lactis	TGATTAGGTATGTGGTGTACGGATT	McEachern and Blackburn 1994

The sequence corresponding to the template region of the *S. cerevisiae* telomerase RNA, ACCACACCCACACA, is also found in the *Saccharomyces* species *S. paradoxus*, *S. cariocanus*, *S. mikatae*, *S. kudriavzevii*, *S. bayanus*, and *S. pastorianus* (Dandjinou et al. 2004).

Even though some motifs of the *TERT* proteins are well conserved throughout evolution, there may be substantial variation not only in the primary sequence but also in the regulation and constitution of the holoenzyme. For example, while *S. cerevisiae* telomerase is a homodimer, with each *TERT* protein complexed to its own separate telomerase RNA, the human holoenzyme contains two catalytic *TERT* proteins in complex with a single RNA molecule (Chen et al. 2000).

Even very closely related yeast species show an unexpected high degree of divergence in the telomerase RNA (*TER*) genes (Tzfati et al. 2000), and the isolation of *TLC1* genes from these species by traditional PCR cloning and hybridization techniques proved cumbersome and had a low success rate. Therefore, the study of *TLC1* genes has benefited considerably from the special sequencing efforts for several very closely related yeast species (Feldmann 2000; Cliften et al. 2003; Kellis et al. 2003). Such projects have facilitated the identification of *TLC1* homologues, and the elucidation of secondary structures and functional elements.

A picture is emerging of an RNA where only a few discrete domains are needed for the enzymatic function of telomerase (Cech 2004; Zappulla and Cech 2004). Primary sequence can be highly divergent but there is, nonetheless, substantial conservation of the secondary folding structure over a wide range of species (Romero and Blackburn 1991; Chen et al. 2000; Tzfati et al. 2003).

The molecular covariation of *TERT* and *TER* suggests a bioinformatics approach in which one can screen new putative *TER* genes by their ability to form functional hairpin loops (Dandjinou et al. 2004; Zappulla and Cech 2004). Interestingly, it may be that the *TERT* protein recognises a sequence-independent secondary structure in the RNA, since a stem-loop, which lacks bulges in the duplex is demonstrated to mediate the protein-RNA interaction (Chappell and Lundblad 2004). Complementation tests confirmed that the newly identified *TER* genes were functional homologues, and experiments in which specific *TER* regions were swapped between species were helpful in determining the functionality of hairpin structures where covariation could be seen. Analyses such as these have been made possible by a range of computational tools, including ClustalW for sequences alignment, mFold for RNA folding predictions, and phylogenetic analyses aided by the algorithm X2s, which identifies compensatory base changes in aligned sequences. The future bioinformatics analysis of RNA genes will clearly require the development of software combining multiple approaches: probing primary sequence and tertiary structure simultaneously to identify functional RNA structures in rapidly evolving RNA genes.

5 Other telosome proteins

Telomerase is only one of the many protein components of the 'telosome' (a term describing the whole nucleoprotein structure capping chromosome ends). An increasing number of different proteins have been found to bind specifically to the ends of chromosomes. In addition to building up a protective cap, which keeps the telomere from triggering cellular rescue mechanisms, these proteins play very specific and individual roles in the regulation of telomere length. They may do so by facilitating the recruitment of telomerase to its substrate, the 3' overhang, or they may instead inhibit the access of telomerase. Some of the proteins in the telosome structure are DNA-binding proteins, binding either the single-stranded overhang or the double-stranded telomere repeat array, while other telosome proteins are assembled into the structure by interacting with those DNA-binding proteins.

Among the telomere-binding proteins found along the length of the telomere are the *RAP1*, RIF, and SIR proteins. Rap1p is one of the most well-studied and has shown to be extremely multifunctional. It binds double-stranded telomeric repeats, and telomere length is determined at least in part by a mechanism of Rap1p counting (Marcand et al. 1997; Brevet et al. 2003). It also binds numerous other locations and recognises a somewhat broad range of consensus sequences (for review see Wahlin and Cohn 2002b). Rap1p binding sites appear to be barriers against the spread of silencing chromatin, but it also seems to create silencing chromatin at telomeres and at silent mating-type loci (Morse 2000). Rap1p also serves as a transcriptional activator at many promoters, especially for ribosomal protein genes (Lieb et al. 2001). Binding appears, in at least some cases, to be regulated by phosphorylation (Tsang et al. 1990). The protein contains a BRCT domain commonly found in cell-cycle checkpoint proteins.

The telomere-associated proteins Rif1p and Rif2p (*RAP1* Interacting Factors) are recruited to the telomere by binding to the Rap1p C-terminal domain. Loss of Rif1p and Rif2p leads to telomere elongation, so they evidently participate in the control of telomeric length and also in establishing silent chromatin. The C-terminal domain of Rap1p is also bound by Sir3p and Sir4p (Silencing Information Regulators). Some mutant alleles of *SIR4* have the unusual property of prolonging cell lifespan (Strahl-Bolsinger et al. 1997); in these mutants Sir4p and Sir3p associate not with the telomeres but with the nucleolus, and may somehow suppress the accumulation of extrachromosomal rDNA circles; known to produce aging-like effects in budding yeast (Park et al. 1999). Sir4p may be phosphorylated by Cdc28p, the catalytic heart of the primary cell-cycle-regulating cyclin-dependent kinase. Sir3p and Sir4p bind to histones (Hecht et al. 1995). The *SIR3* gene seems to have arisen from a duplication of *ORC1* (encoding the main subunit of the origin recognition complex) as part of the whole genome duplication event (Kellis et al. 2004). The duplicate that gave rise to *SIR3* shows signs of having undergone accelerated evolution before developing its new function.

The telomere-associated *SIR2* protein, Sir2p, associates with telomeres and silent mating-type loci like the other two, but it also regulates silencing of rDNA loci (Strahl-Bolsinger et al. 1997). Sir2p contains a domain that is extraordinarily well conserved across all kingdoms of life, even in bacteriophages, and there are multiple homologues of the gene (*HST1-4*) in *S. cerevisiae* alone (Brachmann et al. 1995).

The Ku proteins, Yku70p and Yku80p, form a heterodimer which binds to certain nucleotide structures common to telomeres and to DNA breaks, chiefly the junction between single and double-stranded DNA (Mimori and Hardin 1986; Falzon et al. 1993). This dimer plays a role in the non-homologous end-joining (NHEJ) mechanism of DNA repair, bringing loose ends together. The Ku heterodimer plays a role in regulating chromosome length, and may help to recruit telomerase RNA to the end (Stellwagen et al. 2003). It is highly conserved in evolution, from prokaryotes to eukaryotes, but chiefly forms homodimers in the former and heterodimers in the latter (Downs and Jackson 2004). There is evidence that these proteins play an unexpected role in the ultrastructure of telomeric DNA in the nucleus, with deletions appearing to result in a decoupling of telomeres from the nuclear periphery, where they normally cluster into foci; following this disruption, telomeric and subtelomeric genes that are ordinarily silenced become derepressed (Laroche et al. 1998).

The telomere-specific protein Cdc13p binds to the single-stranded TG-rich overhang of the telomere. It is an essential protein, which has been demonstrated to recruit telomerase through binding of Est1p and Est3p. It also binds Pol1p, the major subunit of DNA polymerase alpha (Qi and Zakian 2000). When Cdc13p is disabled, ssDNA is generated towards the centromere and the long exposed TG-rich tail triggers cell arrest (Booth et al. 2001). Other telomere-interacting proteins are Tel2p, which also binds the single-strand overhang and is essential, Stn1p, Ten1p and Stm1p. Stm1p, though primarily a cytoplasmic protein, binds to Cdc13p, and also to G-quadruplex DNA and to Y' elements (see Section 8), as well as to ribosomal components (Van Dyke et al. 2004). It is rapidly degraded by proteasomes and plays a role in triggering cell death; deletion can actually increase cell survival under normal growth conditions (Ligr et al. 2001).

What then does a comparative approach contribute to the understanding of these proteins and of telomere biology in general? Rap1p homologues have been isolated from a number of budding yeasts and beyond. The DNA binding specificity of these homologues has evidently been conserved to a considerable degree, which suggests that Rap1p binding is a major selective force operating on budding yeast telomeric DNA sequences (Wahlin and Cohn 2002a; Wahlin et al. 2003). However, the human and *Schiz. pombe* Rap1p orthologues do not bind DNA directly. Instead, they are recruited to the telomere by their respective DNA-binding protein partners, Taz1p and Trf2p.

In contrast, Cdc13p, though conserved amongst budding yeasts (Mitton-Fry et al. 2004), is not found in any recognisable form in higher eukaryotes. Instead in various distantly related eukaryotes there are functionally analogous proteins, different in sequence but nonetheless containing an OB-fold in their DNA-binding domains (Mitton-Fry et al. 2002). The OB-fold is a conserved structural element used for oligonucleotide, oligosaccharide, and oligopeptide binding, and the presence of this structural similarity indicates that Cdc13p shares a common ancestor with other telomere end-binding proteins. The profile-profile comparison algorithm COMPASS was used to probe for distant homology among telomeric OB-fold domains; the Cdc13p homologues identified by BLAST of genomic sequence databases were added in this analysis. The results support the contention that all OB-fold telomeric end-binding proteins arise from a common origin (Theobald et al. 2003).

6 Telomere length

The total length of the telomeric repeat array often varies throughout the cell cycle and between different telomeres in the same cell, and between different cells in the same organism or colony. Even the mean length appears to vary substantially from species to species: less than 100 bp in certain ciliates to several kb in humans and more than 100 kb in certain mouse species (for reviews see Greider 1998; McEachern et al. 2000). By these standards the telomeric repeat arrays in fungi are relatively short. In *Podospora*, the mean length is ~200bp (Javerzat et al. 1993), in *Aspergillus* ~100-125bp (Bhattacharyya and Blackburn 1997; Kusumoto et al. 2003), and similarly 130-175 bp in *Candida tropicalis* (McEachern and Blackburn 1994). Among other yeasts *S. castellii, S. dairenensis, S. exiguus, S. kluyveri, Candida albicans*, and several additional *Candida* species there is a mean length comparable to that of *S. cerevisiae,* around 300-500bp (McEachern and Blackburn 1994; Cohn et al. 1998). However, even within *S. paradoxus* strains, the mean length from different isolates ranges from 150 to 750 bp (Liti, unpublished data). Telomere function is evidently not disrupted even by substantial length variation, but telomerase activity is certainly regulated at multiple levels.

The expression of the telomerase catalytic component is controlled both by positive and negative transcriptional regulators. In humans, and probably many other eukaryotes, the enzyme is also regulated through tissue-specific alternate splicing (Ulaner et al. 2001). Moreover, the ability of telomerase to access its 3' single-stranded substrate is also regulated by positive and negative factors (Evans and Lundblad 2000). Several of the proteins constituting telomeric chromatin are implicated in regulation at this level, and the postulated model for this regulation involves the three-dimensional folding of that chromatin (Vega et al. 2003). As already mentioned, there is evidence that a Rap1p-counting mechanism is used to gauge the length of telomeres and thereby regulate it (Marcand et al. 1997).

A recent systematic and comprehensive screen of *S. cerevisiae* deletion mutants to identify those with abnormal telomere length produced ~150 candidates, one third of which produce abnormally long telomeres while the rest led to telomere shortening (Askree et al. 2004). These genes spanned a huge range of roles, painting an alarmingly complex picture of telomere length regulation. There is within this subset a relative overabundance of nucleotide processing factors and chromatin modification factors (both perhaps to be expected) and also of factors involved in intracellular trafficking, presumably due to interference with the normal turnover of telomeric proteins (Askree et al. 2004).

7 Strand bias and telomeric DNA conformation

It is striking that so many telomeres, across so many phyla, are TG-rich in the 3'-ending strand. G-rich DNA may adopt forms other than the usual B-form double-helix; there is evidence that it can adopt a Z-form conformation more readily (Walmsley et al. 1983; Reich et al. 1993) as well as forming a G-quadruplex (Liu et al. 1995), or hairpin structure. There is evidence that such structures do form at telomeres (Henderson et al. 1987; Sundquist and Klug 1989), Rap1p promotes the formation of G-quadruplex DNA in vitro (Giraldo and Rhodes 1994) and there is also biochemical evidence (Salazar et al. 1996) to support the hypothesis that secondary structures of this kind in telomeric DNA play a role in the (ATP and GTP-independent) translocation of telomerase along the single-strand overhang for the next round of reverse transcription (Shippen-Lentz and Blackburn 1990). Stm1p, as already mentioned, binds G quadruplex DNA (Hayashi and Murakami 2002).

Certainly, a functional screen of random *S. cerevisiae* template RNA sequences showed that CA-rich templates were strongly favoured (Forstemann et al. 2003). By contrast, *Drosophila* telomeres, which are maintained by a mechanism other than telomerase, are not TG-rich (Louis 2002).

8 Telomere-associated elements

In almost every eukaryote studied there are, behind the telomere proper, 'telomere-associated sequences' (TASs). Such TASs may be common to many or all chromosome ends, and they often possess some repeat structure.

In *S. cerevisiae* three types of TAS were originally identified through hybridization experiments: an X element, a '131' sequence, and a Y element (Chan and Tye 1983). At the time, Chan and Tye suggested that the "X sequences may actually be composed of a number of smaller repetitive elements and each of the homologous X sequences may contain a subset of these smaller repetitive elements"; they were correct and the early definitions have indeed been refined. The X element is now regarded as a set of smaller elements (Louis et al. 1994): a "core X" sequence and four types of pseudo-repetitive elements called X element combinatorial repeats (XCRs) or subtelomeric repeats (STRs). In individual telomeres, the core X sequence may be accompanied by none, a subset of, or all of the XCR types (Fig. 2). The 131 and Y elements were quickly combined and are now referred to collectively as a Y' element (Walmsley et al. 1984).

The term XCR, recently adopted by the *Saccharomyces* Genome Database, should be preferred over the original term STR (subtelomeric repeat). The latter is a confusing acronym since it is more widely used to refer to simple tandem repeats (which these elements manifestly are not) and because it perhaps implies more regularity than actually exists.

The core X sequence is found at all chromosome ends in the sequence of *S. cerevisiae* S288C, albeit fairly weakly conserved and varying in length between 200 and 470 bp. Almost all core X elements contain an ARS consensus sequence and a binding site for Abf1p (ARS-binding factor 1). Core X elements improve the segregation efficiency of plasmids carrying telomere repeats (Enomoto et al. 1994). Since core X elements are found at all ends, they are considered to delimit the end of the 'telomeric region' (a term used to collectively describe telomeric repeats and telomere-associated sequences) and the start of the subtelomeric domain.

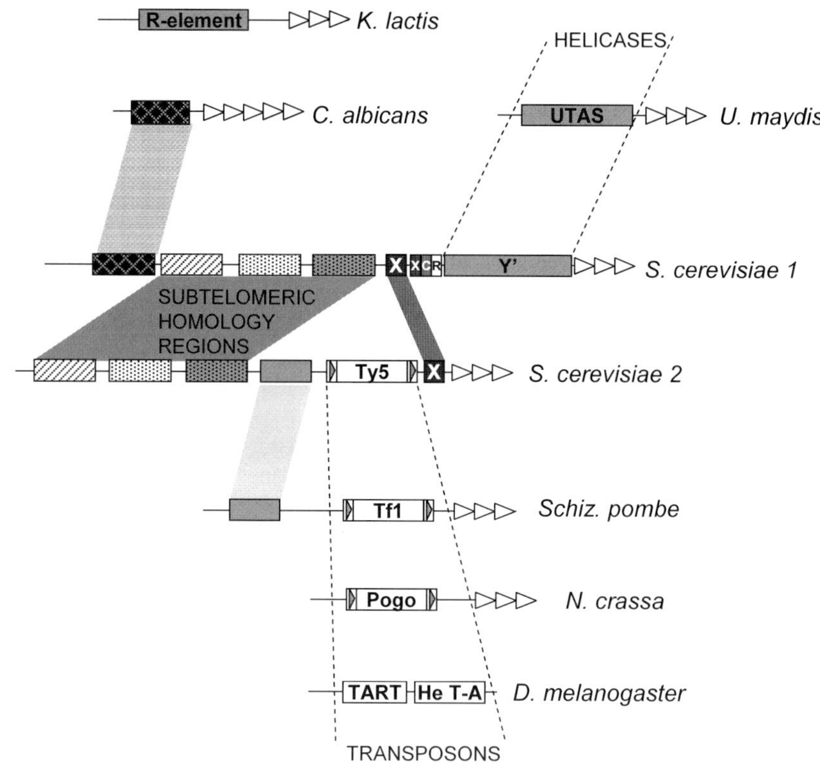

Fig. 2. General telomeric and sub-telomeric structure. Triangles designate telomeric repeats. Grey boxes indicate sequence homology. Two hypothetical *S. cerevisiae* telomeres are depicted and subtelomeric homology regions (SHR) are represented. Different degrees of sequence homology are also present across species. Dashed lines indicate elements having structural similarity but not sequence homology.

XCRs, when present, lie between the core X element and the telomeric repeats or adjacent Y' element. Four pseudo-repeat types have been tentatively identified and assigned the letters A to D, but it should be noted that the distinctions and boundaries between them are hazy at best. Different combinations of these pseudo-repeat units are present at different ends (so that XCR lengths range from 100-370bp) though order is conserved.

XCR-A, the most proximal to telomeric repeats but missing from 11 chromosome ends, contains complete and/or degenerate DNA binding sites for Tbf1p, an essential protein with some homology to *Schiz. pombe* Taz1p, and to the human telomere-binding proteins *TRF1* and *TRF2*. Intriguingly, its standard binding site (TTAGGG) is the same sequence as the telomeric repeat in various other fungi, and in humans (Flint et al. 1997; see Table 1). In *S. cerevisiae*, Tbf1p binding would appear to block the spread of transcription-silencing chromatin (Fourel et al. 1999; Koering et al. 2000). In *Schiz. pombe*, Taz1p is involved in chromatin silencing and various other telomeric functions as well. Elucidation of the function of its homologues in other fungi such as *Aspergillus*, *Neurospora*, or *Cryptococcus* (which still have Tbf1p binding sites in their telomeric repeats) may reveal how the role of this protein has evolved.

In two telomeres of the sequenced strain (IX-L & X-L) the boundary between the XCRs and neighbouring Y' element is overwritten by a recent insertion, the majority of which is identical in sequence to a group I self-splicing intron from the mitochondrial cytochrome gene BI4 (Louis and Haber 1991). It is likely that the insertion occurred only once but that the telomeric region containing it was subsequently duplicated.

Y' elements are a different breed of telomere-associated element: they have the qualities of a mobile genetic element that inserts into telomeric repeats. In the sequence of *S. cerevisiae* S288C they are present in about half the telomeres, in some cases as multiple tandem copies that may or may not be separated by telomeric repeats. There appears to be no correlation between Y' copy number and other features of the chromosome or chromosome end in question, and the particular configuration of Y' elements seems to vary considerably from species to species and even within a given species (Button and Astell 1986; Zakian and Blanton 1988; Jager and Philippsen 1989; Louis and Haber 1990a, 1990b, 1992; Liti et al. 2005).

The sequenced *S. cerevisiae* Y' elements are highly conserved at either end, but there are various deletions in the middle of some elements. They have been subclassified into long (6.7kb) and short (5.2kb) varieties (Chan and Tye 1983; Louis and Haber 1992; Louis et al. 1994; Yamada et al. 1998). Although sequence data shows that there is in fact a smooth range of intermediates from 4.7 to 6.9kb, there is certainly a distinct shared set of deleted regions in the short form. High frequency of recombinational exchange has been measured between the Y' elements (Louis and Haber 1990a; Louis et al. 1994), however both long and short forms of Y' are maintained, which suggests that recombination is not random.

All Y' elements contain potential ORFs. While some of these are short and encode proteins that resemble no other (and may therefore not be genuine or functional) there are 7-13 very similar lengthy Y' ORFs in S288C that encode a protein with some of the features of RNA helicases. These genes can be highly transcribed when normal telomerase activity is disrupted, and the proteins do indeed have helicase activity (Yamada et al. 1998). Y' transcripts have also been detected during meiosis in normal cells (Louis 1995). Some versions of these ORFs contain short introns. Y' elements also contain a small number of TTAGGG Tbf1p-binding motifs (Louis and Borts 1995).

Interestingly, Y' elements have been detected as extrachromasomal circular elements (Horowitz and Haber 1985); they contain autonomous replication sequences (ARSs) and may therefore be able to survive and propagate independently. This suggests a mechanism for their dispersal through excision, duplication and insertion via recombination. It should be noted that Y' elements lack the features of known mobile genetic elements in yeast (authors unpublished analysis): no sequence homology has been detected with any yeast transposons, killer viruses, linear plasmids, or the 2μm plasmid, nor do Y' elements share analogous features such as the terminal repeats of transposons and linear plasmids or the small direct repeats involved in the partitioning of 2μm plasmids. There is, however, some weak homology between the Y' helicase and translation initiation factors Tif1p and Tif2p, and to other DEAD-box family RNA helicases. Efforts to discern the origin of these elements, through further comparative studies, are ongoing.

Y' elements do have one known function in budding yeast, but that function operates only in exceptional circumstances and may be nothing more than a coincidentally useful by-product of its selfish transposon-like behaviour. In yeast populations where telomerase function has been lost, most cells undergo progressive telomere loss until the cell cycle permanently arrests, a phenomenon known as senescence. However, in a large enough population, some cells can overcome the loss of functional telomerase by using alternative strategies to rebuild their telomeres. In *S. cerevisiae*, two main ALT (Alternative Lengthening of Telomeres) mechanisms have been observed, dubbed 'type I' and 'type II', both relying on homologous recombination, functional Rad52p and other DNA repair proteins (Lundblad and Blackburn 1993; McEachern and Blackburn 1995; Chen et al. 2001; Lundblad 2002). In type I cells, generally the first survivors to appear in a senescing population, Y' elements have somehow proliferated into long tandem arrays at all chromosome ends. This appears to be a 'quick fix' in as much as type I survivors eventually senesce and recover through the type II pathway, in which the telomeric repeats themselves are suddenly and rapidly amplified to ~20 times their usual length.

Once the sequence of a TAS is known, its presence or absence at particular ends can be experimentally determined by separating chromosomes on CHEF gels followed by digestion with *NotI* and Southern blotting with a suitable probe. Using this method, Y' elements have been detected in other species of the *Saccharomyces* sensu stricto group but not in the most distant member *S. bayanus*, and not, so far, in more distant relatives (Liti et al. 2005). Copy number may be determined with real-time PCR.

Because telomeric and subtelomeric sequence is challenging to acquire and assemble, there are few other fungi in which telomere-associated sequences have been identified. Recent whole genome shotgun sequencing of the ascomycetous yeasts *S. paradoxus*, *S. mikatae*, *S. kudriavzevii*, *S. bayanus*, *S. castellii*, *S. kluyveri*, and *Kluyveromyces waltii* (Cliften et al. 2001; Kellis et al. 2003, 2004) produces sequence with BLAST homology to fragments of various *S. cerevisiae* TASs, but these fragments cannot yet be assigned any order or position (authors analysis).

A dedicated effort to sequence such regions in *Kluyveromyces lactis* has detected a completely novel telomere-associated sequence just internal to the telomeric repeats (which, at 25bp, are unusually long). This 'R element', which is present at almost all chromosome ends, is 1.5–2 kb long, extremely (~70%) purine rich in the 3'-ending strand, and contains some repetitive segments (including a TTTTTTCA repeat unit at the end adjacent to the telomeric repeats) but few convincing ORFs. These R elements were then detected by hybridization in some but not all of the other *Kluyveromyces* strains checked (Nickles and McEachern 2004).

In any given species, the presence of specific sequence elements next to many or all telomeres indicates that these sequences have some function, but on the other hand there are often (as we shall discuss in Section 9) lengthy regions of homology shared between chromosome ends, and some duplicated sequence elements might simply be remnants of the process that gave rise to these. Some TASs contain ORFs; others do not. Some are transposon-like (see the Pogo element in *Neurospora crassa*, described in the next section), while others are not. Some have a discernable repeat structure, while others do not. The comparative study of these elements is in its infancy, but it is reasonable to expect that a more precise, descriptive and meaningful nomenclature will emerge.

9 Subtelomeric homology regions, multiple gene families, and adaptation

It had been known for some time that there were many duplicated genes at the ends of chromosomes: as *S. cerevisiae* S288C was sequenced, some large blocks of homology, shared between two or more chromosome ends, were noted (e.g. Bowman et al. 1997; Jacq et al. 1997). Once the sequence was complete, many more subtelomeric 'cluster homology regions' (CHRs) could be identified (Louis and Becker, unpublished analysis). Some of these cover over 25kb of a chromosome end. Others are smaller but still extensive islands of homology, which do not extend all the way to the telomere. A few CHRs are noticeably more degenerate and presumably more ancient. The prevalence of these duplications indicates that subtelomeric regions experience an elevated number of ectopic recombination events, and the high similarity in many of these regions demonstrates that they arose through recombination in the recent past.

Any organism must balance the need for genome stability with the need for adaptability. A more variable genome allows a population to adapt much more rapidly to environmental change, but will also give rise to a greater incidence of inviability and infertility amongst offspring, because vital genes are also more prone to deleterious mutation. 'Subtelomeric plasticity' circumvents this catch-22 dilemma. Essential genes can be partitioned away in 'safe' areas of the genome, while non-essential but adaptive genes can be left at chromosome ends, subject to rapid recombination, duplication and mutation (Louis 1995).

It is striking how many of the duplicated genes in subtelomeric CHRs are involved in adaptive but essentially disposable processes, particularly secondary metabolism, toxin resistance and cell-to-cell interaction. Many of these genes have been identified and genetically mapped in various yeasts, but cannot necessarily be physically mapped onto the sequence of S288C because that strain has lost or rearranged the ORFs. Others do correlate with a specific sequence but mutation in S288C has left only a pseudogene; proof, if proof was needed, of the extent of subtelomeric polymorphism between strains.

In the *SUC* gene family, encoding sucrose-hydrolyzing invertases, at least six members have been genetically identified in budding yeasts. All are subtelomeric

(Carlson et al. 1985) but only one, *SUC2* (YIL162W), has been physically mapped on the S288C sequence. The *MAL* genes (maltose fermentation) occur in more complex multigene loci, with an activator gene, a maltose permease, and the catalytic enzyme maltase, all adjacent to one another (sometimes misleadingly referred to as a regulon). At least five such loci have been identified (Chow et al. 1989), but only two of these appear in S288C sequence: MAL1 (YGR288W, YGR289C and YGR292W) on chromosome VII-L and MAL3 (YBR297W, YBR298C and YBR299W) on II-R. The activator proteins at both loci appear to be non-functional in S288C. Also apparently forming multigene units, which are duplicated at multiple chromosome ends, are the *MEL* genes, encoding proteins such as alpha-galactosidase, responsible for galactose and melibiose metabolism and regulated by the well-known *GAL* proteins. None of these can be physically mapped in S288C; only a few *Saccharomyces* species have the ability to ferment these sugars (Liljestrom 1985; Naumov et al. 1991, 1995). The three subtelomeric *ERR* genes are divergent from but still strongly homologous to the two *ENO* (enolase) genes, from which they presumably claim descent (Pryde et al. 1995). There appears to be only one *ERR* gene in *S. paradoxus*, and no copies have been found, either by hybridization or blast search, in other available yeast sequences (Liti et al. 2005).

There are many more examples: thiamine synthesis (*THI*) genes (Hohmann and Meacock 1998) and aryl-alcohol dehydrogenase (*AAD*) genes can be found side-by-side in four subtelomeres (IV-L, VI-L, X-R and XIV-L), though only one of the *AAD* genes is considered fully functional and capable of responding to oxidative stress (Delneri et al. 1999a, 1999b). Many, though not all members of the following gene families have arisen via subtelomeric duplication: *ADH* (NADPH-dependent alcohol dehydrogenases), *PAU* (seripauperins; active during alcoholic fermentation but little understood), and *COS* (membrane proteins involved in salt resistance (Mitsui et al. 2004).

Another well-known family of subtelomeric genes arising from duplication are the *FLO* genes 1, 5, 9, and 10, which encode lectin-like cell-surface proteins controlling flocculation (Teunissen et al. 1995). In addition there are several other *FLO*-like pseudogenes. Flocculation, the clumping together of cells, is a characteristic that was deliberately and systematically bred out of the S288C lineage as experimentally undesirable (Mortimer and Johnston 1986), so the S288C genes are likely to differ substantially from wild type.

The role of cell-surface proteins in adaptability is most obvious in parasitic organisms: they must evade host defences and their virulence often depends on their binding to host cells. In these cells such 'contingency genes' are often found in subtelomeric domains, with subtelomeric chromatin silencing regulating which are expressed. This is the case not only in protozoan pathogens such as *Plasmodium falciparum* (Hernandez-Rivas et al. 1997) and *Trypanosoma brucei* (Chaves et al. 1999; Becker et al. 2004) but in fungal pathogens too.

Candida glabrata is an opportunistic human pathogen much more closely related to *S. cerevisiae* than to its better-known namesake *Candida albicans*. The *EPA* gene family in this organism resembles the *S. cerevisiae FLO* genes mentioned above, and is also subtelomeric. Though these genes appear to usually be silenced, they do contribute to the binding of pathogen to host cells in vitro and thus play some role in virulence (De Las Penas et al. 2003).

The simple oomycetous fungus *Phytophthora infestans* causes potato late blight and was responsible for the Irish Potato Famine of the 1840s (May and Ristaino 2004). Such pathogens cannot only produce virulence factors but also 'avirulence' factors, which, oddly, trigger host defences and block infection (often through the self-destruction of infected plant tissue). The avirulence factors are often host-strain specific and, although the molecular basis for these interactions is still poorly understood, a number of *P. infestans* avirulence genes have been genetically mapped to subtelomeres (van der Lee et al. 2001).

In the rice blast ascomycetous fungus *Magnaporthe grisea* (Gao et al. 2002) and the corn smut basidiomycetous fungus *Ustilago maydis* (Sanchez-Alonso and Guzman 1998), there are subtelomeric families of helicases (as there is, of course, in the Y' elements of *S. cerevisiae*). Unlike the Y' helicases however, these are significantly homologous to the *E. coli* RecQ helicase and to *S. cerevisiae* Sgs1p. The latter, which strongly localises to the nucleolus and potentially plays a role in rDNA processing, is also known to preferentially unwind G-quadruplex DNA (Sun et al. 1999). The exact role of these helicases, and the reason for their recurrence as subtelomeric gene families in widely divergent strains of fungi, are two great unsolved mysteries in telomere biology.

As previously mentioned, Y' elements may constitute a novel kind of mobile genetic element. In *S. cerevisiae* as well as in other fungi, there are other more typical transposons, some of which seem to have a preference for telomeric or subtelomeric regions. Four of the five types of transposon in *S. cerevisiae* normally insert near tRNA genes, but Ty5 preferentially integrates into the silent chromatin of subtelomeres and mating loci (Zou et al. 1996; Zou and Voytas 1997). It would appear that accumulated mutations have left the *S. cerevisiae* Ty5 incapable of further transposition, however, in *S. paradoxus* this element is still active and can be found in high copy numbers in certain strains (Zou et al. 1995). Ty5-like transposons have also been found, by sequence homology, in *S. exiguus*, *Pichia angusta*, *Debaryomyces hansenii* and *C. albicans* (Neuveglise et al. 2002), but it has not been established whether these related transposons retain their preference for telomeres and silent mating loci.

In *Schiz. pombe*, although the subtelomeres remain poorly elucidated, a single Tf1 LTR has been found in the right end of chromosome I (Hunt et al. 2001). In the filamentous fungus *Neurospora crassa*, a telomere-associated element dubbed Pogo was identified in sequence from one chromosome end (Schechtman 1990). Lying next to the terminal TTAGGG repeats, the Pogo element contains some direct and inverted repeats which lend the element a structure reminiscent of a transposable element. There is a 207aa ORF in the Pogo element, which lacks any significant homology to any other characterised protein (apart from some slight homology to one obscure bacterial tRNA synthetase) but is very similar to a second hypothetical ORF in the same organism. A BLAST search against the contigs from the recent *N. crassa* Whole Genome Shotgun project (Galagan et al. 2003) reveals several regions that are homologous to the Pogo element, but on the basis of exonuclease sensitivity (Bal31 digestion) the copy present at the V-R telomere appears to be the only one present in a subtelomeric region. Thus, it might be premature to label this a telomere-associated element when its presence at that location is in fact mere happenstance.

Comparative genomics has revealed at least 11 subtelomeric reciprocal translocations across the *Saccharomyces* sensu stricto complex (Kellis et al. 2003), and these appear to have originated because of ectopic recombination between transposons, tRNA arrays and duplicated genes (Fischer et al. 2000). Duplications therefore, particularly subtelomeric ones, make further duplications and translocations more likely.

10 Nuclear architecture: influence of spatial positioning on genomic dynamics

Telomere biology, in fungi and elsewhere, can seem like a bewildering collection of seemingly disparate phenomena. There are the unusually high rates of ectopic recombination leading to subtelomeric plasticity. There is the patchy silencing of genes at chromosome ends, related in some way to the chromatin-mediated silencing of other genomic locations such as the two mating cassettes HML and HMR in chromosome III, and the long rDNA array in chromosome XII. There are hints of unusual DNA conformations, and complicated relations between telomeres and DNA repair. But telomeres are not normally distributed at random within the nucleus (certainly in *S. cerevisiae*, and doubtless in other organisms too) and the different facets of telomere biology may eventually be drawn together by a deeper understanding of the non-random architecture of chromosome ends within the nucleus.

Fluorescence in situ hybridization and live cell imaging both show a perinuclear arrangement of telomeres and generally continuous movement along the nuclear periphery. The peripheral clustering of subtelomeric regions affects the interaction between interstitial and subtelomeric regions. Ectopic recombination between alleles occurs at much lower rates if one allele is at a chromosome end and the other is not, suggesting the presence of some form of recombinational barrier between the subtelomeric and interstitial loci consistently with the chromosome ends being physically sequestered away from the rest of the genome. As already mentioned, some genes required for maintaining the nuclear architecture (e.g. Yku70/80) are also responsible for maintaining this recombination barrier (Laroche et al. 1998). However, neither *KU* nor *SIR* proteins have a membrane domain, which indicates that there are other proteins involved in the anchoring of telomeres to the nuclear membrane (possibly Esc1p; Taddei and Gasser 2004).

Silencing at telomeres is patchy in both spatial and temporal terms. The telomere position effect (TPE; reviewed in Tham and Zakian 2002) was firstly described when a marker, *ADE2*, was inserted near newly synthesised telomeric repeats (Gottschling et al. 1990). *S. cerevisiae* with a telomeric *ADE2* gene generated red colonies (the phenotype of ade2-) with white sectors (the phenotype of *ADE2*+), or vice versa, indicating that the transcriptional state at this telomere was reversible (and might be linked to rapid changes in gene expression in different growth conditions). As the distance between marker and telomere increased, this variegated expression pattern gradually decreased. A number of genes were implicated in this effect, some of them specific to the telomeric silencing (*RAP1*, YKU70/80) while other (e.g. the *SIR* genes) were also involved in the suppression of the two silent mating type cassettes. It should be noted at this point that recent genome sequencing of related hemiascomycetes *K. lactis* and *Debaryomyces hansenii* has revealed silent mating type cassettes similar to those of *S. cerevisiae* (Dujon et al. 2004).

These initial studies of TPE involved truncated telomere ends (the repetitive nature of the subtelomeric DNA making it difficult to target a specific sequence with a marker gene), but studies of silencing in native ends have produced a different picture of TPE (Fourel et al. 1999; Pryde and Louis 1999). Firstly, the repression along the proximity of the telomere is not continuous but punctate: there are zones of strong repression, coinciding with the core X element, and others with very little repression (such as in Y' elements). What is more, some native ends show no repression at all, though the differences between repressive and non-repressive ends have yet to be determined. Finally, the gene regulation of the repression is different between truncated and native ends. For example, the mutations orc2-1 and orc5-1 do not affect silencing at native ends but do abolish TPE at truncated ends.

The results using native telomere ends are significant in that they support a new structural loop-like model of *S. cerevisiae* telomeres (Pryde and Louis 1999), in which telomeric repeats and their associated proteins (e.g. *RAP1* and SIR proteins) fold back and physically associate with the core X element and its associated proteins. Y' elements, sitting in this loop, distance themselves from the 'knot' and thereby escape silencing. This model has yet to be confirmed, and the issue of whether the perinuclear position of telomeres is sufficient for TPE remains controversial, but the debate highlights the importance of nuclear architecture in understanding telomere biology.

11 The future revealed by comparative genomics of telomeres

Yeast is a useful organism in many fermentative production processes of food or beverages. Some of these processes rely on the natural occurrence of yeast populations that are present on the fruit. Depending on the geographical location, the population of yeast present may differ. Furthermore, within the same geographical region the population may differ from year to year due to weather conditions. Since different yeast species will differ in their metabolism and the production of by-products, such a variation in population profile makes it hard to predict the final quality of the product. Therefore, to ensure a uniform quality of the product, modern wine producers are controlling the fermentation by seeding the fruit juice with large inoculates of specific yeast mixtures. However, some wine producers are still depending on a naturally occurring flora of yeasts, since the large spectrum of different yeast strains contribute with biochemical characteristics responsible for very specific flavours. Here it is hard to determine all the different types of yeast species present and how they contribute to the final product. This is quite a challenge, since there is a need to determine which strains are active in all the different phases of the fermentation process. Cell morphology is not a suitable character to use when aiming for the determination of which yeast species is present in a sample. Instead, a species-specific bar code could be used. The telomeric DNA could be used for this purpose, since the telomeric DNA sequences are highly variable among yeast species. Even relatively closely related yeast species show some specific differences in the sequence (Cohn et al. 1998). The telomeric DNA could therefore be used as an identification tag for yeast species: a 'telomere fingerprint'. The screening procedure could be developed as a hybridization of a chip array on which all the different known telomeric sequences are represented. This kind of telomere sequence analysis may also be envisioned as a useful tool to identify yeast species in patients infected by yeast pathogens.

Assaying ORF deletion libraries for effects on telomere length may be a useful method for identifying all genes that are important for telomere biology (Askree et al. 2004). However, as with other genes, telomere maintenance genes may be masked due to the conditions or the particular strain backgrounds used in the study. This points out the need to not only increase the conditional variations but also extend such analyses to other non-conventional yeast species. Under the same conditions, a particular genetic network may be masked in one species but not the other.

The inclusion of telomeric sequences in the whole genome sequencing efforts will provide us with important and useful information. Of course, the telomeric sequence data would bring important knowledge for telomere biology research, about the evolution of telomere structure and maintenance. The sequencing of closely related species, and even of different strain isolates, may provide interesting information about population substructures and differences in telomere maintenance among yeasts. Furthermore, the telomeric regions will provide essential information on general genome evolution. Because subtelomeric regions play a specific evolutionary role, as generation sites for new genes, a telomere-including genomic sequencing program would give crucial information regarding this phenomenon. For this purpose, the analysis of closely related species will benefit the delineation of the evolutionary pathway paved by gene duplication and a subsequent divergence of function. With this knowledge at hand, maybe we will learn how to understand not only the past of a genome but its future as well.

Long may the eukaryotic genomic sequencing efforts continue, but hopefully with an awareness of the importance of the telomeric and subtelomeric regions. Procedures to accurately sequence and assemble these, chromosome-walking, for example, should now become an established, valued, and routine part of every such project. In simple fungi at least the relatively short telomeres should not pose much problem for the sequence retrieval and analysis; it is the subtelomeric regions, which will prove difficult because of their considerable length, variability, and duplication. It would be worth the trouble.

Acknowledgements

We thank Ed Louis for critical discussions on the manuscript. D.B.H. Barton is supported by BBSRC (to E. Louis), G. Liti is supported by The Wellcome Trust (to E. Louis).

References

Askree SH, Yehuda T, Smolikov S, Gurevich R, Hawk J, Coker C, Krauskopf A, Kupiec M, McEachern MJ (2004) A genome-wide screen for *Saccharomyces cerevisiae* deletion mutants that affect telomere length. Proc Natl Acad Sci USA 101:8658-8663

Beattie TL, Zhou W, Robinson MO, Harrington L (1998) Reconstitution of human telomerase activity in vitro. Curr Biol 8:177-180

Becker M, Aitcheson N, Byles E, Wickstead B, Louis E, Rudenko G (2004) Isolation of the repertoire of VSG expression site containing telomeres of *Trypanosoma brucei* 427 using transformation-associated recombination in yeast. Genome Res 14:2319-2329

Bhattacharyya A, Blackburn EH (1997) *Aspergillus nidulans* maintains short telomeres throughout development. Nucleic Acids Res 25:1426-1431

Booth C, Griffith E, Brady G, Lydall D (2001) Quantitative amplification of single-stranded DNA (QAOS) demonstrates that cdc13-1 mutants generate ssDNA in a telomere to centromere direction. Nucleic Acids Res 29:4414-4422

Bowman S, Churcher C, Badcock K, Brown D, Chillingworth T, Connor R, Dedman K, Devlin K, Gentles S, Hamlin N, Hunt S, Jagels K, Lye G, Moule S, Odell C, Pearson D, Rajandream M, Rice P, Skelton J, Walsh S, Whitehead S, Barrell B (1997) The nucleotide sequence of *Saccharomyces cerevisiae* chromosome XIII. Nature 387:90-93

Brachmann CB, Sherman JM, Devine SE, Cameron EE, Pillus L, Boeke JD (1995) The *SIR2* gene family, conserved from bacteria to humans, functions in silencing, cell cycle progression, and chromosome stability. Genes Dev 9:2888-2902

Brevet V, Berthiau AS, Civitelli L, Donini P, Schramke V, Geli V, Ascenzioni F, Gilson E (2003) The number of vertebrate repeats can be regulated at yeast telomeres by Rap1-independent mechanisms. EMBO J 22:1697-1706

Button LL, Astell CR (1986) The *Saccharomyces cerevisiae* chromosome III left telomere has a type-X, but not a type-Y', ARS Region. Mol Cell Biol 6:1352-1356

Carlson M, Celenza JL, Eng FJ (1985) Evolution of the dispersed SUC gene family of *Saccharomyces* by rearrangements of chromosome telomeres. Mol Cell Biol 5:2894-2902

Cech TR (2004) Beginning to understand the end of the chromosome. Cell 116:273-279

Chan CS, Tye BK (1983) A family of *Saccharomyces cerevisiae* repetitive autonomously replicating sequences that have very similar genomic environments. J Mol Biol 168:505-523

Chappell AS, Lundblad V (2004) Structural elements required for association of the *Saccharomyces cerevisiae* telomerase RNA with the Est2 reverse transcriptase. Mol Cell Biol 24:7720-7736

Chaves I, Rudenko G, Dirks-Mulder A, Cross M, Borst P (1999) Control of variant surface glycoprotein gene-expression sites in *Trypanosoma brucei*. EMBO J 18:4846-4855

Chen JL, Blasco MA, Greider CW (2000) Secondary structure of vertebrate telomerase RNA. Cell 100:503-514

Chen Q, Ijpma A, Greider CW (2001) Two survivor pathways that allow growth in the absence of telomerase are generated by distinct telomere recombination events. Mol Cell Biol 21:1819-1827

Chow TH, Sollitti P, Marmur J (1989) Structure of the multigene family of MAL loci in *Saccharomyces*. Mol Gen Genet 217:60-69

Cliften P, Sudarsanam P, Desikan A, Fulton L, Fulton B, Majors J, Waterston R, Cohen BA, Johnston M (2003) Finding functional features in *Saccharomyces* genomes by phylogenetic footprinting. Science 301:71-76

Cliften PF, Hillier LW, Fulton L, Graves T, Miner T, Gish WR, Waterston RH, Johnston M (2001) Surveying *Saccharomyces* genomes to identify functional elements by comparative DNA sequence analysis. Genome Res 11:1175-1186

Cohn M, Blackburn EH (1995) Telomerase in yeast. Science 269:396-400

Cohn M, McEachern MJ, Blackburn EH (1998) Telomeric sequence diversity within the genus *Saccharomyces*. Curr Genet 33:83-91

Coleman MJ, McHale MT, Arnau J, Watson A, Oliver RP (1993) Cloning and characterisation of telomeric DNA from *Cladosporium fulvum*. Gene 132:67-73

Dandjinou AT, Levesque N, Larose S, Lucier JF, Abou Elela S, Wellinger RJ (2004) A phylogenetically based secondary structure for the yeast telomerase RNA. Curr Biol 14:1148-1158

De Las Penas A, Pan SJ, Castano I, Alder J, Cregg R, Cormack BP (2003) Virulence-related surface glycoproteins in the yeast pathogen *Candida glabrata* are encoded in subtelomeric clusters and subject to *RAP1-* and SIR-dependent transcriptional silencing. Genes Dev 17:2245-2258

Delneri D, Gardner DC, Bruschi CV, Oliver SG (1999a) Disruption of seven hypothetical aryl alcohol dehydrogenase genes from *Saccharomyces cerevisiae* and construction of a multiple knock-out strain. Yeast 15:1681-1689

Delneri D, Gardner DC, Oliver SG (1999b) Analysis of the seven-member AAD gene set demonstrates that genetic redundancy in yeast may be more apparent than real. Genetics 153:1591-1600

Downs JA, Jackson SP (2004) A means to a DNA end: the many roles of Ku. Nat Rev Mol Cell Biol 5:367-378

Dujon B, Sherman D, Fischer G, Durrens P, Casaregola S, Lafontaine I, De Montigny J, Marck C, Neuveglise C, Talla E, Goffard N, Frangeul L, Aigle M, Anthouard V, Babour A, Barbe V, Barnay S, Blanchin S, Beckerich JM, Beyne E, Bleykasten C, Boisrame A, Boyer J, Cattolico L, Confanioleri F, De Daruvar A, Despons L, Fabre E, Fairhead C, Ferry-Dumazet H, Groppi A, Hantraye F, Hennequin C, Jauniaux N, Joyet P, Kachouri R, Kerrest A, Koszul R, Lemaire M, Lesur I, Ma L, Muller H, Nicaud JM, Nikolski M, Oztas S, Ozier-Kalogeropoulos O, Pellenz S, Potier S, Richard GF, Straub ML, Suleau A, Swennen D, Tekaia F, Wesolowski-Louvel M, Westhof E, Wirth B, Zeniou-Meyer M, Zivanovic I, Bolotin-Fukuhara M, Thierry A, Bouchier C, Caudron B, Scarpelli C, Gaillardin C, Weissenbach J, Wincker P, Souciet JL (2004) Genome evolution in yeasts. Nature 430:35-44

Edman JC (1992) Isolation of telomerelike sequences from *Cryptococcus neoformans* and their use in high-efficiency transformation. Mol Cell Biol 12:2777-2783

Enomoto S, Longtine MS, Berman J (1994) Enhancement of telomere-plasmid segregation by the X-telomere associated sequence in *Saccharomyces cerevisiae* involves *SIR2*, *SIR3*, *SIR4* and ABF1. Genetics 136:757-767

Evans SK, Lundblad V (2000) Positive and negative regulation of telomerase access to the telomere. J Cell Sci 113:3357-3364

Fajkus J, Zentgraf U (2002) Structure and maintenance of chromosome ends in plants. In: Krupp G, Parwaresch R (eds) Telomeres and Telomerases: Cancer and Biology. Landes Bioscience, pp314-331

Falzon M, Fewell JW, Kuff EL (1993) EBP-80, a transcription factor closely resembling the human autoantigen Ku, recognizes single- to double-strand transitions in DNA. J Biol Chem 268:10546-10552

Farman ML, Leong SA (1995) Genetic and physical mapping of telomeres in the rice blast fungus, *Magnaporthe grisea*. Genetics 140:479-492

Feldmann H (2000) Genolevures- a novel approach to 'evolutionary genomics'. FEBS Lett 487:1-2

Fischer G, James SA, Roberts IN, Oliver SG, Louis EJ (2000) Chromosomal evolution in *Saccharomyces*. Nature 405:451-454

Flint J, Bates GP, Clark K, Dorman A, Willingham D, Roe BA, Micklem G, Higgs DR, Louis EJ (1997) Sequence comparison of human and yeast telomeres identifies structurally distinct subtelomeric domains. Hum Mol Genet 6:1305-1313

Forstemann K, Lingner J (2001) Molecular basis for telomere repeat divergence in budding yeast. Mol Cell Biol 21:7277-7286

Forstemann K, Zaug AJ, Cech TR, Lingner J (2003) Yeast telomerase is specialized for C/A-rich RNA templates. Nucleic Acids Res 31:1646-1655

Fourel G, Revardel E, Koering CE, Gilson E (1999) Cohabitation of insulators and silencing elements in yeast subtelomeric regions. EMBO J 18:2522-2537

Frydrychova R, Grossmann P, Trubac P, Vitkova M, Marec F (2004) Phylogenetic distribution of TTAGG telomeric repeats in insects. Genome 47:163-178

Galagan JE, Calvo SE, Borkovich KA, Selker EU, Read ND, Jaffe D, FitzHugh W, Ma LJ, Smirnov S, Purcell S, Rehman B, Elkins T, Engels R, Wang S, Nielsen CB, Butler J, Endrizzi M, Qui D, Ianakiev P, Bell-Pedersen D, Nelson MA, Werner-Washburne M, Selitrennikoff CP, Kinsey JA, Braun EL, Zelter A, Schulte U, Kothe GO, Jedd G, Mewes W, Staben C, Marcotte E, Greenberg D, Roy A, Foley K, Naylor J, Stange-Thomann N, Barrett R, Gnerre S, Kamal M, Kamvysselis M, Mauceli E, Bielke C, Rudd S, Frishman D, Krystofova S, Rasmussen C, Metzenberg RL, Perkins DD, Kroken S, Cogoni C, Macino G, Catcheside D, Li W, Pratt RJ, Osmani SA, DeSouza CP, Glass L, Orbach MJ, Berglund JA, Voelker R, Yarden O, Plamann M, Seiler S, Dunlap J, Radford A, Aramayo R, Natvig DO, Alex LA, Mannhaupt G, Ebbole DJ, Freitag M, Paulsen I, Sachs MS, Lander ES, Nusbaum C, Birren B (2003) The genome sequence of the filamentous fungus *Neurospora crassa*. Nature 422:859-868

Gao W, Khang CH, Park SY, Lee YH, Kang S (2002) Evolution and organization of a highly dynamic, subtelomeric helicase gene family in the rice blast fungus *Magnaporthe grisea*. Genetics 162:103-112

Giraldo R, Rhodes D (1994) The yeast telomere-binding protein *RAP1* binds to and promotes the formation of DNA quadruplexes in telomeric DNA. EMBO J 13:2411-2420

Gottschling DE, Aparicio OM, Billington BL, Zakian VA (1990) Position effect at *S. cerevisiae* telomeres: reversible repression of Pol II transcription. Cell 63:751-762

Greider CW (1998) Telomeres and senescence: the history, the experiment, the future. Curr Biol 8:R178-R181

Guzman PA, Sanchez JG (1994) Characterization of telomeric regions from *Ustilago maydis*. Microbiology 140 (Pt 3):551-557

Hayashi N, Murakami S (2002) STM1, a gene which encodes a guanine quadruplex binding protein, interacts with *CDC13* in *Saccharomyces cerevisiae*. Mol Genet Genomics 267:806-813

Hecht A, Laroche T, Strahl-Bolsinger S, Gasser SM, Grunstein M (1995) Histone H3 and H4 N-termini interact with *SIR3* and *SIR4* proteins: a molecular model for the formation of heterochromatin in yeast. Cell 80:583-592

Henderson E, Hardin CC, Walk SK, Tinoco I Jr, Blackburn EH (1987) Telomeric DNA oligonucleotides form novel intramolecular structures containing guanine-guanine base pairs. Cell 51:899-908

Hernandez-Rivas R, Mattei D, Sterkers Y, Peterson DS, Wellems TE, Scherf A (1997) Expressed var genes are found in *Plasmodium falciparum* subtelomeric regions. Mol Cell Biol 17:604-611

Hohmann S, Meacock PA (1998) Thiamin metabolism and thiamin diphosphate-dependent enzymes in the yeast *Saccharomyces cerevisiae*: genetic regulation. Biochim Biophys Acta 1385:201-219

Horowitz H, Haber JE (1985) Identification of autonomously replicating circular subtelomeric-Y' elements in *Saccharomyces cerevisiae*. Mol Cell Biol 5:2369-2380

Hunt C, Moore K, Xiang Z, Hurst SM, McDougall RC, Rajandream MA, Barrell BG, Gwilliam R, Wood V, Lyne MH, Aves SJ (2001) Subtelomeric sequence from the right arm of *Schizosaccharomyces pombe* chromosome I contains seven permease genes. Yeast 18:355-361

Jacq C, Alt-Morbe J, Andre B, Arnold W, Bahr A, Ballesta JP, Bargues M, Baron L, Becker A, Biteau N, Blocker H, Blugeon C, Boskovic J, Brandt P, Bruckner M, Buitrago MJ, Coster F, Delaveau T, del Rey F, Dujon B, Eide LG, Garcia-Cantalejo JM, Goffeau A, Gomez-Peris A, Zaccaria P, et al. (1997) The nucleotide sequence of *Saccharomyces cerevisiae* chromosome IV. Nature 387:75-78

Jager D, Philippsen P (1989) Many yeast chromosomes lack the telomere-specific Y' sequence. Mol Cell Biol 9:5754-5757

Javerzat JP, Bhattacherjee V, Barreau C (1993) Isolation of telomeric DNA from the filamentous fungus *Podospora anserina* and construction of a self-replicating linear plasmid showing high transformation frequency. Nucleic Acids Res 21:497-504

Kellis M, Birren BW, Lander ES (2004) Proof and evolutionary analysis of ancient genome duplication in the yeast *Saccharomyces cerevisiae*. Nature 428:617-624

Kellis M, Patterson N, Endrizzi M, Birren B, Lander ES (2003) Sequencing and comparison of yeast species to identify genes and regulatory elements. Nature 423:241-254

Koering CE, Fourel G, Binet-Brasselet E, Laroche T, Klein F, Gilson E (2000) Identification of high affinity Tbf1p-binding sites within the budding yeast genome. Nucleic Acids Res 28:2519-2526

Kruk PA, Rampino NJ, Bohr VA (1995) DNA damage and repair in telomeres: relation to aging. Proc Natl Acad Sci USA 92:258-262

Kusumoto KI, Suzuki S, Kashiwagi Y (2003) Telomeric repeat sequence of *Aspergillus oryzae* consists of dodeca-nucleotides. Appl Microbiol Biot 61:247-251

Larionov V, Kouprina N, Graves J, Chen XN, Korenberg JR, Resnick MA (1996) Specific cloning of human DNA as yeast artificial chromosomes by transformation-associated recombination. Proc Natl Acad Sci USA 93:491-496

Laroche T, Martin SG, Gotta M, Gorham HC, Pryde FE, Louis EJ, Gasser SM (1998) Mutation of yeast Ku genes disrupts the subnuclear organization of telomeres. Curr Biol 8:653-656

Larrivee M, LeBel C, Wellinger RJ (2004) The generation of proper constitutive G-tails on yeast telomeres is dependent on the MRX complex. Genes Dev 18:1391-1396

Lieb JD, Liu X, Botstein D, Brown PO (2001). Promoter-specific binding of Rap1 revealed by genome-wide maps of protein-DNA association. Nat Genet 28: 327-334

Ligr M, Velten I, Frohlich E, Madeo F, Ledig M, Frohlich KU, Wolf DH, Hilt W (2001) The proteasomal substrate Stm1 participates in apoptosis-like cell death in yeast. Mol Biol Cell 12:2422-2432

Liljestrom PL (1985) The nucleotide sequence of the yeast *MEL1* gene. Nucleic Acids Res 13:7257-7268

Lin J-J, Zakian VA (1995) An in vitro assay for *Saccharomyces* telomerase requires *EST1*. Cell 81:1127-1135

Lingner J, Hughes TR, Shevchenko A, Mann M, Lundblad V, Cech TR (1997) Reverse transcriptase motifs in the catalytic subunit of telomerase. Science 276:561-567

Link AJ, Olson MV (1991) Physical map of the *Saccharomyces cerevisiae* genome at 110-kilobase resolution. Genetics 127:681-698

Liti G, Peruffo A, James SA, Roberts IN, Louis EJ (2005) Inferences of evolutionary relationships from a population survey of LTR-retrotransposons and telomeric associated sequences in the *Saccharomyces* sensu stricto complex. Yeast 22:177-192

Liu Z, Lee A, Gilbert W (1995) Gene disruption of a G4-DNA-dependent nuclease in yeast leads to cellular senescence and telomere shortening. Proc Natl Acad Sci USA 92:6002-6006

Louis EJ (1995) The chromosome ends of *Saccharomyces cerevisiae*. Yeast 11:1553-1573

Louis EJ (2002) Are *Drosophila* telomeres an exception or the rule? Genome Biol 3:reviews0007.1-0007.6

Louis EJ, Borts RH (1995) Complete set of marked telomeres in *Saccharomyces cerevisiae* for physical mapping and cloning. Genetics 139:125-136

Louis EJ, Haber JE (1990a) Mitotic recombination among subtelomeric Y' repeats in *Saccharomyces cerevisiae*. Genetics 124:547-559

Louis EJ, Haber JE (1990b) The subtelomeric Y' repeat family in *Saccharomyces cerevisiae*: an experimental system for repeated sequence evolution. Genetics 124:533-545

Louis EJ, Haber JE (1991) Evolutionarily recent transfer of a group I mitochondrial intron to telomere regions in *Saccharomyces cerevisiae*. Curr Genet 20:411-415

Louis EJ, Haber JE (1992) The structure and evolution of subtelomeric-Y repeats in *Saccharomyces cerevisiae*. Genetics 131:559-574

Louis EJ, Naumova ES, Lee A, Naumov G, Haber JE (1994) The chromosome end in yeast: its mosaic nature and influence on recombinational dynamics. Genetics 136:789-802

Lue NF, Peng Y (1997) Identification and characterization of a telomerase activity from *Schizosaccharomyces pombe*. Nucleic Acids Res 25:4331-4337

Lundblad V (2002) Telomere maintenance without telomerase. Oncogene 21:522-531

Lundblad V, Blackburn EH (1993) An alternative pathway for yeast telomere maintenance rescues est1- senescence. Cell 73:347-360

Marcand S, Gilson E, Shore D (1997) A protein-counting mechanism for telomere length regulation in yeast. Science 275:986-990

Matsumoto T, Fukui K, Niwa O, Sugawara N, Szostak JW, Yanagida M (1987) Identification of healed terminal DNA fragments in linear minichromosomes of *Schizosaccharomyces pombe*. Mol Cell Biol 7:4424-4430

May KJ, Ristaino JB (2004) Identity of the mtDNA haplotype(s) of *Phytophthora infestans* in historical specimens from the Irish potato famine. Mycol Res 108:471-479

McEachern MJ, Hicks JB (1993) Unusually large telomeric repeats in the yeast *Candida albicans*. Mol Cell Biol 13:551-560

McEachern MJ, Blackburn EH (1994) A conserved sequence motif within the exceptionally diverse telomeric sequences of budding yeasts. Proc Natl Acad Sci USA 91:3453-3457

McEachern MJ, Blackburn EH (1995) Runaway telomere elongation caused by telomerase RNA gene mutations. Nature 376:403-409

McEachern MJ, Krauskopf A, Blackburn EH (2000) Telomeres and their control. Annu Rev Genet 34:331-358

Metz AM, Love RA, Strobel GA, Long DM (2001) Two telomerase reverse transcriptases (*TERTs*) expressed in *Candida albicans*. Biotechnol Appl Bioc 34:47-54

Mimori T, Hardin JA (1986) Mechanism of interaction between Ku protein and DNA. J Biol Chem 261:10375-10379

Mitsui K, Ochi F, Nakamura N, Doi Y, Inoue H, Kanazawa H (2004) A novel membrane protein capable of binding the Na+/H+ antiporter (Nha1p) enhances the salinity-resistant cell growth of *Saccharomyces cerevisiae*. J Biol Chem 279:12438-12447

Mitton-Fry RM, Anderson EM, Hughes TR, Lundblad V, Wuttke DS (2002) Conserved structure for single-stranded telomeric DNA recognition. Science 296:145-147

Mitton-Fry RM, Anderson EM, Theobald DL, Glustrom LW, Wuttke DS (2004) Structural basis for telomeric single-stranded DNA recognition by yeast Cdc13. J Mol Biol 338:241-255

Morse RH (2000) RAP, RAP, open up! New wrinkles for *RAP1* in yeast. Trends Genet 16:51-53

Mortimer RK, Johnston JR (1986) Genealogy of principal strains of the yeast genetic stock center. Genetics 113:35-43

Nakamura TM, Morin GM, Chapman KB, Weinrich SL, Andrews WH, Lingner J, Harley CB, Cech TR (1997) Telomerase catalytic subunit homologs from fission yeast and human. Science 277:955-959

Naumov G, Naumova E, Turakainen H, Suominen P, Korhola M (1991) Polymeric genes *MEL8*, *MEL9* and *MEL10*--new members of alpha-galactosidase gene family in *Saccharomyces cerevisiae*. Curr Genet 20:269-276

Naumov GI, Naumova ES, Louis EJ (1995) Genetic mapping of the alpha-galactosidase *MEL* gene family on right and left telomeres of *Saccharomyces cerevisiae*. Yeast 11:481-483

Neidle S, Parkinson G (2002) Telomere maintenance as a target for anticancer drug discovery. Nat Rev Drug Discov 1:383-393

Neuveglise C, Feldmann H, Bon E, Gaillardin C, Casaregola S (2002) Genomic evolution of the long terminal repeat retrotransposons in hemiascomycetous yeasts. Genome Res 12:930-943

Nickles K, McEachern MJ (2004) Characterization of *Kluyveromyces lactis* subtelomeric sequences including a distal element with strong purine/pyrimidine strand bias. Yeast 21:813-830

Noskov VN, Kouprina N, Leem SH, Ouspenski I, Barrett JC, Larionov V (2003) A general cloning system to selectively isolate any eukaryotic or prokaryotic genomic region in yeast. BMC Genomics 4:16

Olovnikov AM (1971) Principle of marginotomy in template synthesis of polynucleotides. Dokl Akad Nauk SSSR 201:1496-1499

Olovnikov AM (1973) A theory of marginotomy. The incomplete copying of template margin in enzymic synthesis of polynucleotides and biological significance of the phenomenon. J Theor Biol 41:181-190

Park PU, Defossez PA, Guarente L (1999) Effects of mutations in DNA repair genes on formation of ribosomal DNA circles and life span in *Saccharomyces cerevisiae*. Mol Cell Biol 19:3848-3856

Powell WA, Kistler HC (1990) In vivo rearrangement of foreign DNA by *Fusarium oxysporum* produces linear self-replicating plasmids. J Bacteriol 172:3163-3171

Pryde FE, Huckle TC, Louis EJ (1995) Sequence-analysis of the right end of chromosome XV in *Saccharomyces cerevisiae*: an insight into the structural and functional significance of sub-telomeric repeat sequences. Yeast 11:371-382

Pryde FE, Louis EJ (1999) Limitations of silencing at native yeast telomeres. EMBO J 18:2538-2550

Qi H, Zakian VA (2000) The *Saccharomyces* telomere-binding protein Cdc13p interacts with both the catalytic subunit of DNA polymerase alpha and the telomerase-associated est1 protein. Genes Dev 14:1777-1788

Reich Z, Friedman P, Scolnik Y, Sussman JL, Minsky A (1993) On the metastability of left-handed DNA motifs. Biochemistry 32:2116-2119

Riethman HC, Moyzis RK, Meyne J, Burke DT, Olson MV (1989) Cloning human telomeric DNA fragments into *Saccharomyces cerevisiae* using a yeast-artificial-chromosome vector. Proc Natl Acad Sci USA 86:6240-6244

Riethman HC, Xiang Z, Paul S, Morse E, Hu XL, Flint J, Chi HC, Grady DL, Moyzis RK (2001) Integration of telomere sequences with the draft human genome sequence. Nature 409:948-951

Romero DP, Blackburn EH (1991) A conserved secondary structure for telomerase RNA. Cell 67:343-353

Salazar M, Thompson BD, Kerwin SM, Hurley LH (1996) Thermally induced DNA.RNA hybrid to G-quadruplex transitions: possible implications for telomere synthesis by telomerase. Biochemistry 35:16110-16115

Sanchez-Alonso P, Guzman P (1998) Organization of chromosome ends in *Ustilago maydis*. RecQ-like helicase motifs at telomeric regions. Genetics 148:1043-1054

Schechtman MG (1990) Characterization of telomere DNA from *Neurospora crassa*. Gene 88:159-165

Shampay J, Szostak JW, Blackburn EH (1984) DNA sequences of telomeres maintained in yeast. Nature 310:154-157

Shampay J, Blackburn EH (1988) Generation of telomere-length heterogeneity in *Saccharomyces cerevisiae*. Proc Natl Acad Sci USA 85:534-538

Shippen-Lentz D, Blackburn EH (1990) Functional evidence for an RNA template in telomerase. Science 247:546-552

Steiner BR, Hidaka K, Futcher B (1996) Association of the Est1 protein with telomerase activity in yeast. Proc Natl Acad Sci USA 93:2817-2821

Stellwagen AE, Haimberger ZW, Veatch JR, Gottschling DE (2003) Ku interacts with telomerase RNA to promote telomere addition at native and broken chromosome ends. Genes Dev 17:2384-2395

Strahl-Bolsinger S, Hecht A, Luo K, Grunstein M (1997) *SIR2* and *SIR4* interactions differ in core and extended telomeric heterochromatin in yeast. Genes Dev 11:83-93

Sundquist WI, Klug A (1989) Telomeric DNA dimerizes by formation of guanine tetrads between hairpin loops. Nature 342:825-829

Taddei A, Gasser SM (2004) Multiple pathways for telomere tethering: functional implications of subnuclear position for heterochromatin formation. Biochim Biophys Acta 1677:120-128

Teunissen AW, van den Berg JA, Steensma HY (1995) Transcriptional regulation of flocculation genes in *Saccharomyces cerevisiae*. Yeast 11:435-446

Tham WH, Zakian VA (2002) Transcriptional silencing at *Saccharomyces* telomeres: implications for other organisms. Oncogene 21:512-521

Theobald DL, Cervantes RB, Lundblad V, Wuttke DS (2003) Homology among telomeric end-protection proteins. Structure (Camb) 11:1049-1050

Tsang JS, Henry YA, Chambers A, Kingsman AJ, Kingsman SM (1990) Phosphorylation influences the binding of the yeast *RAP1* protein to the upstream activating sequence of the *PGK* gene. Nucleic Acids Res 18:7331-7337

Tzfati Y, Fulton TB, Roy J, Blackburn EH (2000) Template boundary in a yeast telomerase specified by RNA structure. Science 288:863-867

Tzfati Y, Knight Z, Roy J, Blackburn EH (2003) A novel pseudoknot element is essential for the action of a yeast telomerase. Genes Dev 17:1779-1788

Ulaner GA, Hu JF, Vu TH, Giudice LC, Hoffman AR (2001) Tissue-specific alternate splicing of human telomerase reverse transcriptase (*hTERT*) influences telomere lengths during human development. Int J Cancer 91:644-649

Underwood AP, Louis EJ, Borts RH, Stringer JR, Wakefield AE (1996) *Pneumocystis carinii* telomere repeats are composed of TTAGGG and the subtelomeric sequence contains a gene encoding the major surface glycoprotein. Mol Microbiol 19:273-281

Underwood AP, Louis EJ, Borts RH, Wakefield AE (1994) A technique for cloning the telomeres and subtelomeric regions from *Pneumocystis carinii*. J Eukaryot Microbiol 41:113S

van der Lee T, Robold A, Testa A, van 't Klooster JW, Govers F (2001) Mapping of avirulence genes in *Phytophthora infestans* with amplified fragment length polymorphism markers selected by bulked segregant analysis. Genetics 157:949-956

Van Dyke MW, Nelson LD, Weilbaecher RG, Mehta DV (2004) Stm1p, a G4 quadruplex and purine motif triplex nucleic acid-binding protein, interacts with ribosomes and subtelomeric Y' DNA in *Saccharomyces cerevisiae*. J Biol Chem 279:24323-24333

Vega LR, Mateyak MK, Zakian VA (2003) Getting to the end: telomerase access in yeast and humans. Nat Rev Mol Cell Biol 4:948-959

Virta-Pearlman V, Morris DK, Lundblad V (1996) Est1 has the properties of a single-stranded telomere end-binding protein. Genes Dev 10:3094-3104

Wahlin J, Cohn M (2002a) Analysis of the *RAP1* protein binding to homogeneous telomeric repeats in *Saccharomyces castellii*. Yeast 19:241-256

Wahlin J, Cohn M (2002b) *RAP1* binding and length regulation of yeast telomeres. In: Krupp G, Parwaresch R (eds) Telomeres and Telomerases: Cancer and Biology. Landes Bioscience, pp259-281

Wahlin J, Rosen M, Cohn M (2003) DNA binding and telomere length regulation of yeast *RAP1* homologues. J Mol Biol 332:821-833

Walmsley RM, Szostak JW, Petes TD (1983) Is there left-handed DNA at the ends of yeast chromosomes? Nature 302:84-86

Walmsley RW, Chan CSM, Tye BK, Petes TD (1984) Unusual DNA-sequences associated with the ends of yeast chromosomes. Nature 310:157-160

Watson JD (1972) Origin of concatemeric T7 DNA. Nat New Biol 239:197-201

Wellinger RJ, Wolf AJ, Zakian VA (1993) *Saccharomyces* telomeres acquire single-strand TG_{1-3} tails late in S phase. Cell 72:51-60

Woods JP, Goldman WE (1992) In vivo generation of linear plasmids with addition of telomeric sequences by *Histoplasma capsulatum*. Mol Microbiol 6:3603-3610

Wright JH, Gottschling DE, Zakian VA (1992) *Saccharomyces* telomeres assume a non-nucleosomal chromatin structure. Genes Dev 6:197-210

Yamada M, Hayatsu N, Matsuura A, Ishikawa F (1998) Y'-Help1, a DNA helicase encoded by the yeast subtelomeric Y' element, is induced in survivors defective for telomerase. J Biol Chem 273:33360-33366

Zakian VA (1985) Nuclear structure. Taken with a grain of salt. Nature 314:223-224

Zakian VA, Blanton HM (1988) Distribution of telomere-associated sequences on natural chromosomes in *Saccharomyces cerevisiae*. Mol Cell Biol 8:2257-2260

Zappulla DC, Cech TR (2004) Yeast telomerase RNA: A flexible scaffold for protein subunits. Proc Natl Acad Sci USA 101:10024-10029

Zhou J, Hidaka K, Futcher B (2000) The Est1 subunit of yeast telomerase binds the Tlc1 telomerase RNA. Mol Cell Biol 20:1947-1955

Zou S, Kim JM, Voytas DF (1996) The *Saccharomyces* retrotransposon Ty5 influences the organization of chromosome ends. Nucleic Acids Res 24:4825-4831

Zou S, Voytas DF (1997) Silent chromatin determines target preference of the *Saccharomyces* retrotransposon Ty5. Proc Natl Acad Sci USA 94:7412-7416

Zou S, Wright DA, Voytas DF (1995) The *Saccharomyces* Ty5 retrotransposon family is associated with origins of DNA replication at the telomeres and the silent mating locus HMR. Proc Natl Acad Sci USA 92:920-924

Barton, David B. H.
Institute of Genetics, University of Nottingham, Queen's Medical Centre, Nottingham NG7 2UH, UK

Cohn, Marita
Department of Cell and Organism Biology, Molecular Genetics, Lund University, Sölvegatan 35, S-223 62 Lund, Sweden
marita.cohn@cob.lu.se

Liti, Gianni
Institute of Genetics, University of Nottingham, Queen's Medical Centre, Nottingham NG7 2UH, UK

Employing protein size in the functional analysis of orthologous proteins, as illustrated with the yeast HOG pathway

Marcus Krantz and Stefan Hohmann

Abstract

Comparative genomics has provided us with a new handle on the interpretation of protein sequences. The sequencing of numerous fungal genomes has offered the possibility to compare genomes over a degree of evolutionary divergence. More importantly, the large number of sequenced genomes provides for a statistical analysis of orthologues, which allows analysis of protein size in addition to primary sequence. Like protein structure, size is expected to be conserved between orthologues. As significant deviations in size are indicative of altered protein domain structure, we propose size analysis as a supplementary tool for interspecies sequence comparison. Herein, we summarize a comparative analysis of the yeast HOG pathway and highlight the use of protein size analysis in identification and comparison of orthologues, and for evaluation of sequencing and annotation. Such comparisons were used to define differences in the signalling pathway architecture between species, such as absent components or components with altered domain structure. Furthermore, they provide a powerful complement to functional protein analysis, such as discovery of novel, and evaluation of characterised, functional domains, as well as evaluation of potential sites for posttranslational modification. The fungal genome sequences provide a unique resource that should facilitate the functional analysis of any fungal proteins.

1 Introduction

One of the major challenges of the post genomic era is the interpretation of protein sequences. This far, our limited comprehension of the protein blueprints has restricted us to comparative analyses in our attempts to deduce function from primary sequence. The increased availability of genomes, however, has opened new avenues for such analyses. By comparing orthologous and syntenic features, we gain a power in analysis not available by simple paralogous comparisons. The most obvious advantage is improved annotation, as functional ORFs are expected to be conserved between species, and share start and stop positions (Cliften et al. 2003; Kellis et al. 2003). The second is in the analysis of regulation, as promoter comparisons should allow the isolation of functional and, therefore, conserved

Topics in Current Genetics, Vol. 15
P. Sunnerhagen, J. Piškur (Eds.): Comparative Genomics
DOI 10.1007/4735_113 / Published online: 18 November 2005
© Springer-Verlag Berlin Heidelberg 2005

elements (Cliften et al. 2003; Kellis et al. 2003). In this essay, we focus on a third possibility, namely functional analysis of orthologous proteins, which by definition shares the same function in the same context, and, inferentially, the same functional domains.

2 Orthologue identification

This strategy of analysis presumes that the proteins compared share the same function and functional domains, and, consequently, relies on the correct identification of orthologues. Essentially, three tools can be used for this identification: (1) primary sequence, (2) synteny, and (3) protein structure. Primary sequence is the primary method of identification of putative orthologues, which, when used reciprocally, can, if not identify an unambiguous candidate, at least narrow down the number of candidates substantially. Synteny, which is a tool available in more closely related species, is an indication of a common ancestry and, consequently, a strong indication of functional relationship. In cases where species are too diverged to have syntenic homologues, synteny can still be employed indirectly by using two syntenic homologues for separate blast searches and limit the result to those proteins identified from both starting points. Finally, protein domain structure should be conserved between orthologues. While protein domain structure can be analysed by searching for conserved domains, this is both laborious and limited to already characterised domains. A more analytically convenient translation of protein domain structure is size, which should be conserved together with domain structure. Although the intrinsic degree of variation will depend on the protein in question, a significant deviation from the average size indicates that either the protein is not the orthologue, or it has acquired or lost functional domains, in which case it is no longer a true orthologue. An interesting variant on this theme is the multienzymatic proteins that occur in eukaryote metabolic pathways (Davidson et al. 1993), but these should be detectable already in the reciprocal blast searches. Similarly, limited structural rearrangements between components of a pathway should be characterised by a reciprocal size deviation in the two components. Alternatively, significant deviation in size can be indicative of mis-annotation or sequencing errors, which would otherwise obscure the analysis. The issue of size will be discussed thoroughly below.

2.1 The yeast HOG pathway as an example

This method was employed to characterize the HOG pathway of *Saccharomyces cerevisiae* in nineteen other fungal species (Krantz et al. 2005b). The High Osmolarity Glycerol pathway is, as its name suggests, monitoring turgor and responding to increases in extracellular osmolarity. This well characterized pathway contains both a phosphotransfer system (Sln1-Ypd1-Ssk1) and a MAPK module (Ssk2/22 or Ste11, Pbs2, and Hog1). Without elaborating on the pathway, which has been

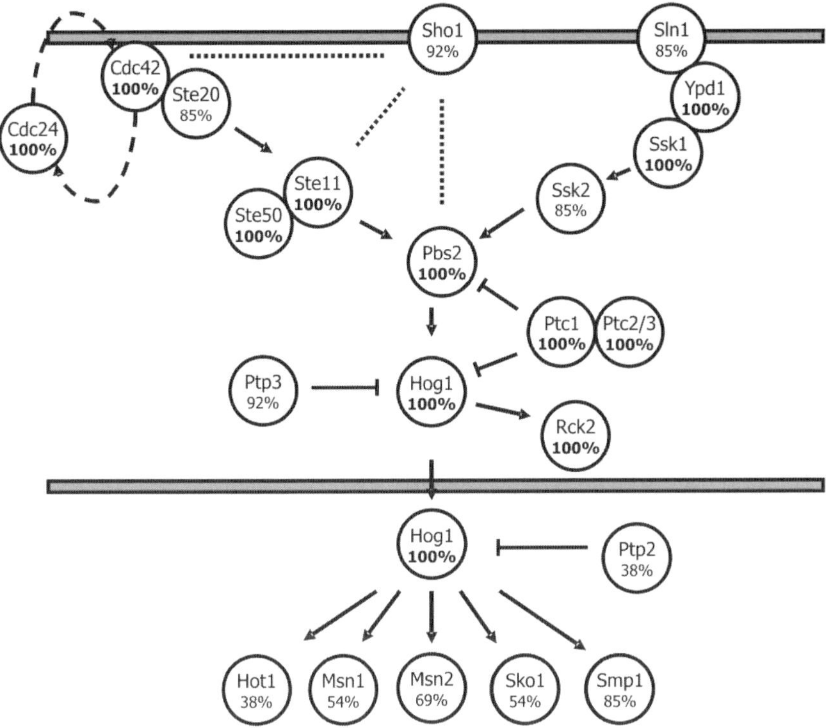

Fig. 1. The HOG pathway. Numbers denote the frequency of identifiable orthologues in thirteen non-*Saccharomyces* species. Note that the core pathway is all but ubiquitous and readily identified. In contrast, transcription factors are too diverged to be easily identifiable. Pathway input is different in *Sz. pombe* and *U. maydis*, which both lack any transmembrane histidine kinases, i.e. Sln1 orthologues. A Sho1 orthologue is present in each organism but *Sz. pombe*, although it is linked to the HOG pathway in a subset of species only. Two organisms scored as lacking Ssk2 homologues due to corrupt sequences, presumably due to sequence and/or annotation errors. In at least one of these cases, it has been shown to be functional (Furukawa et al. 2005).

described elsewhere in detail (Hohmann 2002), it should be noted that it consists of a MAPK core, which can be activated by either one of two upstream branches. One of these consists of the phosphotransfer module, whose primary input is the Sln1 transmembrane histidine kinase and the presumed osmosensor. The second branch is called the Sho-branch, after the Sho1 membrane protein. The input of this branch is not known, but its activation relies on components shared with the mating and pseudohyphal differentiation pathways. The output of the HOG pathway encompasses cytoplasmic targets such as the Rck proteins as well as nuclear transcription factors (Fig. 1).

The core components are ubiquitous and readily identifiable, while the downstream targets are much harder to identify due to poor sequence conservation. The most interesting differences are at the level of pathway input, where two species lack a transmembrane histidine kinase, i.e. an Sln1 orthologue. In addition, several species lack the domain of Pbs2 connecting the MAPK core to the Sho-branch, which consequently does not connect to the HOG pathway in these species. This has been demonstrated for *A. nidulans* (Furukawa et al. 2005). The branch and its components, however, are conserved in these species, consistent with its role in other signal transduction pathways. Intriguingly, *A. nidulans* Sho1 is conserved sufficiently well to complement the *S. cerevisiae sho1Δ* mutant, although this function appear to be absent in *A. nidulans*.

The emerging picture is that of a consensus osmosensing pathway, consisting of the histidine kinase module (Sln1-Ypd1-Ssk1) linked to a MAPK module (Ssk2/22-Pbs2-Hog1). Apart from the input, which appears to be fundamentally different in *Schizosaccharomyces pombe* and *Ustilago maydis*, the only component not identified in all species is Ssk2, which is apparently lacking in *Candida glabrata* and *Aspargillus nidulans*. Homologous sequences are present in both organisms, but the sequence from *C. glabrata* contains a frameshift and is not annotated as a functional protein. The sequence from *A. nidulans* is annotated as fused to a fungal transporter, but has been shown to be functional (Furukawa et al. 2005). This suggests that both cases are due to annotation and sequencing errors.

In contrast to the Sln1 branch, the Sho1 branch is only connected to the osmosensing pathway in a subset of organisms and, therefore, probably does not primarily function in osmosensing. Yet, this "original" function is similar enough to allow the Sho1 proteins to fulfil the osmosensing function even when derived from species were this link is absent. Apparently, the stimulus perceived as osmotic stress is very similar, if not identical, to that resulting in pseudohyphal growth. It is also noteworthy that the structure of the Sho1 orthologues is highly conserved (four TMs, linker region, SH3 domain), although alignment only reveal six residues outside the SH3 domain that are conserved across all species. Obviously, the structural requirements are much more stringent than primary sequence. This would be consistent with its role as a scaffold, as protein-protein interaction domains often show limited sequence conservation while structural conservation is crucial (Krylov et al. 2003).

It is important to stress that an orthologous function may well be encoded by more than one paralogue in some species. Although the genetic drift makes true redundancies unstable, they may occur temporarily, and could potentially be stabilized by divergent expression rather than function. Yet, as will be shown below, isogenes that have been retained tend to diverge in size, indicative of divergence in function.

3 Functional analysis

The attempt to deduce function from primary sequence information is as yet limited to sequence comparison. While paralogous comparisons are limited to well conserved motifs, such as enzymatic domains or highly conserved interaction domains, orthologous comparisons can be employed in the analysis of more weakly conserved interaction domains as well as potential targets for posttranslational modifications. Alignments of orthologues should allow the dissection of such proteins domain by domain.

Protein function is not only defined by its biochemical activity, but also by its interaction partners. An excellent example is the Ste11 MAPKKK in *S. cerevisiae*, which activates two MAPKK in three different signalling pathways. Yet, the output is specific for a given input, as the downstream connectivity is defined by the mode and mechanism of activation. Thus, an orthologous function requires both the same enzymatic activity, and the same interaction and regulation domains. While e.g. catalytic domains show a high degree of sequence conservation interaction or linker domains may show a high degree of sequence variation. At the same time, those are expected to show a rather low variation in size. Hence, orthologous proteins are expected to display conserved size, in addition to primary sequence and domain structure.

Orthologue comparisons have also been used to evaluate known and search for new domains of importance in HOG pathway proteins (Krantz et al. 2005a). Alignments revealed motifs which, although presently uncharacterised, are sufficiently well conserved to suggest functional importance. In several proteins, the multiple alignments highlighted short sequences which were highly conserved across most species. As an example, three highly conserved motifs were found in the N-terminal part of Sko1. The first is located close to the N-terminus and all but perfectly conserved in all ten orthologues but lacks characterised function. The second is slightly less well conserved and interspaces two of the Hog1 phosphorylation sites. It may constitute a recognition signal whose activity is modulated by its phosphorylation status. The third one is the largest, consisting of seventeen well conserved residues, again without any known function. The identification of these conserved sequences provides little information on their function, but suggests them as subjects for further experimental studies.

Along the same line of reasoning, characterised domains are expected to display a certain level of conservation both in sequence and in position. For example, the Sho1 and Nbp2 binding motifs in Pbs2 are relatively poorly conserved with respect to sequence and/or position. It may be that these domains are rather flexible in their structure and/or position, but it may also be that they have not been thoroughly characterised. The lack of convincing conservation suggests that they should be revisited experimentally.

In addition to identify novel and evaluate already characterised motifs, orthologue alignments can be used to determine the exact location of e.g. transmembrane domain. For instance, there was some variation between the location of predicted transmembrane domains in both Sho1 and Sln1, and in the latter even some discrepancies as to the number of them. In contrast to individual predictions, the multiple alignments provided a consensus approach to transmembrane domain predictions. Transmembrane domains have a profound impact on protein structure and would be expected to be positionally conserved although their primary sequence may have diverged.

Conversely, protein modification motifs, such as phosphorylation sites, are small but require perfect conservation. Unlike the transmembrane domains, they can be clearly defined but occur frequently by chance alone. Still, they can readily be evaluated by comparative genomics, on the basic assumption that important sites will be positionally conserved. The study of the HOG pathway discussed examples of both well conserved Hog1 target sites (in Sko1) and predicted sites lacking function (in Hot1). In the Hot1 case, none of five putative phosphorylation sites is conserved across the *Saccharomyces* family. Consistently, experimental analysis has shown that neither of them is important for *in vivo* function, although they are phosphorylated (Alepuz et al. 2003). In contrast, the Sko1 phosphorylation sites are conserved across all ten orthologues and located in a highly conserved region. They have been shown to be important experimentally (Proft and Struhl 2002).

Finally, examination of the transmembrane histidine kinase, Sln1, revealed a conserved sequence of interspaced hydrophobic regions in the extracellular domain, which is reminiscent of a leucine zipper. This region has been shown to be important for protein function and required for dimerisation, which has been suggested to govern the activation status of Sln1. Indeed, replacement of the first transmembrane domain and parts of the extracellular domain with a leucine zipper renders the protein constitutively active (Ostrander and Gorman 1999). This, together with the length of the domain, suggests that it is responsible for dimerisation and that it can possibly interact over different registers. Conceivably, the degree of interaction directly influences the structure of the intracellular parts of the dimers, providing a possible mechanism for osmosensing.

4 Protein size

Comparative genomics does not only allow analysis of the conservation of primary sequence, but also of protein size. By definition, orthologous proteins share function and consequently functional domains, which presumably do require structures of similar size. Indeed, orthologues have been shown to display a much lower size variation than paralogues (Wang et al. 2005). In fact, orthologue sizes are similar in yeast and man, although protein size on average is larger in man (Wang et al. 2005). Similarly, the syntenic orthologues in *S. cerevisiae* and *Ashbya gossypii* tend to be conserved in size while syntenic non-ORF regions have dramatic differences in size (Riccarda Rischatsch and Peter Philippsen, personal communication). Not surprisingly, orthologue size proved rather well conserved across fungal species, and sometimes considerably more so than the primary sequence. An example of this is Sho1, with a size coefficient of variation of 9.3% across sixteen fungal species. This is comparable to Hog1, with a coefficient of variation of 9.1% across the same species, although the average degree of identity is 80% in Hog1 as compared to 41% in Sho1. Certain protein domains provide even more pronounced examples. In Sln1, the size of the spacer between the HATPase domain and the ATP binding domain is perfectly conserved across yeasts (Krantz et al. 2005a). Sequence conservation is on average 64% identity, but drops as low as 29%. Only 11 out of 48 residues are conserved across all yeast

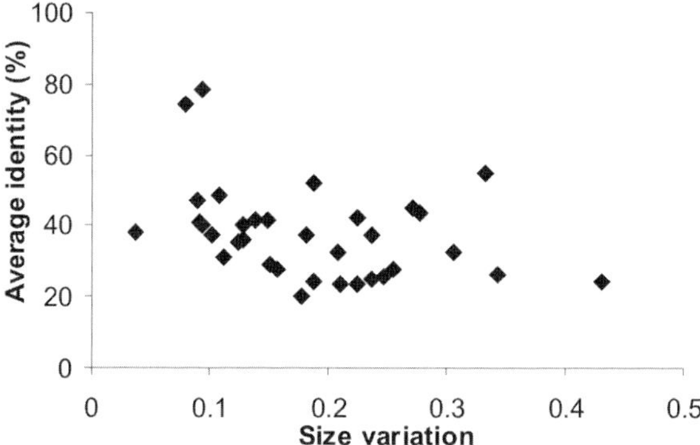

Fig. 2. The correlation between conservation of protein size and primary sequence. Average identity scores plotted against the coefficient of variation of the protein size for each set of orthologues. The correlation between the two is surprisingly low (R^2=0.13), although the most highly conserved proteins, i.e. Cdc42 and Hog1, are well conserved in both respects. Figure reproduced from Krantz et al. 2005b.

species. Apparently, the constraints on sequence and structure/size are, at least partially, different, and it has been reported that sequence conservation may be less of an issue in protein-protein interaction domains, whereas structure and thus size is critical (Krylov et al. 2003). A significant deviation in protein size thus predicts that the protein has lost or acquired additional functions/interactions, in which case it is no longer a true orthologue. We propose size analysis as a supplementary tool for interspecies sequence comparison, and illustrate this possibility on the HOG pathway components of *S. cerevisiae*.

The variance in size shows a surprisingly low correlation to the average degree of sequence conservation (Fig. 2). This suggests that size and primary sequence are under different selective pressure, thus providing an additional handle on orthologue identification and functional analysis. Furthermore, size analysis can reveal misannotations or sequencing errors, as illustrated below with the HOG pathway and the *Saccharomyces* species (Table 1). Finally, the majority of the presumably redundant isogenes included in this study show a significant difference in size, advocating their divergence in function.

The implementation of protein size in the analysis of fungal orthologues is biased towards the *Saccharomyces* species, as these closely related species represent one third of the total number of subjects. Still, as the analysis is relative, the perspective should have no significant impact. As a consequence, however, few significant differences are found in these species, and these differences are likely due to sequencing and annotation errors rather than biological variation.

Table 1. Most outliers from the *Saccharomyces* species are artefacts of misannotations or sequencing errors

Protein	Species	Cause of rejection	
Bem4	*S. castelii*	A(F)	
Ste11	*S. bayanus*	S	
Sln1	*S. kluyveri*	F(t)	
Rck1	*S. kudriavzevii*	A(F)	
Rck1	*S. bayanus*	?	Remaining: ±0.56%
Gic1	*S. bayanus*	A	
Gic1	*S. castelii*	?	Remaining: ±0.55%
Gic2	*S. kudriavzevii*	A, F(t)	
Msn2	*S. bayanus*	S	
Ptp2	*S. mikatae*	S	
Rtn2	*S. paradoxus*	A	

A total of 40 sets of orthologues were examined for outliers with a significant difference in size (compare Fig. 3). Each orthologue thus rejected in any of the *Saccharomyces* species is listed together with the apparent cause of rejection. Nine out of eleven proteins suffer from one or more of the following apparent defects: A(F); the ORF contains an early frameshift mutation and the protein is thus annotated from an alternative, later, start, S; The DNA sequence corresponding to the *S. cerevisiae* start is missing and the protein is consequently annotated from a later start, A: The protein is annotated from an alternative start although the start in *S. cerevisiae* is conserved and F(t); The protein contains a frameshift resulting in a truncation. The two proteins labelled "?" have no apparent sequence error and may constitute the false positives. They are the only remaining non-*Saccharomyces sensu strictu* species and the remaining population is extremely homogenous as regards size.

4.1 Protein size as an analytical tool

Implementing size as an analytical tool in orthologue comparison is straightforward. The basic assumption is that in any given population of orthologues, size should be approximately normally distributed around a mean value. In the next step, the size of each orthologue is compared to a confidence interval based on the remaining population. The comparison is based on the t-distribution and provides a probability for the protein to be part of that population. A significant difference here suggests that the protein has lost or received domains, and, consequently, is no true orthologue. The analysis can be obscured if the original population consists of subpopulations. Indeed, in the analysis of the HOG pathway, it turned out that protein sizes were often significantly different between filamentous fungi and yeasts, and these were consequently considered as separate populations.

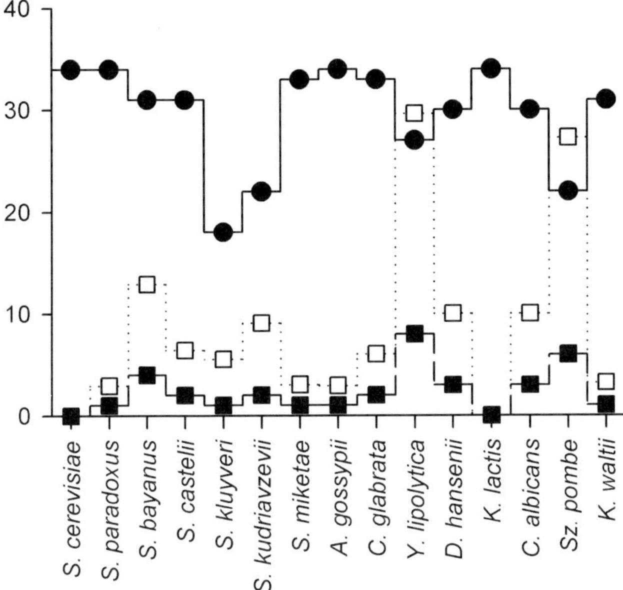

Fig. 3. Frequency of outliers as compared to number of identified function. Outliers are proteins with a significant deviation in size. Identified functions are the number of proteins present, but isogenes score as one if one or both are present. The frequency of outliers is the number of outliers divided with the number of identified functions. Each of these is plotted as a function of species. The highest frequency of outliers can be observed in *Y. lipolytica* and *Sz. pombe*, while the least number of identified functions are in *Saccharomyces kluyveri* and *Saccharomyces kudriavzevii* (here indicative of sequencing quality). In the case of *Y. lipolytica* and *Sz. pombe*, the number of identified functions drops, but probably reflects biological differences or too diverged sequences rather than sequence quality. Compare Figure 4. (•) Number of identified functions (max 34). (■) Number of outliers. (□) Frequency of outliers (%). A total of 40 proteins, 28 unique and six pairs of isogenes, were included in the comparison. The twenty four proteins indicated in Figure 1 and sixteen other proteins potentially connected to the Hog-pathway (Krantz et al. 2005b).

Revisiting this analysis for yeasts only, and applying the less stringent normal distribution as base for the analysis, yields a total of thirty-five outliers. Although the number of false positives potentially increases, this can be controlled by looking at the outliers in the *Saccharomyces* species. Of these outliers (a total of eleven when a normal distribution was used instead of a t-distribution), three lack the sequence corresponding to the *S. cerevisiae* start, three more are annotated from an alternative start (although the one from *S. cerevisiae* is conserved), two have a truncated C-terminus due to frameshifts and two have alternative starts due to early frameshifts. Only two sequences contain no obvious potential errors, and may constitute false positives. On the other hand, these are also the only (remaining) non-*Saccharomyces sensu strictu* species in these groups, and the remaining

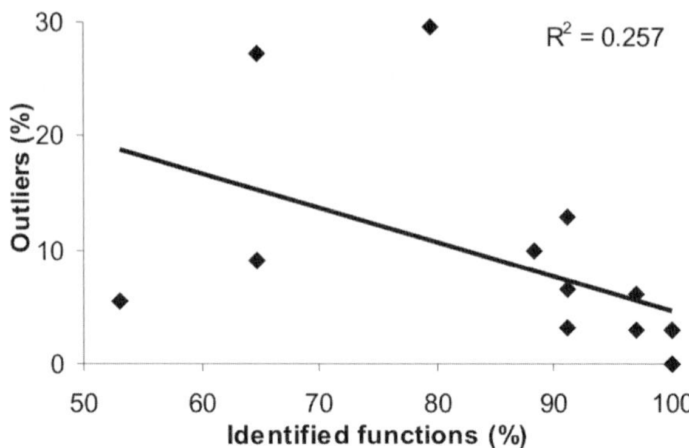

Fig. 4. The frequency of outliers vs. the frequency of identified functions. Each spot corresponds to one species. Species without outliers and 100% identified functions (*S. cerevisiae* and *Kluyveromyces lactis*) are located in the lower right corner. The two "upper" species are *Sz. pombe* and *Y. lipolytica*, which have a high degree of outliers, while the two spots in the lower left are indicative of low sequence coverage (*S. kluyveri* and *S. kudriavzevii*). Excluding these two low coverage genomes, the linear correlation coefficient, R^2, rises from 0.26 to 0.81. If these proteins are representative for the genome, it highlights a clear relationship between the two kinds of divergence; loss of orthologues and altered protein structures.

proteins are exceptionally homologous with regards to size (coefficient of variation of 0.56% and 0.55%, respectively).

Yet, if we consider these two as false positives, and the frequency of these and of the sequencing errors/annotation mistakes to be representative, the expected false positive rate is 0.29 and 1.29 per species, respectively. That is, even with this criterion for significant size difference, most of the outliers in other yeasts cannot be explained either as false positives or as sequence/annotation inconsistencies, as the frequency of outliers is considerably higher in most other yeast species (Fig. 3 and 4). If related to the percentage of identified functions (i.e. proteins, isogenes counting as one), the difference becomes even more pronounced, advocating that this is indeed indicative of biological divergence. Consistent with the implications on functional divergence, the highest frequency of outliers is found in *Yarrowia lipolytica* and *Sz. pombe*, which are evolutionary most distinct from *S. cerevisiae*.

Interestingly, a majority of isogene pairs shows a significant size difference, which is indicative of a diverged function. Only two pairs fail to show a significant difference in size, one of which cannot be statistically evaluated due to too small sample size. As a comparison, all seven of the pairs known to be diverged show a significant size difference (Table 2). Although the study includes too few protein pairs to draw general conclusions, it appears that isogenes that have been retained have acquired structural differences. It seems likely that many proteins presumed to be redundant in fact have acquired such alterations of structure and thus function.

Table 2. Isogenes tend to have diverged in size

Protein 1	Protein 2	p-value	confounders (rejected)
Rtn1	Rtn2	2.9*10E-7	*S. paradoxus* Rtn2 too large (p=8.6*10E-5), A
Ssk2	Ssk22	N/A	*Sz. pombe* homologues cluster together - separate duplication
Ssk2	Ste11	6.5*10E-6	*S. bayanus* Ste11 too small (p=4.1*10E-5), S
Ssk22	Ste11	5.8*10E-4	
Msn2	Msn4	6.7*10E-4	*S. bayanus* Msn2 too small (p=6.7*10E-4), S *C. glabrata* orthologues dubious designation
Ptc1	Ptc2	4.6*10E-5	
Ptc1	Ptc3	6.0*10E-9	
Ptc2	Ptc3	0.62	*Sz. pombe* homologues cluster together - separate duplication
Gic1	Gic2	3.8*10E-5	*S. kudriavzevii* Gic2 too small (p=3.8*10E-5), A, F(t)
Rck1	Rck2	3.1*10E-4	
Hot1	Msn2	4.5*10E-9	
Ptp2	Ptp3	0.007	*Sz. pombe* homologues cluster together - separate duplication
Cla4	Ste20	5.7*10E-5	*Sz. pombe* Cla4 too small (p=0.0002)

All paralogous proteins included in the study were compared pair wise to determine if they were significantly different in size. Confounders, i.e. outliers and isogenes apparently duplicated separately, were eliminated before comparisons as noted. Thereafter, the size differences between paralogous proteins were analysed with paired t-tests. The paralogous protein pairs known to have developed diverged functions are indicated in bold, and seven out of seven of them are significantly different in size. Of the isogenes, which are presumed to be redundant, four out of six pairs display a significant size difference and one of the remaining pairs cannot be tested due to small remaining sample size. Annotation is as per Table 1. N/A; probability cannot be calculated due to sample size. The *Sz. pombe* orthologues are excluded from three of the isogene comparisons as they cluster together by themselves, indicative of separate duplication.

5 Conclusions

Comparative genomics is a valuable addition to the study of any system. It can be employed to predict functional domains, as well as relevance of putative phosphorylation sites. In addition to phosphorylation sites, it should be equally efficient in predicting e.g. functional glycosylation or acetylation sites and should prove a powerful complement to analysis on protein modification and its importance. Apparently, phosphorylation of a residue does not always correlate to the requirement for phosphorylation of that residue (Alepuz et al. 2003). Furthermore, the architecture of the system studied may be derived from orthologue identification alone. Not only may architectural differences be due to absence of orthologous components, but also due to structural differences between the corresponding proteins, which may be reflected in protein size. The information gained on pathway architecture is most sensitive to the sequence and annotation quality, as functions not properly annotated will invariably be scored as absent or altered. Consequently, erroneous sequencing and annotation will have a high penetrance in architectural comparisons, calling for high coverage and high quality genome sequencing. Unlike system architecture, analysis of protein function is rather insensitive to misannotation as such proteins are simply excluded from multiple comparisons.

Acknowledgements

We acknowledge the contribution of Jonas Warringer in conceptual discussions and critical reading of the manuscript. This work was supported by the QUASI EC-funded project (contract LSHG-CT2003-530203). SH was a recipient of a research position from the Swedish Research Council.

References

Alepuz PM, de Nadal E, Zapater M, Ammerer G, Posas F (2003) Osmostress-induced transcription by Hot1 depends on a Hog1-mediated recruitment of the RNA Pol II. EMBO J 22:2433-2442

Cliften P, Sudarsanam P, Desikan A, Fulton L, Fulton B, Majors J, Waterston R, Cohen BA, Johnston M (2003) Finding functional features in *Saccharomyces* genomes by phylogenetic footprinting. Science 301:71-76

Davidson JN, Chen KC, Jamison RS, Musmanno LA, Kern CB (1993) The evolutionary history of the first three enzymes in pyrimidine biosynthesis. Bioessays 15:157-164

Furukawa K, Hoshi Y, Maeda T, Nakajima T, Abe K (2005) *Aspergillus nidulans* HOG pathway is activated only by two-component signalling pathway in response to osmotic stress. Mol Microbiol 56:1246-1261

Hohmann S (2002) Osmotic stress signaling and osmoadaptation in yeasts. Microbiol Mol Biol Rev 66:300-372

Kellis M, Patterson N, Endrizzi M, Birren B, Lander ES (2003) Sequencing and comparison of yeast species to identify genes and regulatory elements. Nature 423:241

Krantz M, Becit E, Hohmann S (2005a) Comparative analysis of HOG pathway proteins to generate hypotheses for functional analysis. Curr Genet (in press)

Krantz M, Becit E, Hohmann S (2005b) Comparative genomics of the HOG signalling system in fungi. Curr Genet (in press)

Krylov DM, Wolf YI, Rogozin IB, Koonin EV (2003) Gene loss, protein sequence divergence, gene dispensability, expression level, and interactivity are correlated in eukaryotic evolution. Genome Res 13:2229-2235

Ostrander DB, Gorman JA (1999) The extracellular domain of the *Saccharomyces cerevisiae* Sln1p membrane osmolarity sensor is necessary for kinase activity. J Bacteriol 181:2527-2534

Proft M, Struhl K (2002) Hog1 kinase converts the Sko1-Cyc8-Tup1 repressor complex into an activator that recruits SAGA and SWI/SNF in response to osmotic stress. Mol Cell 9:1307-1317

Wang D, Hsieh M, Li WH (2005) A general tendency for conservation of protein length across eukaryotic kingdoms. Mol Biol Evol 22:142-147

Hohmann, Stefan
Department for Cell and Molecular Biology, Göteborg University, Box 462, S-40530 Göteborg, Sweden
hohmann@gmm.gu.se

Krantz, Marcus
Department for Cell and Molecular Biology, Göteborg University, Box 462, S-40530 Göteborg, Sweden. Present address: The Systems Biology Institute, 6-31-15 Jingumae M31 6A, Shibuya, Tokyo 150-0001, Japan

Lager brewing yeast

Yukiko Kodama, Morten C. Kielland-Brandt, Jørgen Hansen

Abstract

Lager brewing yeast is a group of closely related strains of *Saccharomyces pastorianus/S. carlsbergensis* used for lager beer production all over the world, making it one of the most important industrial yeasts. The pure cultivation of yeast was established in the early 1880's with immediate practical success for lager brewing yeast. However, almost a century would elapse before its genetics could be approached in detail, despite the development of the genetics of *Saccharomyces cerevisiae*, starting in the 1930's. During the last few decades, the complex nature of the genome of lager brewing yeast was elucidated, showing that it is a hybrid between *Saccharomyces cerevisiae* and another *Saccharomyces* species.

Here we review current knowledge on genetics and genomics of lager brewing yeast and introduce the most updated information about its whole genome sequence. These studies throw further light on the complex chromosomal structure of this yeast. They may also open the door for the elucidation of how inter-species hybrids maintain their chromosomes.

1 Introduction

1.1 Brewing yeast - history and biotechnology

Beer is one of the most popular and most highly consumed alcoholic beverages in the world. Although many types of beer exist, the large majority of beer types may be classified into two major types, ale and lager, reflecting the yeast used as well as the fermentation conditions. To produce an ale type of beer, the fermentation is carried out using a "top-fermenting" yeast (ale yeast) at temperatures from 20°C to 25°C, followed only by a short period of aging, or none at all. In contrast, lager beers are produced by using "bottom-fermenting" yeast strains (lager brewing yeast) at lower temperatures, from 8°C to 15°C, and then subjected to a long (often a few weeks) low-temperature period of maturation (aging, "lagering"). Within these main beer categories, diversity is still high. Both ales and lagers can have colours ranging from pale to black, with all intermediate nuances. Both can have an alcohol content ranging from a few to more than ten percent by volume. Both can be modestly or very strongly flavoured. Historically, ale brewing pre-dates lager brewing by several hundred years, maybe even by thousands of years. It is generally accepted that beer brewing was known as early as 3000 BC from Meso-

Topics in Current Genetics, Vol. 15
P. Sunnerhagen, J. Piškur (Eds.): Comparative Genomics
DOI 10.1007/b106370 / Published online: 7 January 2005

potamia and perhaps even earlier from Egypt (for a review of beer brewing history see Corran 1975). Beer brewing in Europe is much newer, and the technique was most likely acquired from the Middle East independently by Germanic and Celtic tribes around the 1st century AD (Corran 1975). On the British Isles, the tradition of ale brewing was taken to perfection, whereas, in large parts of mainland Europe, ale brewing was transformed into lager brewing.

One of the most important technological milestones in beer brewing was the pure culturing of brewing yeast; in fact it constituted a shift of paradigms for the early 20th century brewers. Until then, wort was invariably inoculated with spent yeast of the preceding fermentation. In 1883, Emil Chr. Hansen of the Carlsberg Laboratory devised a method for using single-cell cultures of yeast in beer production (Hansen 1883). The use of pure yeast cultures was quickly adopted worldwide, first in lager and subsequently in ale brewing. While mixed cultures of several yeast strains are being used quite widely in ale brewing, spontaneous fermentation is used on an industrial scale only in few cases (e.g. Belgian *lambic* brewing). At present, lager beer constitutes more than 90% of the global beer production, and therefore, research has tended to concentrate on lager brewing yeast.

The first lager brewing yeast that was pure-cultured was initially called "Bottom Fermenting Strain # I", but later named *Saccharomyces carlsbergensis* (Hansen 1908). It is kept as the *S. carlsbergensis* type strain CBS1513 (IFO11023), and probably most lager brewing yeasts used today are closely related to this strain. Ale brewing yeasts, on the other hand, seem to constitute a broader variety of *Saccharomyces* strains, some of which are closely related to standard laboratory strains of *S. cerevisiae* (Pedersen 1986a). Modern lager brewing yeast strains and the *S. carlsbergensis* type strain are recognised as part of the *S. pastorianus* group (Vaughan-Martini and Martini 1987). They were previously included in the taxon *Saccharomyces cerevisiae* (Yarrow 1984), after several changes in naming (e.g. Rainieri et al. 2003).

1.2 Genetic structure of lager brewing yeast

1.2.1 The hybrid genome

From early on, it was evident that lager brewing yeast was very different from other brewing yeast. For one thing, the lower temperature at which growth is optimal indicated that this yeast was different. It also became evident that lager brewing yeast strains did not (or only with great difficulty) produce meiotic offspring. Much work has been put into studying the genetic set-up of this yeast, and it is now well established that it is indeed a polyploid species hybrid formed from *S. cerevisiae* and a closely related *Saccharomyces* species.

This conclusion was reached through molecular studies (Southern hybridisation analysis, etc.), as well as transmission genetic analyses to be reviewed below. Quite extensive Southern analysis of genomic DNA from lager brewing yeast has been carried out, employing DNA probes originating from cloned *S. cerevisiae* genes or lager brewing yeast genes. For example, *BAP2* (chr. II) (Kodama et al.

2001), *HIS4* (chr. III) (Nilsson-Tillgren et al. 1981), *LEU2* (chr. III) (Pedersen 1985), *MAT* (chr. III), *HML* (chr. III), *HMR* (chr. III), and *SUP-RL1* (chr. III) (Holmberg 1982), *ILV1* (chr. V), *CAN1* (chr. V), *CYC7* (chr. V), and *URA3* (chr. V) (Nilsson-Tillgren et al. 1986), *MXR1* (chr. V) (Hansen 1999), *ILV3* (chr. X) and *CYC1* (chr. X) (Casey 1986a, 1986b), *ILV5* (chr. XII) and *ILV2* (chr. XIII) (Petersen et al. 1987), *MET2* (chr. XIV) (Hansen and Kielland-Brandt 1994), and *ATF1* (chr. XV) (Fujii et al. 1994) genes have been investigated. In almost all cases, two divergent types of the gene in question were found, of which one invariably exhibited a restriction and hybridisation pattern identical or almost identical to that found in the corresponding *S. cerevisiae* gene, while the other showed divergent patterns. The former is often referred to as the *S. cerevisiae*-like type (variously described as the Sc-, cer- or -CE type) and the other the *S. pastorianus*-(Sp-), lager- (Lg-), non-*cerevisiae*- (non-Sc-), or *S. carlsbergensis*-specific (-CA) type. In the present text we will use the denotations Sc- and non-Sc-, respectively.

The finding of two types of genes is consistent with the possibility that lager brewing yeast contains two types of chromosome such as Sc- and non-Sc-type chromosomes. In fact, the essence of this idea preceded much of the molecular work and was based on pioneering genetic studies of the nuclear genome of lager brewing yeast, employing the technique of *kar*-mediated single-chromosome transfer. Thus, several lager brewing yeast chromosomes were transferred individually into genetically well defined strains of *S. cerevisiae*, after which these chromosomes were genetically characterised. Hence, the transfer of chromosomes III (Nilsson-Tillgren et al. 1981), V (Nilsson-Tillgren et al. 1986), VII (Nilsson-Tillgren unpublished; see Kielland-Brandt et al. 1995), X (Casey 1986b), XII and XIII (Petersen et al. 1987) from lager brewing yeast to *S. cerevisiae* was accomplished. In some cases, so-called substitution strains could be created, such as strains in which the original *S. cerevisiae* chromosome was lost, meaning that in each of these cases the transferred chromosome was a functional equivalent of its *S. cerevisiae* counterpart, at least for the essential genes. Now, meiotic crossing-over between the different chromosomes was assayed in diploid crosses of the substitution strains with standard *S. cerevisiae* strains. The chromosomes derived from lager brewing yeast were found to be of three types: i) **homologous** chromosomes, which recombined normally with their *S. cerevisiae* counterparts, ii) so-called **homoeologous** chromosomes, which rarely recombined with their *S. cerevisiae* counterparts, and iii) **mosaic** chromosomes that were composed of homologous and homoeologous segments.

The hybrid nature of lager brewing yeast has also been confirmed and characterised by hybridisation of radioactive probes to chromosome-sized DNA separated by pulsed-field electrophoresis (Casey 1986b; Tamai et al. 1998; Yamagishi and Ogata 1999), as has the mosaic structure of several chromosomes, by genomic DNA hybridisation to *S. cerevisiae* gene arrays, as described below.

Southern hybridisation experiments indicated at an early point that the lager brewing yeast Sc-type of any given gene is identical, or practically identical, to the corresponding *S. cerevisiae* gene (e.g. Holmberg 1982; Nilsson-Tillgren et al. 1986; Petersen et al. 1987), a notion which was confirmed by partial or full sequencing of a few lager brewing yeast Sc-genes (Hansen and Kielland-Brandt

1994; Fujii et al. 1996; Børsting et al. 1997; Johannesen and Hansen 2002). Several lager brewing yeast-specific (non-Sc) genes were also studied at the nucleotide level already a decade or more ago, and when compared to the equivalent *S. cerevisiae* genes, ORF sequences of Sc-type and non-Sc-type genes were invariably found to be highly related and always of the same length, while non-coding sequences show much less general identity. Thus, the non-Sc-*ILV1* and the Sc-*ILV1* genes are 86% identical, whereas, the deduced polypeptide sequences are 96% identical (Gjermansen 1991). The same comparisons of the *ILV2* (Gjermansen 1991), *MET2* (Hansen and Kielland-Brandt 1994), *MET10* (Hansen et al. 1994), and more recently *ACB1* (Børsting et al. 1997), *HIS4* (Porter et al. 1996), *HO* (Tamai et al. 2000), *MET14* (Johannesen and Hansen 2002), *MXR1* (Hansen et al. 2002), *BAP2* (Kodama et al. 2001), and *ATF1* (Fujii et al. 1996) genes give numbers of 78-88% (nt) and 76-97% (aa), respectively.

The studies described above clearly showed that lager brewing yeast is a species hybrid, but they did not reveal much about the ploidy of this yeast. This question has been approached for the *ILV2* locus in a lager brewing yeast strain (Gjermansen et al. 1988; Kielland-Brandt et al. 1989). As part of a study of the consequences of eliminating *ILV2* gene function in lager brewing yeast, deletions of the two wild type alleles were constructed *in vitro*. Using the two-step gene replacement technique (Scherer and Davis 1979) yeasts were obtained which carried one or the other deletion allele of *ILV2* instead of the wild type allele. Southern analysis of these strains showed that the particular lager brewing yeast studied contains two copies of each of the two versions of the *ILV2* gene. Generalisation of this result would indicate this lager brewing yeast to be allotetraploid, but recent studies suggest a pentasomic content of at least the *HIS4*-containing chromosome III regions (Hoffmann 2000). Further, lager brewing yeast chromosomes of three sizes hybridise with a chromosome X probe (Casey 1986b) or with a chromosome III probe (Pedersen 1986b). Thus, even though this lager brewing yeast has a total DNA content approximately corresponding to tetraploidy (Hoffmann 2000), it appears to be irregular in its chromosome set-up and probably aneuploid for some chromosomes, or chromosome regions.

To summarise, lager brewing yeast is an allopolyploid species hybrid, containing parts of two diverged genomes, one derived from *S. cerevisiae* and one that is derived from another *Saccharomyces* yeast. A particular, well-studied strain appears to be largely allotetraploid. Sexual reproduction in this organism is impaired, perhaps due to the mosaic structure of some chromosomes, perhaps due to mutations in genes important for sporulation, etc. Classical genetic analysis with tetrads is, therefore, not possible. A fact with long recognised grave consequences for breeding. One may, however, speculate whether the low efficiency of sexual reproduction has contributed to maintaining a relatively high genetic stability of a yeast population with desirable characteristics for beer production.

1.2.2 Phylogenetic and taxonomic position

While once included in the taxon *Saccharomyces cerevisiae* (Yarrow 1984), and commonly denoted *S. carlsbergensis*, lager brewing yeast is now generally recognised as part of the *S. pastorianus* group (Vaughan-Martini and Martini 1987). As described above, lager brewing yeast seems to be the result of hybridisation between *S. cerevisiae* and another *Saccharomyces* yeast. This hybridisation event, which may well have taken place in nature, must be considered interspecific, since the observed divergence is correlated with a drop in meiotic recombination expected to cause fertility barriers. Several studies have been aimed at identifying the non-*S. cerevisiae* parent. One of the first candidates was another yeast isolated by Hansen (1908), namely "Bottom Fermenting Yeast #II", initially named *Saccharomyces monacensis* (CBS 1503). Southern hybridisation and cloning experiments with a low number of genes, mostly on relatively small chromosomes, indicated that only non-Sc-type genes existed in this yeast (Pedersen 1986a, 1986b; Hansen and Kielland-Brandt 1994; Børsting et al. 1997). It has, however, recently been shown that although deficient in the Sc-version of many genes, strain CBS1503 undoubtedly contains Sc-type genomic DNA (Andersen et al. 2000; Casaregola et al. 2001; Kodama Y, Nakao Y, Nakamura N, Fujimura T, Shirahige K, and Ashikari T: Diversity of chromosomal structure in lager brewing yeast, manuscript in preparation), a notion that has been supported also by AFLP (amplified fragment length polymorphism) studies (de Barros Lopes et al. 2002). Another candidate suggested as the progenitor was the *Saccharomyces bayanus* type strain (CBS 380). This hypothesis was originally based on DNA-DNA reassociation experiments (Vaughan-Martini and Kurtzman 1985), and seemed to gain support from recent findings of identical non-*S. cerevisiae*-like sequences in some lager brewing strains and in the type strain of *S. bayanus* (CBS 380) (Tamai et al. 1998; Yamagishi and Ogata 1999; Kodama et al. 2001; Casaregola et al. 2001). However, strain CBS 380 itself seems to be a hybrid yeast containing two versions of many genes, of which one is identical to the non-Sc type genes of lager brewing yeast and one is about 7% diverged (Casaregola et al. 2001, consistent with data of Pedersen 1986a; Hansen and Kielland-Brandt 1994). The obvious hypothesis to explain all of these observations is that lager brewing yeast represents one hybridisation event, and *S. bayanus* CBS 380 another, having in common the genomic sequences represented by the non-Sc genes, which have still not been found alone in any yeast species (Fig. 1). At the moment, however, it is indicated that the *S. bayanus* isolates IFO539 and IFO1948 might constitute "pure" genetic lines, the only genomic content of which corresponds to the non-Sc genome of lager brewing yeast (Rainieri et al. 2004). It now seems that hybrids and aneuploidy represent a rather common phenomenon in the *Saccharomyces* genus, especially amongst species and isolates used for industrial purposes (Masneuf et al. 1998; de Barros Lopes et al. 2002). The emerging data on the lager brewing yeast will likely be used as a model to understand the genomes of these yeasts.

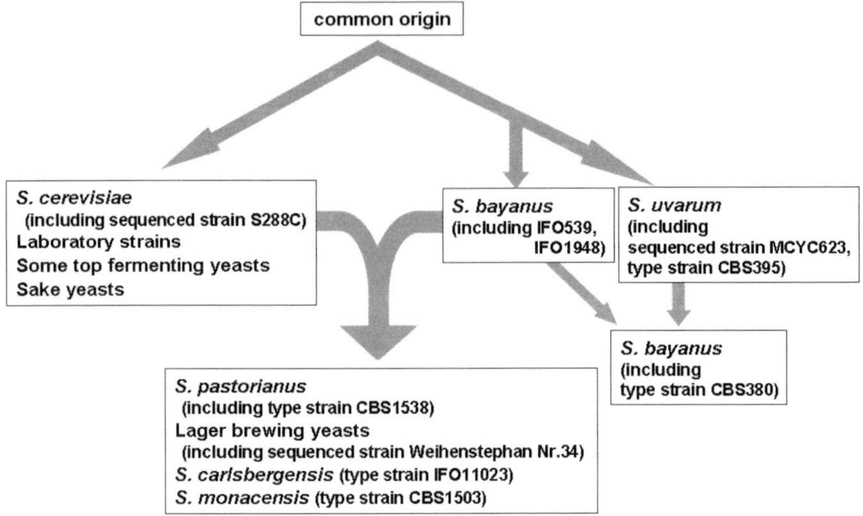

Fig. 1. Relationships among lager brewing yeasts and other *Saccharomyces* species.

1.2.3 Impact on breeding efforts

The complicated genetic nature of lager brewing yeast obviously puts a limit to the means of breeding of this economically very important fungus. At first glance, the deficiency in production of meiotic offspring would seem to obstruct classical breeding efforts, and the hybrid and polyploid nature of the yeast seriously hampers journeys into targeted gene alterations as well as into a selection of recessive mutants. However, in the early 1980's, a method to select for the few viable spores formed by lager brewing yeast and to re-constitute functional brewing yeast from such offspring was devised (Gjermansen and Sigsgaard 1981). First of all, such mating-proficient spores could be used to form a quite heterogeneous population of potential brewing strains, some of which could be better suited to particular brewing conditions, but secondly, these allodiploid or near allodiploid strains (Johannesen and Hansen 2002; Hansen et al. 2002; Hansen and Kielland-Brandt 1996, 1996b; Nilsson-Tillgren et al. 1986; Petersen et al. 1987) could be used for the selection of recessive mutants (Gjermansen 1983). Regarding targeted gene disruption, however, the problem remained that even the allodiploid spore segregants usually contains at least one copy of the non-Sc type of a given gene.

In order to improve yeast characteristics for fast fermentation or high quality beer production, several breeding efforts of lager brewing yeast have been carried out after all (e.g. Hansen and Kielland-Brandt 2003). Gene overexpression is quite feasible, as any expression cassette can be introduced on YEp plasmids or targeted to *S. cerevisiae*-like loci in the lager brewing yeast. Using this technique, it has been attempted, for example, to improve maltose fermentation efficiency (Kodama et al. 1995), to decrease off-flavours such as H_2S (Omura et al. 1995; Bramsted and Hansen 2003) and to increase ester formation (Fujii et al. 1994). As described above, inactivation of a biochemical function by gene disruption in lager brewing yeast is more complicated, and usually at least two copies of each of the Sc- and non-Sc-type of a gene have to be disrupted. Selective markers for nutrient requirements, so commonly used in haploid *S. cerevisiae*, have so far not proven expedient in lager brewing yeast, and therefore, at least four (in the case of tetraploid) types of dominant drug resistant markers, or some kind of marker elimination system to reuse a resistant marker, are necessary. Furthermore, sequence information for both the Sc- and non-Sc-type of a given gene is required, meaning that these gene versions will first have to be isolated and characterised. There are, however, a few examples of such approaches being taken. The *MET10* genes, encoding part of the sulphite reductase (Hansen et al. 1994), were disrupted and resulted in a brewing yeast with a dramatically increased sulphite production (Hansen and Kielland-Brandt 1995, 1996b), and disruption of the *MXR1* genes (encoding methionine sulphoxide reductase) resulted in a decrease in formation of the beer off-flavour dimethyl sulphide (Hansen et al. 2002). For more systematic approaches to breeding, techniques such as global functional gene analysis and transcriptome analysis are highly desirable, but due to the great previous unknown, constituted by half of the lager brewing yeast genome, the brewing geneticists have, until recently, not had access to the true potentials of these techniques. Now, however, times are changing.

2 The whole genome sequence of lager brewing yeast

As mentioned above, the whole genome sequence of the lager brewing yeast, comprising both the Sc- and the non-Sc-genome complement, is required to fully appreciate the behaviour of this yeast in industrial beer production, and to fully exploit the possibilities of constructing "customised" brewing yeast strains. With the rapidly developing DNA sequencing technologies and facilities, such projects are now feasible, and recently the whole genome sequence of one strain of lager brewing yeast, Weihenstephan Nr.34 (34/70), has in fact been obtained (Nakao et al. 2003, sequence to be published on www.suntory.com). A combination of "shotgun sequencing" and sequencing of cosmid libraries was employed to perform a total of 348,001 sequence reads of the genome of this lager brewing yeast, and the obtained nucleotide sequences constitute approximately 160 million base pairs (bp) of DNA, corresponding to a 6.5-fold coverage of the genome. The sequences were assembled into contigs, and a lager brewing yeast/*S. cerevisiae* comparative genomic map was constructed by alignment of the contigs to the *S. cerevisiae* genomic map (*Saccharomyces* Genome Database; SGD) (Fig. 2). In this way, the minimum total size of the lager brewing yeast genome was found to be 23.2 million bp, which is approximately twice the size of the *S. cerevisiae* genome (Table 1). Contigs were classified into two groups: i) Those with a DNA

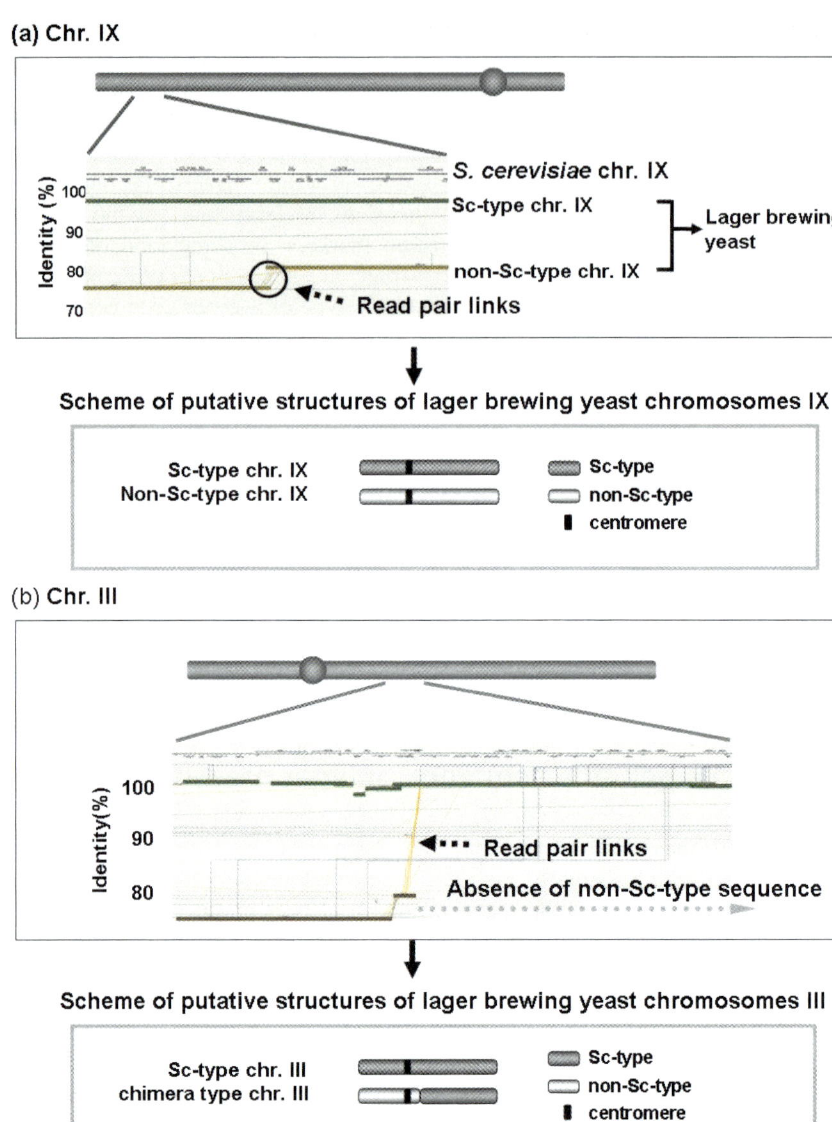

Fig. 2. Partial comparative genomic map of lager brewing yeast vs. *S. cerevisiae*. Based on sequence data to be published (see Nakao et al. 2003).

identity to *S. cerevisiae* of more than 98% (Sc-type), ii) Those with identities around 85% (non-Sc-type). From the lager brewing yeast genome sequence, 6,193 possible Open Reading Frames (ORFs) of the Sc-type were obtained and identified as well as 5,294 ORFs of the non-Sc-type. As of June 2004, almost the same

Table 1. Genome assemblies of *Saccharomyces* species

	S. cerevisiae S288C	S. paradoxus NRRL Y-17217	S. mikatae IFO 815 (CBS 8839)	S. bayanus* MCYC 623	Lager brewing yeast Weihenstephan Nr.34
Sequence coverage (fold)	Finished	7.7	5.9	6.4	6.2
Genome sequence in contig (Mb)	12.16	11.57	11.22	11.32	21.9
Genome length, including gaps (Mb)	12.16	11.75	12.12	11.54	22.11
Percentage of genome in contigs	100	98	93	98	99
Number of scaffolds	16	51	90	100	292
Gaps per 200kb	0	3.2	10.3	4.4	7.7
Average gap length (bp)	0	583	847	679	753

Data for *S. cerevisiae*, *S. paradoxus*, *S. mikatae* and *S. bayanus* are from Kellis et al. (2003). Data for the lager brewing yeast (*S. pastorianus*) are to be published, see Nakao et al. (2003). **S. uvarum* type

number was found in *S. cerevisiae*; SGD lists 6700 ORFs, including 814 dubious ORFs (Nakao et al. 2003). The reason for the lower number of non-Sc-type ORFs is a slight deficit of non-Sc-type DNA, as described below. In almost all cases, the "non-Sc" ORFs are of the same length as their "Sc" counterparts, but a few exceptions have been observed. From the lager brewing yeast/*S. cerevisiae* comparative genomic map (Fig. 2), it can be seen that in large regions of the lager brewing yeast genome the gene synteny (order) is identical to that found in *S. cerevisiae*, with some exceptions reflecting translocations or inversions (see below). The average identities of the lager brewing yeast ORFs to the ORFs found in other *Saccharomyces* species, both at the nucleotide and the amino acid level, are shown in Table 2. The Sc-type ORFs obviously show the highest identity to the *S. cerevisiae* counterparts, while the non-Sc-type ORFs seem most strongly related to the corresponding *S. bayanus (uvarum)* gene sequences, confirming earlier data from single genes (see the introductory sections of this chapter). As mentioned above, the phylogenetic position of *S. bayanus* is complicated because some isolates have a partial hybrid nature, and the particular strain used to obtain the sequence data compared to in Table 2 is a non-hybrid strain (*S. uvarum*) (Kellis et al. 2003), hence, the relatively low identity numbers between these sequences and the lager brewing yeast non-Sc-sequences.

Intriguingly, 20 ORFs present in the non-Sc-type contigs showed no significant identities to any *S. cerevisiae* ORFs, but nonetheless, significantly similar to genes from other organisms registered in GENBANK. The functions of most of these ORFs are unknown, and further investigation is necessary to elucidate their nature.

Table 2. Average sequence identities of lager brewing yeast ORFs to those of other *Saccharomyces* species

		Sc-ORF	Non-Sc-ORF
S. cerevisiae	nt(%)	98.9	84.4
S288C	AA(%)	98.6	83.7
S. paradoxus	nt(%)	89.7	84.1
NRRL Y-17217 (CBS 432)	AA(%)	91.2	83.7
S. mikatae	nt(%)	85.5	83.6
IFO 1815 (CBS 8839)	AA(%)	86.1	82.4
S. bayanus	nt(%)	83.6	91.9
MCYC 623	AA(%)	82.9	92.2

Sc-type and non Sc-type ORFs of lager brewing yeast were subjected to homology searching against the *S. cerevisiae, S. paradoxus, S. mikatae, and S. bayanus* (*S. uvarum*) genome sequence, respectively, and average identities of ORFs at the nucleotide (nt) and amino acid (AA) levels were shown.

However, these findings might indicate that the two yeast species that came together to constitute the lager brewing yeast hybrid were of somewhat different ecotypes. One of these ORFs has been reported as a gene encoding a specific fructose/H^+ symporter in lager brewing yeast (Gonçalves et al. 2000). Active fructose transport is one of the taxonomic markers to distinguish *S. pastorianus* and *S. bayanus* from other *Sacchromyces sensu stricto* species (Rodrigues de Sousa et al. 1995). *S. bayanus* is generally isolated from oenological environments, rich in fructose. Although this sugar does not play a major role in brewing, the gene for the symporter has stayed around in the lager brewing yeast.

In contrast to the protein-encoding regions, the non-Sc-type intergenic regions in the lager brewing yeast are very diverged from the Sc-type counterparts. This may mean that transcriptional regulation of the two gene types differ during the life cycle of the yeast cell. In fact, such differential expression of homoeologues in lager brewing yeast has been reported for the *BAP2* gene (encoding a branched-chain amino acid permease) homoeologues (Kodama et al. 2001), and for *MET2* (encoding homoserine O-acetyl transferase) and *MET14* (encoding adenosylphosphosulphate kinase) (Johannesen and Hansen 2002). Firstly, these findings tell us that the lager brewing yeast is not merely a polyploid with two divergent but similarly functioning genome parts, but is in fact a unique organism with a biological complexity larger than any of the species that took part in its formation. A few large-scale studies of the lager brewing yeast transcriptome, using *S. cerevisiae* DNA micro- or macro-arrays have been performed (Panoutsopoulou et al. 2001; Olesen et al. 2002; James et al. 2003), but these studies obviously could not reveal anything about differential expression, due to preferential hybridisation between the Sc-type fraction of the probe preparations and the *S. cerevisiae* target sequences. Thus, knowledge of the sequence of the non-Sc part of the lager brewing yeast genome is necessary to fully appreciate the behaviour of this yeast in industrial beer production.

To that end, a comprehensive expression analysis during beer fermentation, using massively parallel signature sequencing (Brenner et al. 2000) has recently been performed (Nakao Y, Kodama Y, Fujimura T, Nakamura N and Ashikari T, manuscript in preparation). The expression profiles of almost 1400 homoeologues were investigated, and it was shown that almost half of the lager brewing yeast gene homoeologues showed differential expression. In order to study the phenomenon of differential gene expression in lager brewing yeast in more detail, DNA arrays containing all of the lager brewing yeast ORFs (i.e. both Sc- and non-Sc-types of all genes) have been recently constructed, and transcriptome analysis performed (Nakao Y, Kodama Y, Fujimura T, Nakamura N, and Ashikari T, manuscript in preparation). Data from these studies will surely lead to a rich picture and hopefully an understanding of principles underlying differential expression of homoeologues, not only in lager brewing yeast but in species hybrids in a broader sense.

The mitochondrial genome of lager brewing yeast has also been analysed in detail (Nakao Y, Kodama Y, and Ashikari T, manuscript in preparation). The size of the mitochondrial DNA molecule is around 70kb, which is a little smaller than that of *S. cerevisiae* (85.8kb) (Foury et al. 1998), and almost consistent with the previously reported size of *S. pastorianus* mitochondrial DNA (66.6kb) (Groth et al. 2000). Also the gene order of lager brewing yeast mitochondrial DNA is different from that of *S. cerevisiae* but the same as that of *S. bayanus*. This result is completely consistent with the study by Groth et al. (2000). Furthermore, the intron-exon structures in the *COX1, COB,* and 21s rRNA genes of lager brewing yeast are different from those of *S. cerevisiae*. For example, there is an intron in the 21s rRNA gene of *S. cerevisiae* (Foury et al. 1998), whereas there is none in the lager brewing yeast. The sequence identities of mitochondrial genes from lager brewing yeast and *S. bayanus* are very high (e. g. 100% for the *ATP9* and *COX2* genes) (Groth et al. 2000; Nakao Y, Kodama Y, and Ashikari T, manuscript in preparation), whereas the nucleotide identity of the whole mitochondrial genome of lager brewing yeast to that of *S. cerevisiae* is much lower (65-97%). These results confirm the suggestion (Groth et al. 2000) that lager brewing yeast inherited its mitochondrion from the non-*S. cerevisiae* type ancestor. Thus, lager brewing yeast contains two diverged nuclear genomes and only one mitochondrial genome, and different genetic rules are followed during their inheritance (Piskur 1994).

3 Chromosomal structure

As described above, the mosaic structure of some lager brewing yeast chromosomes was discovered almost twenty years ago. The actual number of recombinational events was, however, not known, and neither were the actual recombination points. Furthermore, none of the existing data provided the answer to several questions regarding the ploidy of lager brewing yeast. How many types exist of each of the chromosomes, and how many copies of each type? Such information is required for establishing the optimal strategies for targeted molecular breeding of this yeast. Recently, these problems have been addressed, partly by analysis of the sequence contigs obtained by the lager brewing yeast whole genome sequence analysis as just described, and partly by hybridisation experiments with *S. cerevisiae* gene arrays and biotin-labelled genomic lager brewing yeast DNA (Kodama Y, Nakao Y, Nakamura N, Fujimura T, Shirahige K and Ashikari T, manuscript in preparation).

Thus, when mapping the contigs of the lager brewing yeast genome to that of *S. cerevisiae,* it was confirmed that there are three kinds of chromosomes in this yeast: Sc-type, non-Sc-type, and various chimerical types, as shown in Fig. 3. The precise structures of the chimera-type chromosomes were determined by the links of forward-reverse shotgun read pairs as shown in Fig. 2. The recombination break points between Sc-type and non-Sc-type chromosomes were also confirmed by PCR using Sc-type and non-Sc-type sequences as primers and subsequent sequencing of PCR fragments. These analyses showed that the lager brewing yeast contains at least eight chimerical chromosomes. For the chromosomes VIII, X, and XI, the situation appears even more complicated, as they come in three types: pure Sc-type, pure non-Sc-type, and chimerical ones. Some of these data are consistent with the previous reports of single chromosome transfer (for a review of genetics of brewing yeasts see Kielland-Brandt et al. 1995), and the quite recent report of competitive comparative genome hybridisation using *S. cerevisiae* DNA microarrays with Cy3-, Cy5-labelled DNA and quantitative real time polymerase chain reaction assays (Bond et al. 2004), whereas some are not. We take this to reflect a diversity of chromosome structure among lager brewing yeasts as described below. Some of the chromosomal breakpoints were found to be inside ORFs, which means that some hybrid ORFs exist. Most of these were classified as "non-Sc" ORFs according to the relatively low nucleotide identities to *S. cerevisiae* ORFs. It is to be anticipated that the further investigation of such hybrid ORFs will be highly rewarding in terms of new knowledge on protein function as well as hybrid speciation. We expect more recombination break points in the telomeric regions, because these regions are subject to frequent rearrangements (Kellis et al. 2003). One example is a telomeric translocation between chr. VIII (Lg-*FLO1*) and chr. IX (YIL169c), which has been reported to cause the conversion from flocculation to non-flocculation in a lager brewing yeast (Sato et al. 2003). Such break points are, however, difficult to identify solely from sequence data.

Furthermore, some translocations between the non-Sc-types of chromosomes II and IV, VIII, and XV must also have taken place, as there are links of forward-reverse shotgun read pairs that connect these chromosomes (Nakao Y, Kodama Y, Nakamura N, Fujimura T, Rainieri S, Ito T, Hattori M, Shiba T, and Ashikari T, manuscript in preparation). These data are consistent with previous results of studies of the chromosomal structure of *S. bayanus* (Ryu et al. 1996), indicating that

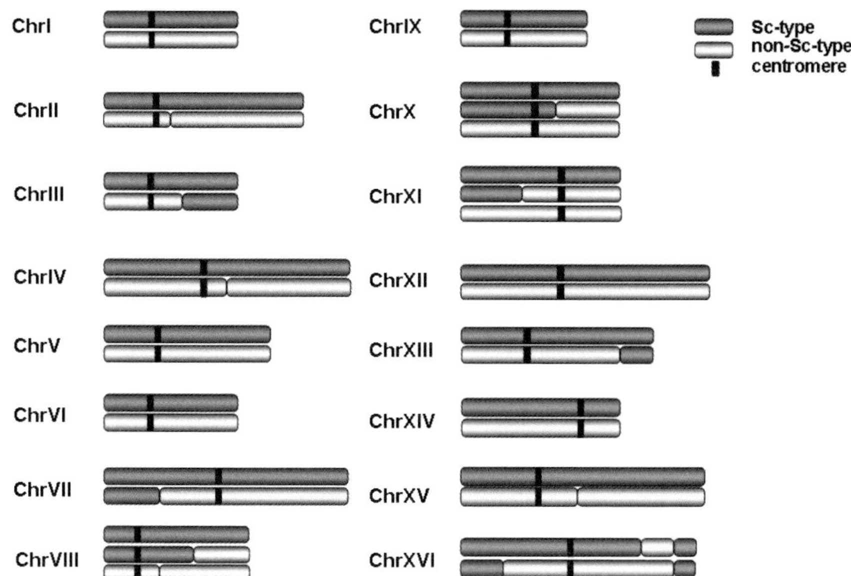

Fig. 3. Putative chromosomal structure of lager brewing yeast strain Weihenstephan Nr.34 (34/70). The breakpoints between "Sc" and "non-Sc" DNA in chromosomes are shown as constrictions (*e.g.* in chromosome III). Constrictions in chromosomes II, IV, VIII and XV denote translocation breakpoints between the non-Sc type chromosomes.

the rearrangements among the non-Sc-type chromosomes have occurred in the non-Sc-type ancestor, followed by the hybridisation with the Sc-type ancestor, and recombination events between Sc-type and non-Sc-type chromosomes. *Saccharomyces* chromosomes in general seem to be quite dynamic and have often been rearranged during their evolutionary history (Langkjaer et al. 2000; Fischer et al. 2000). Characterisation of the *S. pastorianus/carlsbergensis* chromosome breakpoints, creating mosaic chromosomes, confirms these previous observations.

To summarise, we now know the complicated genome of lager brewing yeast in some detail. We have the greater picture of the chromosomal structure, we are aware of the more apparent translocations and recombination events that have taken place, we know the number of types of each chromosome, and we know that on average this yeast has a tetrasomic chromosome content. We still do not know, however, the exact number of copies of each chromosome type, and such knowledge is important for breeding purposes. One possible approach to solve this problem is comparative genomic hybridisation using DNA microarrays containing all of the lager brewing yeast ORFs. The signal intensity of each probe corresponding to each ORF reflects the copy number of each ORF (Kodama Y, Nakao Y, Nakamura N, Fujimura T, Shirahige K, and Ashikari T, manuscript in preparation).

4 Diversity of chromosome structure of lager brewing yeasts and their relatives

Understanding the genomic set-up of one particular brewing yeast obviously leads to the question of generalisation: Do all lager brewing yeast strains have the same set-up? In other words, to what exact extent are these industrial organisms historically related? Some data on this matter are becoming available from studies of comparative genomic hybridisation between genomic DNA of various lager brewing yeast strains and various type strains, with *S. cerevisiae* yeast DNA microarrays (Kodama Y, Nakao Y, Nakamura N, Fujimura T, Shirahige K, and Ashikari T, manuscript in preparation). The techniques used for these investigations were identical to those leading to the results described in the preceding section. Only Sc-type gene fragments hybridise to the *S. cerevisiae* oligonucleotide array, and therefore, the hybridisation signal reflects the relative content of Sc-type DNA to non-Sc-type DNA. A high or exclusive content of non-Sc-type DNA is seen as a negative number in analyses as those depicted in Fig. 4. Several lager brewing strains were analysed with this method, and the chromosome structures were compared (Fig. 5). Even though most of the tested lager brewing yeast strains have the same structure for the shown chromosome (XVI), three show divergent structures. Furthermore, it can be seen that the type strains of *S. pastorianus* (CBS 1538), *S. carlsbergensis* (IFO11023), and *S. monacensis* (CBS 1503) have lost many Sc-type chromosomes, quite in accordance with previous findings for particular genes or chromosome regions (Yamagishi and Ogata 1999; Casaregola et al. 2001). In particular, the type strain of *S. pastorianus* (CBS 1538) has lost eight chromosomes of Sc-type (II, III, IV, VI, VIII, XII, XV, XVI). Thus, these type strains are actually very different from the lager brewing strains that are used for commercial beer production. These results raise interesting questions on the mechanisms and selective forces that control the evolution of hybrid organisms; for example, what is the nature of the determinants for which of the homoeologous genes to keep and which to let go of? Further investigations in this field will certainly yield information on the history of individual genes, and likely also to a general understanding of the forces underlying post-hybridisation speciation processes.

The availability of a DNA array with the whole genome sequence of lager brewing yeast will make possible a detailed comparison of the chromosomal structures of various lager strains. Provided that they still exist, "pure" genetic lines, the only genomic content of which corresponds to the non-Sc-genome of lager brewing yeast, could in principle equally well be identified by the investigation of various *Saccharomyces* strains with this technique.

5 Concluding remarks

From the results of the whole genome analysis, the hybrid nature of lager brewing yeast has been confirmed and clarified in unprecedented detail. For most of the chromosomes both Sc- and non-Sc-like types exist, but also chimerical

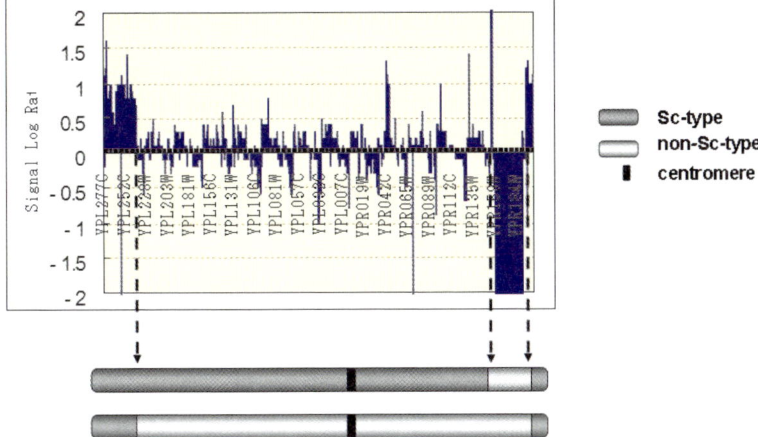

Fig. 4. Dosage of Sc-type sequences on Chr. XVI. Labelled DNA of lager brewing yeast was hybridised to an *S. cerevisiae* gene array, and the signal of each ORF was normalized to that of the haploid strain S288C. The log$_2$ of the resulting ratio was depicted, following the gene order in each chromosome. The non-*S. cerevisiae* type genes do not hybridize to the *S. cerevisiae* array. The points where the signal shows abrupt changes are considered sites of recombination that gave rise to the chimerical chromosome.

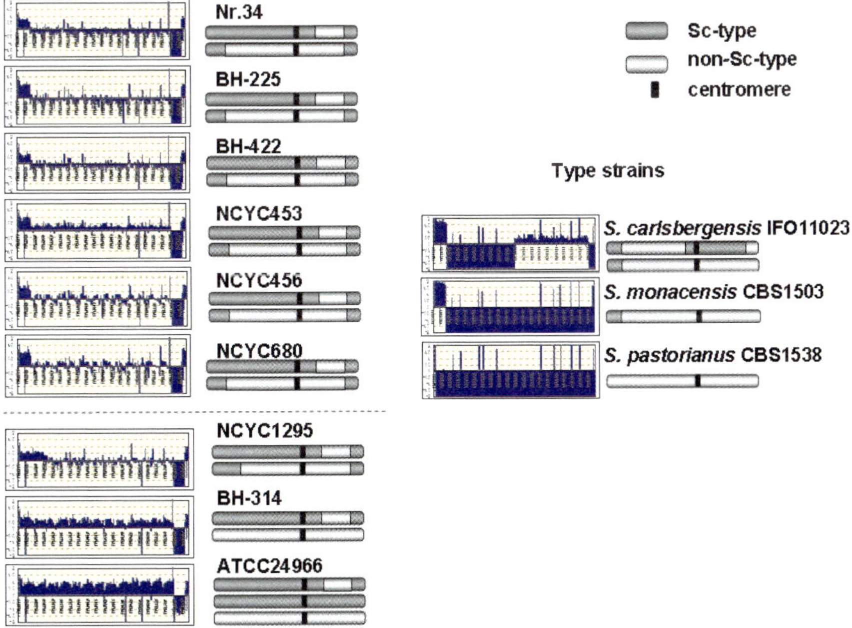

Fig. 5. The structure of various versions of Chr. XVI in several brewing yeasts, as deduced from results of array analysis according to Fig. 4.

chromosomes are present, and some strains even show the complete loss of one or the other chromosome type. Chromosome transmission fidelity following inter-species crosses among *Saccharomyces sensu stricto* species is a matter of interest (Wolfe 2003; Delneri et al. 2003), and lager brewing yeast provides an excellent example of how an inter-species hybrid maintains its chromosomes and how re-combination events between two related, yet speciated, genomes have occurred.

The whole genome sequence of lager brewing yeast is needed to fully exploit the possibilities of constructing customised brewing yeast strains with any combination of desired traits. The whole genome sequence will be available on the website of Suntory Ltd. in the near future (www.suntory.com). The availability of the whole genome sequence will enable the scientific community to carry out comprehensive expression analyses and genome structural analyses of lager brewing yeast strains. Application of such analyses to quality control in beer production and development of new products will be highly beneficial to the brewing industry.

Acknowledgements

We thank Y. Nakao for allowing us to refer to unpublished results. We also thank Dr. T. Ashikari and Dr. S. Rainieri for their helpful suggestions.

References

Andersen TH, Hoffmann L, Grifone R, Nilsson-Tillgren T, Kielland-Brandt MC (2000) Brewing yeast genetics. EBC Monograph 28, Fachverlag Hans Carl, Nürnberg, pp 140-147

De Barros Lopes M, Bellon JR, Shirley NJ, Ganter PF (2002) Evidence for multiple inter-specific hybridization in *Saccharomyces sensu stricto* species. FEMS Yeast Res 1:323-331

Bond U, Neal C, Donnelly D, James TC (2004) Aneuploidy and copy number breakpoints in the genome of lager yeasts mapped by microarray hybridisation. Curr Genet 45:360-370

Bramsted B, Hansen J (2003) Controlling the level of hydrogen sulphide production in lager brewing yeast by the introduction of heterologous enzymatic pathways for cysteine biosynthesis. Proc 29th Conf Eur Brew Conv, Dublin, Chapter 51, pp 554-562

Brenner S, Johnson M, Bridgham J, Golda G, Lloyd DH, Johnson D, Luo S, McCurdy S, Foy M, Ewan M, Roth R, George D, Eletr S, Albrecht G, Vermaas E, Williams SR, Moon K, Burcham T, Pallas M, DuBridge RB, Kirchner J, Fearon K, Mao J, Corcoran K (2000) Gene expression analysis by massively parallel signature sequencing (MPSS) on microbead arrays. Nat Biotechnol 18:630-634

Børsting C, Hummel R, Schultz ER, Rose TM, Pedersen MB, Knudsen J, Kristiansen K (1997) *Saccharomyces carlsbergensis* contains two functional genes encoding the acyl-CoA binding protein, one similar to the *ACB1* gene from *S. cerevisiae* and one identical to the *ACB1* gene from *S. monacensis*. Yeast 13:1409-1421

Casaregola S, Nguyen HV, Lapathitis G, Kotyk A,Gaillardin C (2001) Analysis of the constitution of the beer yeast genome by PCR sequencing and subtelomeric sequence hybridization. Int J Syst Evol Microbiol 51:1607-1618

Casey GP (1986a) Cloning and analysis of two alleles of the *ILV3* gene from *Saccharomyces carlsbergensis*. Carlsberg Res Commun 51:327-341

Casey GP (1986b) Molecular and genetic analysis of chromosomes X in *Saccharomyces carlsbergensis*. Carlsberg Res Commun 51:343-362

Corran HS (1975) A History of Brewing. David & Charles, Newton Abbott, UK

Delneri D, Colson I, Grammenoudi S, Roberts IN, Louis EJ, Oliver SG (2003) Engineering evolution to study speciation in yeasts. Nature 422:68-72

Fischer G, James SA, Roberts IN, Oliver SG, Louis EJ (2000) Chromosomal evolution in *Saccharomyces*. Nature 405:451-454

Foury F, Roganti T, Lecrenier N, Purnelle B (1998) The complete sequence of the mitochondrial genome of *Saccharomyces cerevisiae*. FEBS Lett 440:325-331

Fujii T, Nagasawa N, Iwamatsu A, Bogaki T, Tamai Y, Hamachi M (1994) Molecular cloning, sequence analysis, and expression of the yeast alcohol acetyltransferase gene. Appl Env Microbiol 60:2786-2792

Fujii T, Yoshimoto H, Nagasawa N, Bogaki T, Tamai Y, Hamachi M (1996) Nucleotide sequences of alcohol acetyltransferase genes from lager brewing yeast, *Saccharomyces carlsbergensis*. Yeast 12:593-598

Gjermansen C (1983) Mutagenesis and genetic transformation of meiotic segregants of lager yeast. Carlsberg Res Commun 48:557-565

Gjermansen, C (1991) Comparison of genes in *Saccharomyces cerevisiae* and *Saccharomyces carlsbergensis*. PhD thesis, University of Copenhagen, Copenhagen

Gjermansen C, Sigsgaard P (1981) Construction of a hybrid brewing strain of *Saccharomyces carlsbergensis* by mating of meiotic segregants. Carlsberg Res Commun 46:1-11

Gjermansen C, Nilsson-Tillgren T, Petersen JGL, Kielland-Brandt MC, Sigsgaard P, Holmberg S (1988) Towards diacetyl-less brewers yeast. Influence of *ilv2* and *ilv5* mutations. J Basic Microbiol 28:175-183

Gonçalves P, Rodrigues de Sousa H, Spencer-Martins I (2000) *FSY1*, a novel gene encoding a specific fructose/H$^+$ symporter in the type strain of *Saccharomyces carlsbergensis*. J Bacteriol 182:5628-5630

Groth C, Petersen RF, Piskur J (2000) Diversity in organization and the origin of gene orders in the mitochondrial DNA molecules of the genus *Saccharomyces*. Mol Biol Evol 17:1833-1841

Hansen EC (1883) Recherches sur la physiologie et la morphologie des ferments alcooliques V. Méthodes pour obtenir des cultures pures de *Saccharomyces* et de mikroorganismes analogues. Compt Rend Trav Lab Carlsberg 2:92-105

Hansen EC (1908) Recherches sur la physiologie et la morphologie des ferments alcooliques XIII. Nouvelles études sur des levures de brasserie à fermentation basse. Compt Rend Trav Lab Carlsberg 7:179-217

Hansen J (1999) Inactivation of *MXR1* abolishes formation of dimethyl sulfide from dimethyl sulfoxide in *Saccharomyces cerevisiae*. Appl Env Microbiol 65:3915-3919

Hansen J, Kielland-Brandt MC (1994) *Saccharomyces carlsbergensis* contains two functional *MET2* alleles similar to homologues from *S. cerevisiae* and *S. monacensis*. Gene 140:33-40

Hansen J, Kielland-Brandt MC (1995) Genetic control of sulphite production in brewer's yeast. Proc 25th Congr Eur Brew Conv 1995, Brussels, pp 319-328

Hansen J, Kielland-Brandt MC (1996a) Inactivation of *MET2* in brewer's yeast increases the level of sulphite in beer. J Biotechnol 50:75-87

Hansen J, Kielland-Brandt MC (1996b) Inactivation of *MET10* in brewer's yeast specifically increases SO_2 formation during beer production. Nature Biotechnol 14:1587-1591

Hansen J, Kielland-Brandt MC (2003) Brewer's yeast: genetic structure and targets for improvement. In: H de Winde (ed) Functional Genetics of Industrial Yeasts, Topics in Current Genetics 2, Springer, Berlin, pp 143-170

Hansen J, Cherest H, Kielland-Brandt MC (1994) Two divergent *MET10* genes, one from *Saccharomyces cerevisiae* and one from *Saccharomyces carlsbergensis*, encode the α subunit of sulphite reductase and specify potential binding sites for FAD and NADPH. J Bacteriol 176:6050-6058

Hansen J, Bruun SV, Bech LM, Gjermansen C (2002) Brewing yeast expression of the *MXR1* gene is the major determinant for the content of dimethyl sulphide in beer. FEMS Yeast Res 2:137-149

Hoffman L (2000) The defective sporulation of lager brewing yeast. PhD thesis, University of Copenhagen, Copenhagen

Holmberg S (1982) Genetic differences between *Saccharomyces carlsbergensis* and *S. cerevisiae* II. Restriction endonuclease analysis of genes in chromosome III. Carlsberg Res Commun 47:233-244

James TC, Campbell S, Donnelly D, Bond U (2003) Transcription profile of brewery yeast under fermentation conditions. J Appl Microbiol 94:432-448

Johannesen PF, Hansen J (2002) Differential transcriptional regulation of gene homoeologues in a fungal species hybrid. FEMS Yeast Res 1:315-322

Kellis M, Patterson N, Endrizzi M, Birren B, Lander ES (2003) Sequencing and comparison of yeast species to identify genes and regulatory elements. Nature 423:241-254

Kielland-Brandt MC, Gjermansen C, Nilsson-Tillgren T, Holmberg S (1989) Yeast breeding. Proc 22nd Congr Eur Brew Conv, Zürich, pp 37-47

Kielland-Brandt MC, Nilsson-Tillgren T, Gjermansen C, Holmberg S, Pedersen MB (1995) Genetics of brewing yeasts. In: Wheals AE, Rose AH, Harrison JS (eds) The Yeasts., 2nd edn, Vol 6, Academic Press, London, UK, pp 223-254

Kodama Y, Fukui N, Ashikari T, Shibano Y (1995) Improvement of maltose fermentation efficiency: Constitutive expression of *MAL* genes in brewing yeast. J Am Soc Brew Chem 53:24-29

Kodama Y, Omura F, Ashikari T (2001) Isolation and characterization of a gene specific to lager brewing yeast that encodes a branched-chain amino acid permease. Appl Environ Microbiol 67:3455-3462

Langkjaer RB, Nielsen ML, Daugaard PR, Liu W, Piskur J (2000) Yeast chromosomes have been significantly reshaped during their evolutionary history. J Mol Biol. 304:271-288

Masneuf I, Hansen J, Groth C, Piskur J, Dubourdieu D (1998) New hybrids between *Saccharomyces* sensu stricto yeast species found among wine and cider production strains. Appl Environ Microbiol. 64:3887-3892

Nakao Y, Kodama Y, Nakamura N, Ito T, Hattori M, Shiba T, Ashikari T (2003) Whole genome sequence of a lager brewing yeast. Proc 29th Congr Eur Brew Conv, Dublin, Chapter 48, pp 524-530

Nilsson-Tillgren T, Gjermansen C, Kielland-Brandt MC, Petersen JGL, Holmberg S (1981) Genetic differences between *Saccharomyces carlsbergensis* and *S. cerevisiae*. Analysis of chromosome III by single chromosome transfer. Carlsberg Res Commun 46:65-76

Nilsson-Tillgren T, Gjermansen C, Holmberg S, Petersen JGL, Kielland-Brandt MC (1986) Analysis of chromosome V and the *ILV1* gene from *Saccharomyces carlsbergensis*. Carlsberg Res Commun 51:309-326

Olesen K, Felding T, Gjermansen C, Hansen J (2002) The dynamics of the *Saccharomyces carlsbergensis* brewing yeast transcriptome during a production-scale lager beer fermentation. FEMS Yeast Res 2:563-573

Omura F, Shibano Y, Fukui N, Nakatani K (1995) Reduction of hydrogen sulphide production in brewing yeast by constitutive expression of *MET25* gene. J Am Soc Brew Chem 53:58-62

Panoutsopoulou K, Wu J, Hayes A, Butler P, Oliver SG (2001) Yeast transcriptome analysis during the brewing process. Yeast 18:S300

Pedersen MB (1985) DNA sequence polymorphisms in the genus *Saccharomyces* II Analysis of the genes *RDN1*, *HIS4*, *LEU2* and *Ty* transposable elements in Carlsberg, Tuborg and 22 Bavarian brewing strains. Carlsberg Res Commun 50:263-272

Pedersen MB (1986a) DNA sequence polymorphism in the genus *Saccharomyces* III. Restriction endonuclease fragment patterns of chromosomal regions in brewing and other yeast strains. Carlsberg Res Commun 51:163-183

Pedersen MB (1986b) DNA sequence polymorphism in the genus *Saccharomyces* IV. Homologous chromosomes III in *Saccharomyces bayanus*, *S. carlsbergensis*, and *S. uvarum*. Carlsberg Res Commun 51:185-202

Petersen JGL, Nilsson-Tillgren T, Kielland-Brandt MC, Gjermansen C, Holmberg S (1987) Structural heterozygosis at genes *ILV2* and *ILV5* in *Saccharomyces carlsbergensis*. Curr Genet 12:167-174

Piskur J (1994) Inheritance of the yeast mitochondrial genome. Plasmid. 31:229-241

Porter G, Westmoreland J, Priebe S, Resnick MA (1996) Homologous and homeologous intermolecular gene conversion are not differentially affected by mutations in the DNA damage or the mismatch repair genes *RAD1, RAD50, RAD51, RAD52, RAD54, PMS1,* and *MSH2*. Genetics 143:755-767

Rainieri S, Zambonelli C, Kaneko Y (2003) *Saccharomyces sensu stricto:* Systematics, genetic diversity, and evolution. J Biosci Bioeng 96:1-9

Rainieri S, Kodama Y, Nakao Y, Ashikari T, Mikata K, Kaneko Y (2004) Analysis of the species *Saccharomyces bayanus* and *Saccharomyces pastorianus*; hybrid lines and pure genetic lines. Abstr 10th International Congress for Culture Collections (ICCC), Tsukuba, pp 534-535

Rodrigues de Sousa H, Madeira-Lopes A, Spencer-Martins I (1995) The significance of active fructose transport and maximum temperature for growth in the taxonomy of *Saccharomyces sensu stricto*. Syst Appl Microbiol 18:44-51

Ryu SL, Murooka Y, Kaneko Y (1996) Genomic reorganization between two sibling yeast species, *Saccharomyces bayanus* and *Saccharomyces cerevisiae*. Yeast 12:757-764

Sato M, Yokoi S, Watari J, Takashio M (2003) Model of an inactivation of the Lg-FLO1 gene by translocation of chromosome. Proc 29th Congr Eur Brew Conv, Dublin, Chapter 61, pp 656-668

Scherer S, Davis RW (1979) Replacement of chromosome segments with altered DNA sequences constructed *in vitro*. Proc Natl Acad Sci USA 76:4951-4955

Tamai Y, Momma T, Yoshimoto H, Kaneko Y (1998) Co-existence of two types of chromosome in the bottom fermenting yeast, *Saccharomyces pastorianus*. Yeast 14:923-933

Tamai Y, Tanaka K, Umemoto N, Tomizuka K, Kaneko Y (2000) Diversity of the HO gene encoding an endonuclease for mating type conversion in the bottom fermenting yeast *Saccharomyces pastorianus*. Yeast 16:1335-1343

Vaughan-Martini A, Kurtzman CP (1985) Deoxyribonucleic acid relatedness among species of *Saccharomyces sensu stricto*. Int J Syst Bacteriol 35:508-511

Vaughan-Martini A, Martini A (1987) Three newly delimited species of *Saccharomyces sensu stricto*. Antonie v Leeuwenhoek 53:77-84

Wolfe K (2003) Evolutionary biology: Speciation reversal. Nature 422:25-26

Yamagishi H, Ogata T (1999) Chromosomal structures of bottom fermenting yeasts. Syst Appl Microbiol 22:341-353

Yarrow D (1984) *Saccharomyces* Meyen ex Reess. In: NJW Kreger-van Rij (ed) The Yeasts, a Taxonomic Study., 3rd edn, Elsevier Science Publishers, Amsterdam, pp 379-395

Hansen, Jørgen
    Poalis A/S, Bülowsvej 25, DK-1870 Frederiksberg C, Denmark

Kielland-Brandt, Morten C.
    Carlsberg Laboratory, Gamle Carlsberg Vej 10, DK-2500 Copenhagen Valby, Denmark
    mkb@crc.dk

Kodama, Yukiko
    Suntory Research Center 1-1-1, Wakayamadai, Shimamoto-cho, Mishima-gun, Osaka 618-8503, Japan
    Yukiko_Kodama@suntory.co.jp

Genome evolution: Lessons from Genolevures

Monique Bolotin-Fukuhara, Serge Casaregola, and Michel Aigle

Abstract

In the past years, yeast genome-sequencing programs have been widely developed. Two of them, namely Genolevures I and II, were devoted to the exploration of hemiascomycetous yeasts. The first one covered 13 species with partial random sequencing (0.2-0.4X coverage). The second one led to the complete genome sequence (8-11X coverage) of four species. The overall evolution of genome structures and the phylogeny could already be deduced from the partial sequencing data, while their mechanisms needed the analysis of complete genome. The main results that came out of these projects are that evolution superimposes stochastic discrete events to the continuous mutation flow. Concerning functional aspects, examination of partial sequences revealed only general trends. Here, we performed a detailed analysis of the genes involved in central metabolism and in the anaerobiosis/aerobiosis pathways on the complete genome of three representative yeasts, *Saccharomyces cerevisiae*, *Kluyveromyces lactis*, and *Yarrowia lipolytica*. This comparison revealed subtle and relevant differences in gene content, which explain in a satisfactory way the known physiological specificities of these three species. This analysis outlines the need for high quality and complete genome sequence when comparing biological processes and functions.

1 Introduction

The picture we have of life has changed dramatically since Darwin proposed that life is not a fact but a history. Indeed, it became necessary, not only to describe and to compare living beings and their components, but also to discover the story which has resulted in each specific object. This fundamental feature of living creatures and their components compelled biologists to develop evolution as a specific branch of biology.

For more than a century now, biologists have used evolution to explore different scales of living systems, from global biosphere to simple small metabolic molecules. At each level, the historical features have to be discovered and understood. Among the last and fascinating objects of life that biologists have been able to describe precisely for the last decade are genomes. A genome is, at least theoretically, a perfect object because its description is ideally unique and finite. We consider that the sequence of a genome contains all the information necessary to

Topics in Current Genetics, Vol. 15
P. Sunnerhagen, J. Piškur (Eds.)· Comparative Genomics
DOI 10.1007/b136677 / Published online: 13 April 2005

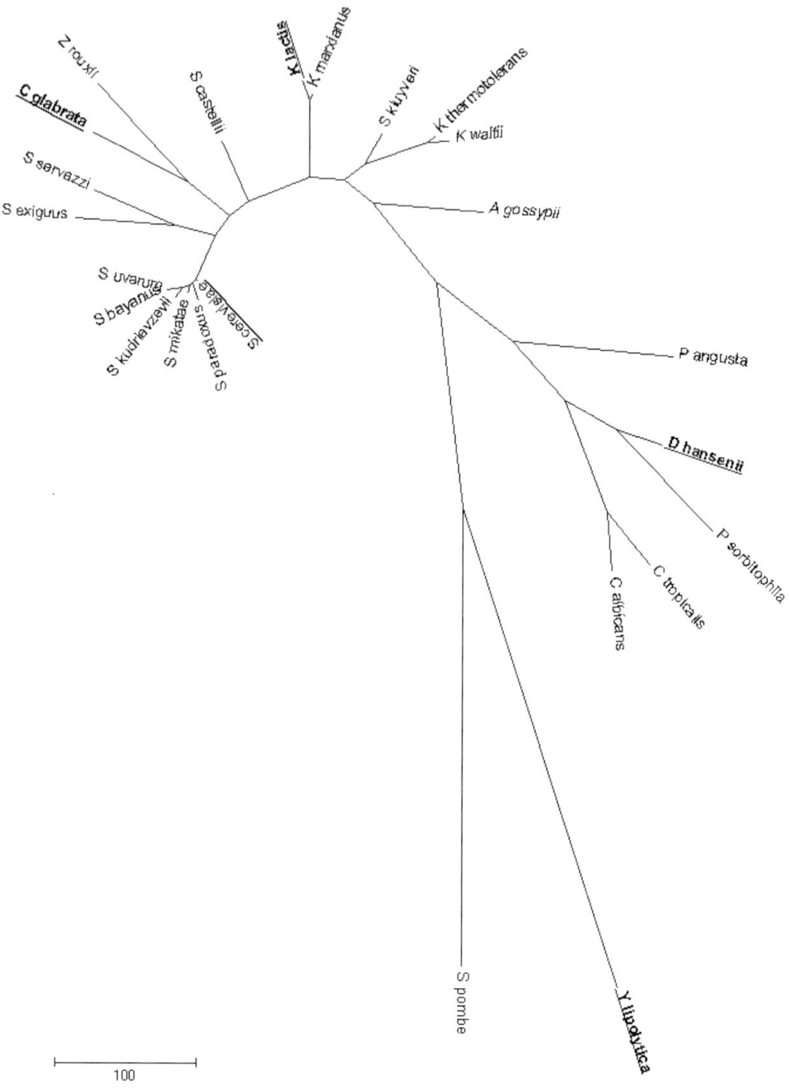

Fig. 1. Evolutionary tree of hemiascomycetous yeasts. The tree is calculated on the basis of the 25S sequences, using the maximum parsimony method (P. Durrens personal communication) In bold, the species analyzed during the Genolevures II program. *S. pombe* is taken as an outer group.

describe it. It is a robust material upon which the work of reconstructing the histories of genomes can now begin.

Less than ten years after the first description of a cellular genome (Fraser et al. 1995), the situation is still in its infancy. The first limitation for eukaryotic cells lies in the material itself. While prokaryotic genome sequences are now available in appreciable numbers and quality, eukaryotic genomes are not. Only some of them are sufficiently finished to be exploited in a global way. Homogeneous, well-defined, high quality sequences are a prerequisite that is difficult to satisfy for work in this field. Such a difficulty can be overcome by an effort in terms of defining strict and homogeneous quality standards. Much has to be done between the sequencing of a genome and the termination of a complete, well-annotated sequence that corresponds to a genuine genome.

The second limitation is our naiveté in front of new biological objects such as genomes. The obvious difficulty is to ask relevant questions. Some of them will give some insights into the history that we are trying to understand. Most of them will just lie sterile because they are not relevant to history. Is the size of a genome an important historical determinant? Is the GC content something that leaves its mark on the whole story? To understand the history of genomes we must learn progressively what are the most pertinent questions to ask, but genome evolution is a recent field and the most we do at this stage is to apply all our tools.

1.1 Genolevures projects

Two projects, named Genolevures I and II, were realized by a network of laboratories, all working on yeasts ("levure" in French). A boundary was defined around the group of hemiascomycetous yeasts (Fig. 1). This group, which used *Saccharomyces cerevisiae* as a reference, proved *a posteriori* to be evolutionary as large as the whole chordate phylum taking protein sequences divergence as a measure. At the same time, other genomes in this group were being sequenced by other laboratories (Cliften et al. 2003; Dietrich et al. 2004; Kellis et al. 2003, 2004).

Two waves of sequences have been achieved in two successive structured efforts. The first consisted in sequencing the genome of thirteen hemiascomycetous yeast species at a very low coverage (0.2 to 0.4 X). The first requirement of this initial phase, which we shall call the exploration phase, was to sequence truly random small pieces (about 1 kb) of the considered genomes, necessary to ensure good representativity of the samples. This has been achieved through careful experimental procedures and *a posteriori* verified by the coherence of several internal parameters. The second requirement of such a partial survey is the adoption of procedures and annotation that guarantee the homogeneity necessary to handle comparable data. All these parameters and most of the results have been published in a special issue (The Génolevures consortium 2000).

A second wave was achieved recently (Dujon et al. 2004). This consisted in the production of four complete annotated hemiascomycetous yeast genomes: *Candida glabrata*, *Kluyveromyces lactis*, *Debaromyces hansenii*, and *Yarrowia lipolytica*. Here, also, a very important requirement was obtaining highly homogeneous data of high quality, assembled from a 10X coverage and a chromosomal map and checked for absence of gaps. In the eukaryotic world, only the genomes *Saccharomyces cerevisiae* (The yeast genome consortium 1997), *Ashbya gossypii*

(Dietrich et al. 2004) and *Schizosaccharomyces pombe* (Wood et al. 2002) reach this quality. Like in the exploration phase, it appears that homogeneity of procedures and quality standards were of primary importance to analysis.

The present review will try to resume in a comprehensive way some of the questions and the answers we obtain through these two projects. Moreover, new aspects will be also proposed as a natural sequel of the work. To avoid repetitive and cumbersome referencing procedure, we suggest that more interested readers consult the two founding publications (Dujon et al. 2004; The Génolevures consortium 2000). All the data are available on the Genolevures site: http://cbi.labri.fr/genolevures

1.2 Life styles

Hemiascomycetous yeasts are unicellular organisms. This first point is significant in the eukaryotic branch of life. The multicellular organisms have obviously heavy constraints for (i) organizing the communication between the cells, (ii) limiting cell multiplication, (iii) maintaining a more or less constant medium in which the cells do not experience abrupt variations. These constraints are not effective in yeasts and the cells are adapted to the opposite situations: little communication limited essentially to sexuality, no positive cell cycle control (only negative controls are known), and well developed mechanisms of environmental adaptation.

How do these facts act on genome evolution? This could be an important clue when a comparison between yeast genomes and the vegetal and animal ones will be done. Because of the common characteristics of hemiascomycetes, we can propose that the constraints linked to unicellular life style are the same for all our genomes.

Nevertheless, if life styles are the same, physiologies are different. *S. cerevisiae* can live in a near anaerobic environment, most of other yeasts cannot. *D. hansenii* can support very high salinity, *Y. lipolytica* is able to use readily paraffin and alkanes as carbon and energy source, *K. lactis* is a frequent host of dairy product, *etc.* (Kurtzman and Fell 1998). Each species has its own specific success in ecological niches. How did these present time specificities arise? Are they an important determinant for genomic evolution? Or, on the contrary, are phenotypic characters, even those as important as aerobic/anaerobic life, the result of small changes, which have no relevant meaning in genomic evolution?

2 Structural aspects

2.1 Sizes

The first obvious characteristic of all hemiascomycetous yeast species is that they have small genomes as compared to many other eukaryotic organisms. Typically the size range is 12-20 Mb. As compared to large eukaryotic genomes which are thousands of megabases, three main features can be seen: drastic reduction of the number of transposon-like sequences, drastic lower number and smaller size of introns, and a modest difference in the number of genes.

2.1.1 Transposons

Two types of transposons have been described, that differ in their propagation mechanism; the retrotransposons that use an RNA intermediate and reverse transcription step make up the class I, whereas the DNA transposons that can propagate as DNA with the help of a transposase constitute the class II. DNA transposons are absent in the "upper" branch of the hemiascomyceteous yeast tree (see Fig. 1, above the *Y. lipolytica* branch), and are in small number (some units per genome) in *Y. lipolytica* (Neuvéglise et al. submitted). This could be due to a stochastic loss of these elements at the divergence between *Y. lipolytica* and the other yeasts in a context of poor multiplication of the transposons. But an invasion of a DNA transposon free ancestor can also be proposed for *Y. lipolytica*.

On the contrary, retrotransposons using RNA-mediated replication are common in yeasts. They are, however, less diverse and tremendously less numerous than they are in large genomes. Another exception exists also in *Y. lipolytica*. This species, which diverged much earlier from the others in our sample, is also the unique yeast bearing active non-LTR retrotransposons. These *Y. lipolytica* elements (Casaregola et al. 2002) are present in about 100 copies, most of them deleted at the 5' end as is usual for this type of elements. The disappearance of the non-LTR retrotransposons from all other yeasts parallels the disappearance of DNA transposons.

Due to the repetition of the LTR-transposon sequences, a relevant analysis has been done from the 13 low coverage sequences of the Genolevures I program (Neuveglise et al. 2002). Only two types of retrotransposons are present in yeast genomes: Ty1/*copia* and Ty3/*gypsy*. *Copia*-like retrotransposons are the more common and most differentiated elements. They are generally present as apparently functional sequences in tens of copies per genome. But numbers of intricate situations exist, like deletions, integration of Ty inside Ty and very frequently dispersed "solo" LTR. These molecular scars result probably from intramolecular crossing-over leading to a "pop out" of the coding part of the Ty. Several exceptions exist in which no *copia*-like elements have been detected: *Kluyveromyces thermotolerans* and *Zygosaccharomyces rouxii* apparently do not harbor these retrotransposons. However, rare occurrence cannot be detected by low sequence coverage. The most relevant is the absence of *copia*-like elements established in *Y. lipolytica* by complete sequence and annotation of its genome. *Gypsy*-like elements are more dispersed in the yeast phylogenetic tree. Present as the unique Ty elements in *Y. lipolytica*, they are found in some genomes and absent in others. We were not able to draw any coherent story considering the link between presence/absence of these elements and the evolution tree. Probably, these elements exist in the ancestral genome and were lost stochastically in the different evolution pathways. As a whole, we can propose that in this branch of life, *gypsy*-like elements were not able to multiply actively. This leads to few copies per genome and frequent stochastic losses in many genomes.

The link between this Ty feature and the size of the genome is not trivial. One view is that because of their low ability to multiply, which can be an intrinsic property of the Tys, yeast genomes simply did not accumulate the transposons, leading in part to the small size of genomes. Rapid elimination of Tys is also a possible mechanism leading to scarcity. But, as compared to higher eukaryotic genomes, the several hundreds of Ty scars found in yeasts are far from the millions of events found in the very large genomes. Another explanation can be that, due to a small size constraint on yeast genomes, only low-replicative Tys survived (when they are not definitively eliminated). This last idea is more in agreement with the fact that other mechanisms, with no apparent link with this low replicative property, tend to reduce genome size. At the scale of the phylum, inside the hemiascomycetes, genome size varies from 20 Mb to 10 Mb. The contribution of transposons to these sizes is low (a few percents at the most).

2.1.2 Introns

Yeasts genomes contain very few and small introns as compared to multicellular eukaryotes in which intronless genes are rare. On the contrary, only between 1% and 13 % of genes have introns in hemiascomycetous yeasts (Bon et al. 2003; Dujon et al. unpublished). The average intron length varies from 100 to 300 nt, and the number of introns in such genes is generally one. A strong bias exists in the position of introns with a marked preference for the 5' part of the gene. In orthologous genes both containing introns, the position of introns is strongly conserved in the different species. This demonstrates a unique and ancient origin of the feature in the whole phylum.

Expectations on this very low occurrence of introns are based on two aspects. The first is a hypothetical mechanism by which introns were massively lost at the origin of the hemiascomycetous branch by a huge retro-transcription phenomenon (Fink 1987). This is in accordance with the 5' position of the residual introns, due to a supposed limited processivity of the Reverse Transcriptase starting from the 3' end of mRNA. The second aspect is the fact that introns are more frequent in specific classes of genes (such as ribosomal protein coding genes or actin) suggesting that their persistence is linked to some functional necessity (Ares et al. 1999; Rodriguez-Navarro et al. 2002). This can be seen as having a counterbalancing effect on a random erasing process.

Another problem is to try to understand if this massive absence is just a stochastic event, which has little meaning in terms of evolution or if, like the Ty proposed hypothesis, it was driven by a necessity of genome size reduction. Here also, the presence of more than 600 introns in the 20 Mb genome of *Y. lipolytica* as well as the large number of genes containing several introns is significant.

2.1.3 Protein encoding genes

Considering that the protein coding sequences represent in most cases about 70% of the total length of a present day yeast genome, their number and size are obviously an important parameter for determining the size of the genome.

The number of protein encoding genes (CDS) ranges between 5,000 and 7,000 in yeast genomes, clearly less than the few tens of thousands found in higher eukaryotes. This trivial conclusion is generally related to the more complex organization of the latter organisms. Nevertheless, if a strong constraint on genome size exists, it will automatically be imposed on CDSs size and/or number, in the limit of functional requirement. Comparing the number and size of CDSs to the genome size within the hemiascomycetes can shed light on the problem. Three situations are clearly identified among the 5 species. *S. cerevisiae*, *C. glabrata*, and *K. lactis* have a 10-12 Mb genome and 5,200 to 5,800 CDSs. Surprisingly, *D. hansenii* with an equivalent genome size, (12 Mb) is estimated to contain more than 6,500 CDSs, up to 1,000 supplementary genes. The augmentation in number of genes is counterbalanced by a reduction in their size. In fact, in *D. hansenii*, the average gene size is 389 codons, as compared to 461-493 for the three other species. This leads to the same coverage of the genome by the CDSs. The last species, *Y. lipolytica*, with a 20 Mb genome, contains 6,700 CDSs with an average size of 476 codons, as the three first species. These numbers lead to a smaller part of the genome occupied by CDSs in this species: 49% versus 70-79% for the others.

In summary, for the very small genomes, the 79% space occupied by CDSs seems to be a maximal limit, the actual number of genes being determined by the size of CDSs inside this limit. For the larger *Y. lipolytica* genome, constraint on the size is not as strong as in other yeasts but the number and size of genes is slightly larger than in small genome species. This can be interpreted as a history with two constraints. First, unicellular free living yeast life style can be achieved with about 6,000 CDSs as was first described for *S. cerevisiae* by (Goffeau et al. 1996) and it is not possible to drastically decrease this number. Secondly, the genome has to be small (10-20 Mb).

The fact that the gene size is reduced to fit in the 12 Mb genome of *D. hansenii* is a strong argument to think that the genome size is actually a strong constraint which weights on the evolution of yeast genomes. A particular event of general tandem duplication of genes (see below) in this species forced the size of the CDSs to be small, to fit with the 12 Mb genome size. In the reciprocal case of *Y. lipolytica*, the genome size constraint was relatively relaxed. The consequences of this less stringent pressure have to be analyzed in detail, but it is already clear that all parameters have been in part relaxed (introns, gene, and Tys).

2.1.4 Conclusions

In conclusion, all the parameters participating in the definition of the size of the genome, as compared to the multicellular eukaryotic organisms, converge on a reduction in size. Transposons disappeared or are rare. Introns disappeared or are small compare to multicellular organism introns. Gene number cannot be compressed much below 6,000. The fight between the diminution of the genome size and the necessity to preserve the 6,000 genes leads to the different architectures that we have described.

If the yeast genome story is as we have just imagined, then the following questions are: why 6,000 genes and why a so small genome? This first question is perhaps a functional one. At this point, the pertinence of the question is itself questionable. Perhaps a good lead is to consider the number of essential genes. But what makes a gene "essential" in nature is something far from being defined.

For the second question (the small size constraint), it has often been suggested that, because yeasts have to multiply quickly to be competitive, DNA must replicate quickly, leading consequently to a reduction in its size. But this apparently coherent explanation may be a naïve one. The S phase occupies only a part of the time necessary to accomplish a cell cycle and the energy necessary to double DNA is certainly negligible as compared to doubling the mass of the cell. The S phase length is also strongly dependent on the number of replicative origins per chromosome. As an example, the segmentation of *Drosophila* eggs is very rapid: one cycle each ten minutes (with a genome ten time larger) as compared to an S phase for *S. cerevisiae* of about 45 minutes (synthetic medium, 30°C). Relevant enough, during egg segmentation, no biomass is synthesized, reducing cell cycle to a simple single division.

In summary, if a small size constraint really seems to act strongly on yeast genome evolution, three features can be identified: (i) the main mechanisms used are more or less the same in all the cases (mainly Ty scarcity and intron loss), (ii) an interplay between number and size of the genes leads to various present day structures, and (iii) the origin and the meaning of the constraint are not obvious.

2.2 Gene duplication

As all genomes, yeast genomes contain a number of paralogous genes. Doubling a gene, an apparently simple operation in the scale of evolution, opens the way to several interesting histories as clearly illustrated by a careful analysis of our sample of yeast genomes. A copy, new or parental, can be lost by deletion, leading to the original situation or to an apparent jump of the gene from one region to the other. A functionally equivalent situation is the destruction, by mutation, of a copy, leaving in place a pseudo gene or a relic, sometimes so divergent from its parent that it is difficult to identify. And finally, the two copies can evolve in diverse functional ways, creating new functions. This last destiny is an economical and powerful way to gain new functions from old ones without the deleterious effect of loosing the original function. Moreover, several successive rounds of duplication can occur. Other specific properties of this general phenomenon like entire genome duplication (Wolfe and Shields 1997) or differential rates of evolution of the various paralogues (Langkjaer et al. 2003) have been analyzed elsewhere. Analyzing the features of gene duplication in the branch of the hemiascomycetes can help us to understand some mechanisms and history of this elegant evolutionary tool.

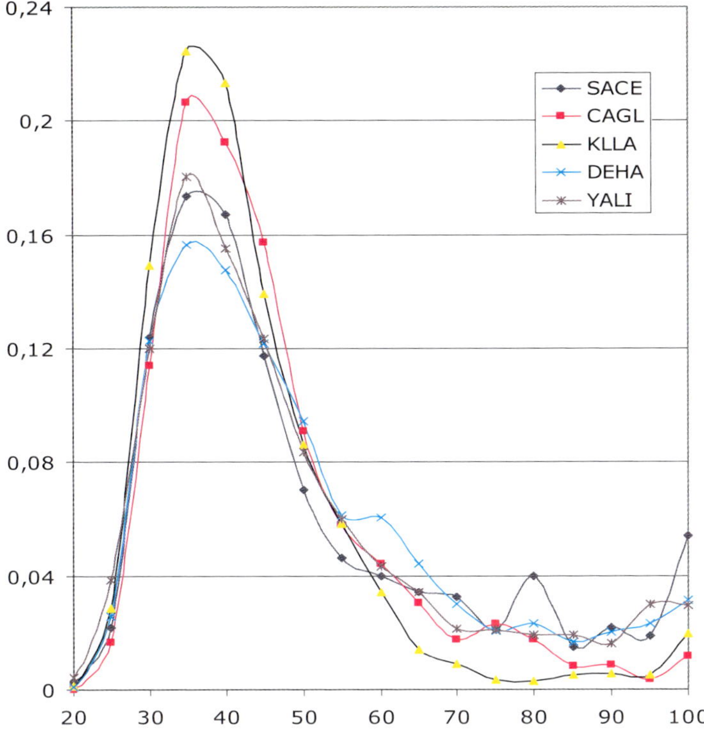

Fig. 2. Distribution of the percentage identity between pairs of paralogous genes in the different yeast species. Taken from Dujon et al. (2004).

2.2.1 Numbers of paralogous genes

First, the repartition of duplicate, triplicate, *etc.* is something approximately constant in the different species. In fact, the number of genes in pair is about three times the number of genes in three copies, these latter being two times more frequent than the genes present in four copies, *etc.* The higher the copy number is, the less often it arises. This apparently trivial situation cannot be explained only by a unique single event like genome doubling. Doubtless, a superposition of stochastic multiplication and deletion events occurred during evolution to lead to this robust image. This phenomenon has been theorized (Tiuryn et al. 1999). Numerous aspects of the phenomenon are not understood: molecular mechanisms, regulation, selection pressures, and can be questioned here.

A first interesting feature is the evolutionary distance between the different copies of paralogous genes. In most species, a mixed distribution appears. The majority of paralogous genes forms a Gaussian-like distribution centered on about 35 % amino acid identity (Fig. 2). This feature is, surprisingly, exactly the same in all the four species we examined. In fact, for *Y. lipolytica* (phylogenetically more distant), *D. hansenii* (which experimented huge duplication waves), or *S. cerevisiae*, the profile of this Gaussian-like curve is identical, centered on the same value. A simple interpretation is that the duplications are all the same age and were produced by a unique ancestral event. Another interpretation is that if different duplication events arose dispersed in time, the evolution of structures, under a functional necessity, reaches a ceiling and stops diverging. If no necessity limits the erasing process, the genes disappeared rapidly. More specific studies, specifically at the level of DNA sequences should shed some light on this question.

A second part of the curve exists or not depending on the genomes. The two extreme situations are found: (i) in *K. lactis* in which very rare genes are outside the curve (20-55% identity); (ii) in *S. cerevisiae* in which about 15% of the paralogous genes present 60-100% identity. The flat profile of this second part of the curve shows that these duplications were dispersed in time, if we consider that evolution rate is statistically the same over the population of genes. Consequently, we can imagine that, after a major important event before hemiascomycetous speciation, several more recent specific duplications occurred in the different species.

Such a possibility is, nevertheless, in strong contradiction with the repartition of the paralogous pairs in the genomes. Several clear-cut situations exist. In *D. hansenii*, 329 pairs of genes are found in tandem, contributing significantly to the total paralogous gene population. In this species, a mechanism leading to local tandem duplication was obviously a major way to obtain paralogous genes. In the four other species, only about 50 pairs of paralogous genes are found in tandem and other duplication mechanisms were probably employed. Careful inspection of the repartition of paralogous genes shows that, in *S. cerevisiae*, most of them are large groups of genes, repeated more or less faithfully in different chromosomes. This must either be the result of large segmental duplications or even of entire genome doubling. Finally, in *Y. lipolytica* most of the paralogous genes are dispersed in the whole genome. It is, thus, difficult to reconcile a unique pre-hemiascomycetous phenomenon, with so different observations, which reflects clearly different mechanisms.

A last feature to consider is the total number of paralogous genes in each species. The present situation, as demonstrated by the presence of various artifacts, is the result of two antagonistic phenomena: duplication and erosion. Independently of the underlying mechanisms, the situation is quantitatively significantly variable from *K. lactis*, which harbors 500 sets of paralogous genes to *D. hansenii* with 900 sets.

A general guide of the histories of duplication/erosion/evolution can be tentatively drawn. At this scale of phylogenic distances, different mechanisms obviously occurred to duplicate small (genes) or long (segments) pieces of DNA. This seems to occur stochastically in the branching points of the tree. On this basis of constant duplication, erosion works, within the confines of physiology and other forces like a strong size constraint. Since this is a continuing story, the present situation is doubtless only a temporary frame of a film and not a final and stable state.

2.2.2 Gene's gain and loss

A very exciting aspect of the Genolevures program was to attempt at answering the old and intriguing question: What makes a species different from another? Is it the nature of the genes (presence/absence) or their quality (differences in the sequences) or even, subtle differences in regulation networks? More realistically, what is the relative weight of each aspect in the answer? In this chapter, we will examine general structural aspects. The meaningful functional aspects will be discussed below.

How shall we go forward? We have at hand five comparable genomes (size, sequences annotation, quality, homogeneity). They are dispersed on different branches of a whole phylum. They belong to species with a common life style, although they adapted to specific ecological niches.

We can first consider the phylum as containing a unique gene pool as compared to other eukaryotic and prokaryotic cells: this ends with the identification of about 2,000 "ascomycete-specific genes" that are not found outside the phylum. These genes are found in the majority of yeast species and are absent outside the phylum. Some of them will be doubtless found outside the hemiascomycetous yeasts when new genomes will be available, but a core of genes will maintain this property. The meaning of such an important part of the yeast genomes (about one third) is obscure. Are they functional genes governing common properties like cell wall or other common life style specificities? Or, on the contrary, are they, for one part, the result of stochastic drift leading to only historical marks of the origin?

Another, more refined analysis is to compare genes between yeast species. About 4,700 families with one or several members were defined in the phylum to avoid strong difficulties in counting paralogous genes. Among them, 2,000 are common to all genomes. These include obviously most of the essential genes involved in basic cell processes such as DNA replication, protein synthesis, *etc.* Very few are missing in only one species. Strangely enough, in this analysis, which uses a cut-off value, *Y. lipolytica* seems to have lost more than 300 families, which is much more than the tens missing in one of all other species. The meaning of this difference is puzzling. Is it due to a simple genetic drift as suggested by the correlation between phylogenetic distance and number of loss? Or is it due to a functional constraint, considering the specificity of numerous aspects of metabolisms of *Y. lipolytica*? Because of the long evolutionary distance between *Y. lipolytica* and the other species, at least a part of these apparent losses are not actual losses, but can be due to the difficulty of finding orthologous genes. In fact, a too divergent orthologous gene could be falsely considered as a loss. In this case, is the hidden orthologous gene always in charge of the same function or a different one?

Reciprocally, if we consider families present in only one specific species, the picture is quite different. *S. cerevisiae, K. lactis,* and *C. glabrata* contain only about 20 specific families, whereas *D. hansenii* and *Y. lipolytica* have about 150 specific families. The evolutionary distance between *Y. lipolytica* and all other yeasts, including *D. hansenii,* is large, as compared to the distance between *D. hansenii* and other yeasts. To date, no correlation exists between phylogenetic distances and the number of species-specific families, we have to conclude that evolutionary time is not the main cause governing the phenomenon. In this case, ecological pressure could be predominant. This is plausible since they have very specific physiologies. *D. hansenii* is a strong osmophilic organism and *Y. lipolytica* is specifically able to use alkanes and other organic compounds.

These two mirror images -missing genes and species specific genes- seem, at this scale, to be determined by the two main evolution forces: to lose by chance, to keep by necessity.

A last point has to be mentioned. Very rare events (if any) of recent acquisition from bacterial genomes to a single yeast species were detected in a careful survey of the five genomes on the basis of conserved amino acid sequence. This approach, which only revealed recent Horizontal Gene Transfer (HGT), may indicate a reduced but continuous gene flow, which could have played a significant role in evolution. HGT concerning genes from other eukaryotes, including yeasts, that so far escaped detection, may also have contributed to this gene flow. Thus, if not totally excluded, horizontal transfer of genes may not be a significant actor of evolution at the genome scale, although it could be very significant in terms of specific functions.

3 Functional aspects

As described previously, yeasts species live in very different environments, from trees and fruits (*S. cerevisiae*), dairy products (*K. lactis*), to environment rich in decaying organic compounds (*Y. lipolytica*) (Kurtzman and Fell 1998). Several possibilities can be advanced to explain how these species have functionally evolved to adapt to their natural environment. A large variation in the number of genes being excluded (they all contain 5,000 to 7,000 ORFs), two mechanisms may account for this diversity: (i) the phenotypic variations that are observed may be due to gene duplication (or more) and function differentiation. Such small and localized changes are most often not detectable in a statistical and unspecific analysis; (ii) these different species exploit the same genes by different regulatory mechanisms - which at large could imply sensory cascades, transduction signal and regulatory networks - to adjust to their specific environment. Such hypothesis relies in great part to the variation of regulatory networks, in which either the number and function of regulators are different or the range of their target genes is different from one organism to another.

The respective weight of these two hypotheses, which are not exclusive, is difficult to estimate. The evolution of regulatory networks is not a mechanism yet seriously substantiated by many data and may be underestimated, while the importance of the other mechanism, proposed many years ago (Ohno 1970) has already been evidenced in many cases (Gu et al. 2004; Papp et al. 2003; Zhang and Kishino 2004; references therein).

The "Genolevures" programs can be used as a powerful tool in which to start examining these hypotheses, because the number of sequences and the yeasts diversity are large. However, the approach taken by the two "Genolevures" programs are different; Genolevures I exploited the species diversity with a relatively small coverage of genomic sequences while Genolevures II was dedicated to the complete nucleotide sequence of a few species. The two programs gave us a unique opportunity to evaluate their respective contribution to the acquisition of knowledge about the possible mechanisms underlying functional evolution of genomes

3.1 Genolevures I revealed that functional classes may be differently represented among species.

3.1.1 Speciation and functional categories

The 49,000 random sequence tags obtained in Genolevures I were submitted to BLASTX and 22,000 were identified as associated to possible ORFs. As a first step, they were compared to the set of *S. cerevisiae* proteins, which in this work was the reference set. These ORFs were then classified into functional categories based on the functional catalog of MIPS (CYGD, http://mips.gsf.de/desc/yeast/). Table 1 summarizes the comparison obtained (Gaillardin et al. 2000). For each species, the frequency of ORFs assigned to a functional class is indicated after a normalization process, which takes into account the number of ORFs detected for each species. In the Genolevures I project, the number of reads were quite different among the different species studied (around 5,000 reads, calculated to correspond to 0.4 times genome coverage for *Y. lipolytica; P. angusta; P. sorbitophila; K. lactis; Z. rouxii; S. bayanus,* and 2,500 reads for *S. exiguus; S. servazii; S. kluyverii; K. thermotolerans; K. marxianus; D. hansenii, and C. tropicalis*). While the ideal value of 1 is not reached as could be expected (different factors can interfere such as bias due to the phylogenetic proximity/distance, different annotators *etc.*) within one column, that is to say, within one species this frequency should be the same. Differences may, therefore, indicate an over- or under-representation of genes belonging to a specific functional class as compared to *S. cerevisiae.* For example, in *Y. lipolytica* one can see a slight over-representation of genes involved in transport facilitation and an under-representation of genes involved in protein synthesis distribution while for *K. lactis* the distribution is much closer to the *S. cerevisiae* distribution. Similarly, an over-representation of the energy class in *C. tropicalis* is also revealed by this analysis.

However, these observations are only trends deduced from sequence similarity and cannot be validated as such. Under-representation may be due to divergence of sequences, which does not allow them to be recognized. In addition, one functional class contains various sub-classes that may have a quite different role in the cell. For example, in the "transport facilitation" class, an over-representation may be due to the raise of one subclass as compared to others (e.g. hexose transport or amino-acids transport) or alternatively to a general raise of all subclasses. These two possible ways to modulate numbers may not have the same biological meaning. A tentative analysis of subclasses has been made in Gaillardin et al. (2000) by

Table 1. Distribution into functional categories of the putative protein fragments obtained from Genolevures I sequences

	Yl	Ct	Dh	Ps	Pa	Km	Kl	Kt	Sk	Zr	Ss	Se	Sb	Sc
Metabolism	1.6	1.7	1.4	1.4	1.4	1.3	1.4	1.2	1.4	1.4	1.2	1.3	1.1	1
Energy	1.3	2.0	1.3	1.4	1.5	1.3	1.3	1.1	1.4	1.4	1.2	1.3	1.1	1
Cell growth, cell division and DNA synthesis	1.2	1.2	1.2	1.4	1.1	1.1	1.3	1.1	1.2	1.2	1.4	1.5	1.2	1
Transcription	1.1	1.0	1.1	1.2	1.1	1.2	1.3	1.2	1.1	1.2	1.3	1.2	1.3	1
Protein synthesis	1.0	1.1	0.9	1.5	1.1	1.1	1.0	1.0	0.8	1.1	1.1	1.1	1.0	1
Protein destination	1.3	1.4	1.2	1.4	1.3	1.1	1.2	1.3	1.4	1.3	1.2	1.4	1.1	1
Transport facilitation	2.0	1.9	2.0	1.5	1.6	1.2	1.4	1.1	1.4	1.3	1.1	1.2	1.2	1
Intracellular transport	1.4	1.4	1.4	1.4	1.3	1.4	1.3	1.1	1.4	1.3	1.2	1.3	1.2	1
Cellular biogenesis	1.6	1.3	1.3	1.5	1.3	1.2	1.2	1.5	1.2	1.3	1.4	1.6	1.2	1
Cellular communication/signal transduction	1.2	1.0	1.2	1.7	1.3	1.1	1.3	1.0	1.3	1.2	1.7	1.5	1.2	1
Cell rescue, defense, cell death and ageing	1.4	1.6	1.3	1.4	1.1	1.0	1.0	1.1	1.2	1.1	1.1	1.1	1.1	1
Ionic homeostasis	1.7	1.8	1.4	1.5	1.3	1.0	1.2	1.3	0.9	1.2	1.1	1.1	1.3	1
Cellular organization	1.3	1.3	1.3	1.4	1.3	1.2	1.2	1.2	1.3	1.3	1.3	1.3	1.2	1
Classification unclear	1.0	1.1	1.6	1.0	1.3	1.0	1.1	1.3	0.8	1.1	1.0	1.1	1.2	1
Unknown function	0.6	0.6	0.7	0.5	0.7	0.7	0.7	0.8	0.7	0.7	0.7	0.7	0.8	1
Total number of genes identified	1235	965	1165	1321	2476	1343	2283	1468	1432	2126	1415	1456	2898	6320

Table 1 was computed from Table 2 of Gaillardin et al. (2000). For each species, the ratio of the fraction of ORFs versus the fraction of *S. cerevisiae* ORFs classified into the same functional category was calculated. Yeasts species are ordered as they are in the phylogenetic tree (from *Y. lipolytica*, the most distant to *S. uvarum*, very close to *S. cerevisiae*). The functional categories are those described at MIPS (at the "Comprehensive yeast genome database", http://mips.gsf.de/desc/yeast/). ORFs can be classified into several functional categories.

Abbreviations to Table 1:
Sc: *Saccharomyces cerevisiae*
Sb: *Saccharomyces bayanus (S. uvarum)*
Se: *Saccharomyces exiguus*
Ss: *Sacharomyces servazii*
Zr: *Zygosaccharomyces rouxii*
Sk: *Saccharomyces kluyveri*
Kt: *Kluyveromyces thermotolerans*
Kl: *Kluyveromyces lactis*
Km: *Kluyveromyces marxianus*
Pa: *Pichia angusta*
Dh: *Debaryomyces hansenii*
Ps: *Pichia sorbitophila*
Ct: *Candida tropicalis*
Yl: *Yarrowia lipolytica*
For more details, see: http://cbi.labri.fr/Genolevures/nomenclature.php3

examination of over-representation only. These data suggest that allantoine and allantoate transporters as well as peroxisomal biogenesis constitute classes with more genes than expected in *Y. lipolytica*. According to this analysis, *Z. rouxii* has more genes than expected for beta-oxidation of fatty acids and mitochondrial biogenesis genes are over-represented in *D. hansenii* and *K. marxianus*. Amino-acid transporters were over-represented in *K. marxianus* and *D. hansenii* but data might not be meaningful since the twenty-five *S. cerevisiae* genes of this functional class were pooled whatever their role at the cellular level (mitochondrial import, vacuolar transport and plasma membrane import) may be. This type of analysis has clearly a limit: in many cases, the figures are small which, in addition to the fact that only partial sequences were analyzed and degree of redundancy is difficult to estimate, renders the conclusions barely significant.

Accordingly, Gaillardin et al. (2000) concluded that changes in major functional classes were associated to global profile difference between species, but that a clear definition of most species according to their profile is biased by phylogenetic distance and gene conservation.

3.1.2 Species-specific and ascomycete-specific genes

One interesting question, which arises from comparative genomics, is the question of species-specific genes. Such genes reflect the acquisition of a specific function, which may have been acquired, either through duplication of an ancestor gene, horizontal transfer from other organisms, or the loss of ancestral functions during evolution in some branches. These genes may be difficult to identify in such a partial sampling since the fact not to find orthologs does not mean that the corresponding gene is not present in another species.

Furthermore, some classes diverge much faster than others, making difficult the identification of orthologs by screening them using only *S. cerevisiae* as a reference. This is especially true for regulatory transcription factors (detailed analysis in Bussereau et al. (2004)) for which constraints exist mostly on the DNA-binding domain. Also of interest is the case of the HAP complex, for which the activator part encoded by the *HAP4* gene in *S. cerevisiae* was considered as species-specific since no orthologs were identified by BLAST in other species. Recent studies, exploiting the identification of a functional homologue in *K. lactis* (Bourgarel et al. 1999) allowed detection of a conserved motif and consequently the identification of functional orthologs in other species (Sybirna et al. 2005 and unpublished data).

While the identification of true species-specific genes is still awaiting a clear answer based on functional tests, the notion of ascomycete-specific genes is more clearly supported as introduced by Malpertuy et al. (2000). Their analysis showed that the fraction of ascomycete-specific genes is far from negligible. Detailed analysis based on amino-acid conservation comparison points to the fact that "asco-genes" tend to be more sensitive to evolutionary pressure than other classes. One of the most interesting and numerous class of "asco-genes" belongs again to the "transcription" functional class and is represented by the Zn2-Cys6 zinc binuclear transcriptional activators which have been found only in fungi (reviewed in Todd and Andrianopoulos 1997). This example indicates more generally that the gene number of different functional classes may be slightly biased by a fraction of "asco-genes" which is more or less important according to the different classes. In the "transcription" functional class, this proportion is quite high.

3.2 The evolution of specific pathways as revealed by Genolevures II

The availability of the complete genome sequence of several yeasts allows us to scrutinize the components of different metabolic pathways. Among the yeasts examined in Genolevures II (*C. glabrata, D. hansenii, K. lactis, and Y. lipolytica*), we focused on two of them, *K. lactis* and *Y. lipolytica*. These two species plus *S. cerevisiae* are well dispersed along the hemiascomycete phylogenetic tree and also represent archetypal cases for the study of central metabolism and anaerobiosis. While *S. cerevisiae* can grow in anaerobiosis and has a fermentation-oriented metabolism, *Y. lipolytica* is exclusively respiratory and *K. lactis* represents an intermediary situation being capable of fermentation and respiration simultaneously (Crabtree-negative). Nevertheless, the latter two are both unable to grow anaerobically.

3.2.1 Transport facilitation

As deduced from Table 1, the extrapolation of data obtained from partial sequences could indicate that the functional category "transport facilitation" contains a slightly larger number of genes in *Y. lipolytica* than expected. *Y. lipolytica* can grow on some amino acids (L-glutamate, L-lysine, and possibly L-proline) as can *K. lactis* (especially on L-lysine) does. Indeed, a closer examination (Wesolowski-Louvel and Bolotin-Fukuhara, unpublished) of Genolevures I data revealed an excess of amino-acids transporters in *K. lactis* (22 in place of the 11 expected in the sample considered). We, therefore, reinvestigated this functional class in more detail and focused on plasma membrane localized amino-acid transporters, necessary to import amino acids. Seventeen transporters (specific either for one amino-acid, to a class of amino-acids or with no specificity) have been characterized in *S. cerevisiae*. While it is difficult in some cases to ascertain the specific correspondence using only sequence comparison rather than phylogenetic studies, it is easy to conclude that there is no expansion of transporter family which could explain growth on some amino-acids, since only 13 amino-acid transporters were identified in *K. lactis*, and 14 in *Y. lipolytica*. In addition, there is no duplication corresponding to a specific amino-acid transporter with the exception of proline in *K. lactis* (two genes). This confirms our conclusions from Genolevures I, that is to say, for functional analysis the data obtained from a random sampling have to be interpreted with extreme caution since: (i) partial sequences may be misleading in terms of functional classification depending upon the part of the sequence examined and (ii) the evaluation of possible duplications is extremely difficult, especially when sequences are very conserved.

3.2.2 Aerobiosis versus anaerobiosis

The assimilatable sugar sources vary greatly according to yeast species, as extensively documented by Barnett (1976), but glucose remains a universal carbon source of yeast. In *S. cerevisiae*, glucose is first directed to the glycolytic fermentation, rather than the respiratory pathways, which are strongly repressed by glucose (Crabtree positive effect). This glucose repression is not found in the two other yeasts examined here. *K. lactis* can ferment and respire on glucose simultaneously, meaning that pyruvate is metabolized partly by fermentation through pyruvate decarboxylase and partly by respiration through pyruvate dehydrogenase (Breunig et al. 2000), a situation also known for other yeasts such as *P. stipitis* (Passoth et al. 1996). Many other types of yeast, such as *Y. lipolytica*, are exclusively respiratory yeasts, unable to ferment (Kurtzman and Fell 1998). In such aerobic yeasts, an important part of glucose seems to go through the pentose phosphate shunt, rather than the classical glycolytic pathway, and the pyruvate produced is consumed by the respiratory metabolism. In many other species of the intermediate status, which possess both fermentation and respiration, the capacity to ferment sugars is often dependent on the simultaneous presence of respiration. In fact, some yeasts (Kluyver effect-positive yeasts) cannot ferment nor grow on certain sugars in the absence of respiration (Fukuhara 2003). This effect is sugar-specific, that is to say, a yeast can be Kluyver effect positive for one sugar and negative for another. However, most identified yeast species can ferment sugars to ethanol and carbon dioxide (Kurtzman and Fell 1998) and can, therefore, grow quite well under oxygen-limited conditions. Only a very few species can grow in the strict absence of oxygen. *S. cerevisiae*, which can grow rapidly under both aerobic and anaerobic conditions (Visser et al. 1990), stands apart from most eukaryotic organisms, hence, its long history of exploitation by our fermentation industries.

Table 2. The *S. cerevisiae* genes involved in anaerobiosis and their putative orthologs in *K. lactis* and *Y. lipolytica*.

Part A: Genes repressed by *ROX1*

Gene name	*K. lactis* ortholog	*Y. lypolitica* ortholog
ERG26/YGL001c	1	1
ERG11/YHR007C	1	1
RTA1/YGR213c	1	2
CYB5/YNL111c	1	1
HEM13/YDR044w	1	1
HEM14/YER014w	1	1
YGR066C	No	2
SPI1/YER150w	No ?	No
SCM4/YGR049w	No	No
CWP1/YKL096w	No	3
RTS3/YGR161c	1	No
FIT2/YOR382w	2	No
YOR306c	1	No
CYC7/YELO39c	No	No
GLC7/YER133w	1	No
FIT3/YOR383c	No?	No?
AAC3/YBR085w	No	3
FET4/YMR319c	1	No
ANB1YJR047c	No?	1
DAN1YJR150c	No	No
HEM13/YDR044w	1	1
COX5b/YIL111w	No	No
OYE2/YHR179w	No	7
LAC1/YKL008c	1?	2
YMR252C	2	No
YHR048W	1	2
FRDS/YEL047C	No	No
SUR2/YDR297w	1	1
SML1/YML058W	No	No
YNR014W	No	No
FRT2/YAL028W	No	No
TIR2/YOR010c	No	No
YGR035C	No	No
YAR028W	No	No
DIP5/YPL265w	No	No
ISU2/YOR226c	No	No
EUG1/YDR518w	No	No
HEM14/YER014w	1	1
PIS1/YPR113w	1	1

Part B : Genes involved in sterol uptake

Gene name	*K. lactis* ortholog	*Y. lypolitica* ortholog
UPC2/YDR213w	No	No
HES1/YOR237w	No	No
DAN1/YJR150c	No	No
DAN3/YBR301w	No	No
DAN4/YJR151c	No	No
TIR4/YOR009w	No	No
YGR131w	No	No
AUS1/YOR011w	No	No
PDR11/YIL013c	No	No
SUT1/YGL162w	No	No

Part C: Genes involved in anaerobiosis independently of *ROX1*

Gene name	*K. lactis* ortholog	*Y. lypolitica* ortholog
YPR149w/NCE2	1	1
YIR019cSTA1	2?	1
YHL042w	No	No
YHR210c	No	No
YHL044w	No	No
YOL015w	No	No
YML083c	No	No
YER011w/TIR1	No	No
YKR053c/YSR3	1?	No
YER188w	No	No
YCL025c/AGP1	?	No
YDL241w	No	No
YBL029w	1?	No
YLR413w	No	No

Part A: Genes that are repressed by *ROX1*
Data are taken from Ter Linde and Steensma (2002). The first column indicates the *S. cerevisiae* gene name (if any) and its generic name. The identification of putative homologues based on BLAST search is described in the second (for *K. lactis*) and third (for *Y. lipolytica*) columns. The number of putative orthologs is provided when they are identified. "No" indicates that (i) no ortholog of *S. cerevisiae* gene was identified or that (ii) only one ortholog was detected while *S. cerevisiae* had two genes, one involved in anaerobiosis and one in aerobiosis. ? means that it was not possible to conclude.
Part B: Genes involved in sterol uptake
Data are taken mostly from Wilcox et al. (2002). Symbols for the second and third columns are the same as for Part A.
Part C: Genes induced in anaerobiosis.
Data are taken from Ter Linde et al. (1999). Legend is the same as for Part B.

In *S. cerevisiae*, a genome-wide transcriptional study in aerobiosis as compared to anaerobiosis has been made (Ter Linde et al. 1999) and revealed only a small number of genes with a clear-cut difference in expression level. This was the case for several of the known targets of *ROX1*, the heme-dependent repressor of hypoxic genes in normoxic conditions, as well as for *ROX1* itself. These data, associated to the work of Ter Linde and Steensma (2002), confirmed *ROX1* as being involved in the transcription of anaerobic genes. In normoxic conditions, *HAP1* senses cellular heme status and increases the expression of aerobic genes in response to oxygen. It also activates the transcription of *ROX1* and consequently, *ROX1* represses hypoxic and specific anaerobiosis genes. This repression is alleviated in hypoxia, a physiological condition in which *HAP1* does not activate *ROX1* anymore. Moreover. *ROX1* plays a role in the regulation of isoforms by oxygen.

We decided to investigate in detail, the genes involved in anaerobiosis in the three species to obtain a comparative genomic picture of the anaerobiosis vs. aerobiosis pathway. Table 2 provides a list of the *S. cerevisiae* genes which according to Ter Linde et al. (1999) and Ter Linde and Steensma (2002) are involved in the anaerobiosis process and searched orthologs in *Y. lipolytica* and *K. lactis*, yeasts which are both unable to grow without oxygen. Some interesting results are commented below.

Since *ROX1* is one of the main actors, we first examined the genes that are dependent upon it. They can be classified in three groups according to Ter Linde and Steensma (2002) from which most of the data interpretation in *S. cerevisiae* are taken from.

The first group is made of genes that encode oxygen-controlled functions. In hypoxic conditions, derepression of these genes may allow for an efficient utilization of the limited oxygen in these conditions. In this group, two main classes are targets of *ROX1*. One class is involved in the oxygen-requiring ergosterol pathway (*ERG11, ERG26, CYB5,* and *RTA1*) either directly or indirectly via transduction signal and regulation. The second class is involved in the oxygen-dependent step of heme biosynthesis (*HEM13* and *HEM14* encoding coproporphyrinogen III oxidase and protoprophyrinogen IX oxidase, respectively). As expected, orthologs of these genes can be identified in *K. lactis* and *Y. lipolytica,* but for the moment no experimental data allow to decide if their expression is effectively dependent upon oxygen tension.

As a consequence of the absence of sterol synthesis in anaerobiosis, *S. cerevisiae* can grow only if ergosterol is added to the medium. This implies the presence of genes involved in sterol uptake. They are described in the Table as a separate section (Part B). Among the genes listed as necessary for sterol uptake in anaerobiosis (Wilcox et al. 2002), three are major components. *SUT1* encodes the transcriptional regulator necessary for uptake in anaerobiosis. This gene is induced in anaerobiosis and has a counterpart, *SUT2*, which is active in aerobiosis. The two others are the sterol transporters *PDR11* and *AUS1*. A striking fact is that the three genes are absent in both *Y. lypolitica* and *K. lactis*.

A second group concerns genes that are necessary for anaerobiosis. In *S. cerevisiae*, these genes are repressed by *ROX1*. An important part of those are encoding anaerobiosis-specific components of the cell wall. In anaerobiosis, the cell switches its cell wall composition by changing the type of mannoproteins (Abramova et al. 2001). Cwp1p and Cwp2p, which are the normal components in normoxy, have to be replaced by the products of the *DAN/TIR* family. The presumed function of *ROX1* would be to avoid the incorporation of the *DAN/TIR* products in aerobic conditions. Since neither *K. lactis* nor *Y. lipolytica* can live in anaerobiosis, one might expect that these genes are unnecessary. A search for orthologs of *S. cerevisiae* genes in these species did not reveal any candidates. The same holds probably true for *SPI1* that, according to its behavior, could belong to the same class. A few genes (*LAC1, SUR2,* and *PIS1*) are involved in sphingolipids synthesis, a process that is necessary for transport of DAN/TIR gene products via their GPI-anchor. Their equivalents exist in the two aerobic species. The product of *EUG1* may play an important role in glycosylation and folding in anaerobiosis. Since another gene is involved in the same process in aerobiosis (*PDI*), it is not surprising that homologues of *EUG1* are absent in *K. lactis* and *Y. lipolytica*

In anaerobiosis, there is a need to keep the redox balance by alternative means than mitochondrial electron transport. An important part is played by the reduction of fumarate to succinate (catalyzed by FRDS1), an activity essential for anaerobic growth (Arikawa et al. 1998). Homologues of this gene could not be identified in the two aerobic species. Finally the situation of iron uptake and mobilization is unclear. *FET4* is the low affinity iron permease. In *S. cerevisiae*, its expression is increased in anaerobiosis since its function is not dependent upon *FET3*, a gene, which encodes a multicopper oxidase and is limiting in anaerobiosis. Its absence in *Y. lipolytica* can be easily explained but its presence in *K. lactis* is less obvious.

Finally the third class of genes, which are regulated by *ROX1*, is composed of one isoform of the pairs, which in *S. cerevisiae* are differentially expressed in anaerobiosis/hypoxia and in aerobiosis. Such a situation is well known for *AAC1/AAC2/AAC3, CYC1/CYC7, COX5a/COX5b, or HYP2/ANB1* (the last form of each group being the one expressed in anaerobiosis). For the same reasons, one can consider the pair *OYE3/OYE2* (coding for the "old yellow enzyme") and *ISU1/ISU2* (serving as scaffold for assembly of Fe/S clusters) as equivalent aerobic/hypoxic gene pairs. The situation is extremely clear in *K. lactis* where all the hypoxic isoforms are probably absent, only one ortholog being identified that may be assumed to correspond to the aerobic form. The situation is more uneven in *Y. lipolytica*; the situation is identical to *K. lactis* for some pairs, only one homologue being identified (*CYC7, COX5,* or *ISU2*) while for the other cases, there is in contrast an increase in the family size (up to 7 copies for the *OYE2/OYE3* family). This may be related to the strong oxidative metabolism of *Y. lipolytica*, but at the moment no simple explanation can be proposed.

While most genes induced in anaerobiosis are controlled by *ROX1*, there are a few genes (mostly of unknown function) which are induced in anaerobiosis and do not seem to be under control of *ROX1* in the two different transcriptome studies described above (Part C). This may be due to a low level of expression variation, which does not allow the retention of these genes in the *ROX1* screen or to the fact that they are regulated by other anaerobiosis regulators. One of them is *UPC2* (described also in Part B; Abramova et al. 2001), which is not found in the two aerobic species. The existence of a second transactivator is suspected (Cohen et al. 2001), but the gene has not yet been identified.

Since *ROX1* seems to be the major regulator of anaerobiosis, the "raison d'être" of its existence in the aerobic yeasts is opened. Searching for homologues, we identified some candidates mostly on the basis of the conservation of the HMG-motif (Fig. 3). Their significance will have to await functional studies currently in progress in our laboratories.

In conclusion, the non-detection of most of the "anaerobic" genes in the two strictly aerobic species is a striking finding. A recent publication (Gojkovic et al. 2004) proposes one reason why some yeasts are strictly aerobic. The fourth enzyme activity of the *de novo* pyrimidine synthesis pathway, dihydroorotate dehydrogenase, is dependent in most yeasts upon active respiratory chain. This is linked to the fact that they possess only the mitochondrial form of the DHODase and lack the cytoplasmic form, which makes them dependent of oxygen. The general situation is probably more complex since, if a unique ortholog of the mitochondrial DHODase can indeed be identified in *Y. lipolytica*, *K. lactis* has the two DHODase forms, but cannot grow anaerobically. Other factors are certainly involved and our analysis shows that sterol uptake is one of them. Experiments are currently in progress to test these hypotheses.

3.2.3 A comparative analysis of central metabolism

The ability to ferment different carbon sources as well as the balance between fermentation and respiration (when both can be achieved) is quite different from one yeast species to another (see Section 3.2.2). Although the pathways of sugar utilization follow the same scheme in all yeasts, important genetic and biochemical variation exist as described in the review by Flores et al. (2000). Having at disposal the complete set of genes encoded by three yeasts (*S. cerevisiae*, *K. lactis*, and *Y. lipolytica*), we compared the presence/absence of genes involved in glycolysis and related pathways. Previous experiments based on the search of *rag* mutants in *K. lactis* ("RAG" means "resistance to antimycin on glucose" and this phenotypic character allowed to select mutants of the glycolytic pathway) have shown that some of the glycolytic steps were encoded by a single gene and not by two as in *S. cerevisiae* and that functional redundancy was absent for many of the glycolytic genes (e.g. see Bianchi et al. 1996). This variability in paralogues number has also been observed in other species (Moller et al. 2004). These observations can now be confirmed/infirmed throughout the complete pathway.

3.2.3.1 The upper part of the glycolysis pathway: storage of carbohydrates. Glucose enters to the cell as glucose-6 phosphate. Part of it can be converted by phosphoglucomutase into Glucose-1P and then to UDP-Glucose. From this, two pathways are possible: UDP-glucose can be converted to glycogen that serves as a polymeric storage form for carbohydrates. Alternatively, it can be converted to trehalose. Table 3 (Part A) gives a comparative list of all genes involved in these pathways. From these data it can be deduced that the high frequency of gene duplication observed in *S. cerevisiae* disappears in the two other yeasts where all these genes are single except for the glycogenin glucosyltransferase. While this is not the only possible hypothesis it is tempting to postulate that storage reactions

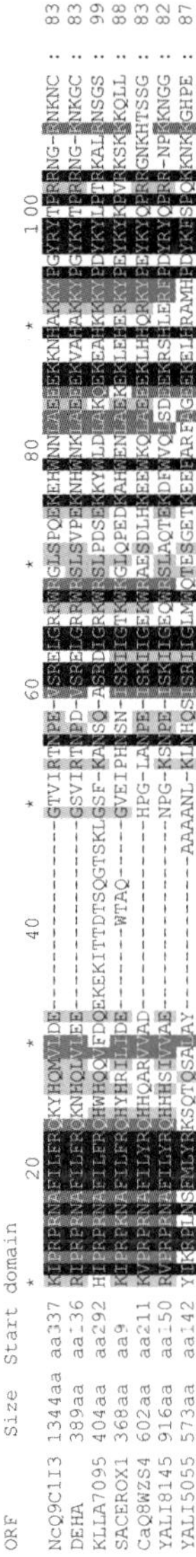

Fig. 3. Search of putative orthologs of *S. cerevisiae ROX1* gene in different species using the alignment of the HMG domain. Sequence alignments were generated using CLUSTAL X (Thompson et al. 1997) and were manually adjusted with Genedoc (http://www.psc.edu/biomed/genedoc). Nc_Q9CI13: *Neurospora crassa*; DEHAE09977g: *Debaryomyces hansenii*; KLLA0B1195g: *Kluyveromyces lactis*; SACE PR065w (*ROX1*): *Saccharomyces cerevisiae*; Ca_Q8WZS4: *Candida albicans*; YALI0B02266g and YALI0D2266g: *Yarrowia lipolytica*. Start domain indicates the start position of the HMG domain within each CDS. The nomenclature of the *Y. lipolytica*, *D. hansenii*, and *K. lactis* CDSs is that of Genolevures (http://cbi.labri.frGenolevures/).

Table 3. Gene number comparison between the three yeast species for carbon metabolism.

Enzyme activity	number of genes in			
	S. cerevisiae		*K. lactis*	*Y. lipolytica*
Part A: Carbohydrate storage				
phosphoglucomutase	2	*PGM1*	1	1
		PGM2		
UDP-glucose pyrophosphorylase	1	*UGP1*	1	1
Glycogen synthase	2	*GSY1*	1	1
		GSY2		
Glycogenin glucosyltransferase	2	*GLG1*	1	2
		GLG2		
1,4-alpha-glucan branching enzyme	1	*GLC3*	1	1
Glycogen phosphorylase	1	*GPH1*	1	1
4-alpha-glucanotransferase	1	*GDB1*	1	1
alpha,alpha-trehalose-phosphate synthase (UDP-forming)	1	*TPS1*	1	1
Trehalose-6-phosphate phosphatase	1	*TPS2*	1	1
alpha,alpha-trehalose-phosphate synthase (UDP-forming):	2	*TPS3*	1	1
regulatory activity		*TSL1*		
alpha,alpha-trehalase	2	*NTH1*	1	1
		NTH2		
trehalase (vacuolar)	1	*ATH1*	1	1
Part B: glycolyse and gluconeogenesis				
glucokinase	2	*HXK1*	1	1
		HXK2		
glucokinase (trehalose pathway)	1	*GLK1*	1	1
glucose-6-phosphate isomerase	1	*PGI1*	1	1
6-phosphofructokinase	2	*PFK1*	1	1
		PFK2		
Fructose-1,6-bisphosphatase (gluconeogenesis)	1	*FBP1*	1	1
fructose-bisphosphate aldolase	1	*FBA*	1	1
triose-phosphate isomerase	1	*TPI1*	1	1+1*
Glyceraldehyde-3-phosphate dehydrogenase	3	*TDH1*	3	1
		TDH2		
		TDH3		
3-phosphoglycerate kinase	1	*PGK1*	1	1
phosphoglycerate mutase	1	*GPM1*	1	?
enolase	5	*ENO1*	1	1
		ENO2		
		ERR1		
		ERR2		
pyruvate kinase	1	*PYK1*	1	1
phosphoenolpyruvate carboxykinase	1	*PCK1*	1	1
Part C: Pentose phosphate pathway				
Glucose-6-phosphate dehydrogenase	1	*ZWF1*	1	1
6-phosphogluconate dehydrogenase	2	*GND1*	1	1
		GND2		
D-ribulose-5-Phosphate 3-epimerase	1	*RPE1*	1	1
Ribose-5-phosphate ketol-isomerase	1	*RKI1*	1	1
Transketolase	2	*TKL1*	2	2
		TKL2		
transaldolase	2	*TAL1*	1	1
xylose reductase	0	*(XYL1)*	1	0
xylitol dehydrogenase	1	*XYL2*	0	1
xylulokinase	1	*XKS1*	1	1

Enzyme activity	number of genes in			
	S. cerevisiae		*K. lactis*	*Y. lipolytica*
Part D: TCA cycle and related pathways				
pyruvate decarboxylase	3	*PDC1*	4	2
		PDC5		
		PDC6		
alcohol dehydrogenases:	4		4*	5
cytosol		*ADH1*		
cytosol		*ADH2*		
mito		*ADH3*		
cytosol		*ADH5*		
aldehyde dehydrogenase	5*	*ALD*	5*	5*
acetyl-coA synthetase	2	*ACS1*	2	1
		ACS2		
pyruvate carboxylase	2	*PYC1*	1	1
		PYC2		
pyruvate dehydrogenase alpha subunit	1	*PDA1*	1	1
dihydrolipoamide acetyltransferase	1	*PDA2*	1	1
pyruvate dehydrogenase (E1 beta)	1	*PDB*	1	1
dihydrolipoamide dehydrogenase	1	*LPD1*	1	1
citrate synthase (major mitochondrial)	2	*CIT1*	1	1
citrate synthase (peroxisomal)		*CIT2*		
citrate synthase (mito)	1	*CIT3*	1	1
aconitase	1	*ACO1*	1	1
aconitate hydratase	1	*XXX*	1	1
isocitrate dehydrogenase(NAD+)	2	*IDH1*	2	2
		IDH2		
alpha-ketoglutarate dehydrogenase	1	*KGD1*	1	2
dihydrolipoyl transsuccinylase	1	*KGD2*	1	3
succinyl-CoA ligase	1	*LSC2*	1	1
succinate dehydrogenase	1	*SDH1*	1	1
succinate dehydrogenase	1	*SDH2*	1	1
succinate dehydrogenase	1	*SDH3*	1	1
succinate dehydrogenase	1	*SDH4*	1	1
fumarase	1	*FUM1*	1	1
malate dehydrogenase (mito)	1	*MDH1*	1	1
malate dehydrogenase (cytosol)	1	*MDH2*	1	1
malate dehydrogenase (peroxisomal)	1	*MDH3*	1	2
isocitrate dehydrogenase (mito)	1	*IDP1*	2*	3
isocitrate dehydrogenase (cytosol)	1	*IDP2*		
isocitrate dehydrogenase (perox)	1	*IDP3*		
Isocitrate lyase	1	*ICL1*	1	1
2-methylisocitrate lyase (mito)	1	*ICL2*	1	1
malate synthase	1	*MLS1*	1*	1

First column lists the enzymatic activities involved in the pathway while the corresponding *S. cerevisiae* genes are listed in the second and third columns. The second column provides the number of genes with the activity and the third the name of the *S. cerevisiae* gene. For each activity, columns 4 and 5 indicate the number of putative orthologs identified in *K. lactis* and *Y. lipolytca*, respectively.

Part A: The carbohydrate storage pathway.

Part B: The glycolysis/gluconeogenesis pathway.

The asterisk corresponds to a second *Y. lipolytica* ortholog of TPI1. This second ORF codes for a product more closely related to bacterial protein and which may have arisen by horizontal transfer (see text).

Part C: The pentose phosphate pathway

Part D: TCA cycle and related pathways

For some activities, that is to say, alcohol dehydrogenase or aldehyde dehydrogenase, the number of genes identified in each species is indicated but no direct correspondence has been attempted between the genes (indicated by an asterisk).

have to be much more active in *S. cerevisiae* than in the two respiratory yeasts. As the energetic yield of glycolysis is drastically smaller than in respiratory metabolism, *S. cerevisiae* could require more storage capacities.

3.2.3.2 The glycolytic pathway and gluconeogenesis. Glycolysis is the conversion of glucose (or in a wider sense of hexoses) to pyruvate and is one of the most conservative processes. Genes and gene products involved in this process have been well characterized in *S. cerevisiae*. In order to get a global picture of the *K. lactis* and *Y. lipolytca* genes of this metabolic pathway, putative orthologs have been searched in the genomes of these two yeasts. They are described in Table 3 (Part B). From it, one can easily deduce that orthologs exist in the two respiratory yeasts but that the tendency to lower gene redundancy, already apparent in the upper part of the pathway, is maintained. The only expansion is observed with the triose isomerase encoded by the TPI1 gene. Two *Y. lipolytica* proteins were found but one is phylogenetically more related to the bacterial TPI than to yeasts (data not shown). Interestingly, this second form is also found in *N. crassa*. For the moment there is no experimental evidence showing that this protein is involved in the glycolysis process.

3.2.3.3 Related pathways. A major role of the pentose phosphate pathway (PPP, Table 3, Part C) is the regeneration of reducing equivalents (NADPH) in its oxidative part and the production of pentoses for biosynthetic reactions (nucleosides and amino-acids syntheses). The PPP is also the catabolic pathway for xylose in naturally xylose-fermenting yeasts. This is not the case of *S. cerevisiae*, but *K. lactis* can do so. The flux going through PPP in *S. cerevisiae* is moderate as shown by the low rate of fermentation of xylulose with respect to glucose fermentation (Walfridsson et al. 1995), probably to reduce overproduction of NADPH. Screening a genomic library of *K. lactis* to rescue the glucose growth defect of the *S. cerevisiae pgi* mutant revealed indeed the presence in this species of a glyceraldehyde-3 phosphate dehydrogenase that accept NADP as cofactor and counterbalances the excess of NADPH (Verho et al. 2002). In *S. cerevisiae*, the gene (xylose reductase, *XYL1*) responsible for the first step of xylose catabolism is absent, but exists in *K. lactis*. The gene (xylitol dehydrogenase, *XYL2*) responsible for the second step is cryptic in *S. cerevisiae* and has to be overexpressed to be functional (Toivari et al. 2004). In *K. lactis*, the best match was found with the gene *SOR1* that in *S. cerevisiae* is a paralogue of *XYL2* and codes for a polyol-dehydrogenase. Its precise function is not known but recent evidence (Toivari et al. 2004) seems to indicate that *SOR1* expression responds in a xylose-specific way. In such case, the two genes *SOR1* and *XYL2* present in *S. cerevisiae* would be reduced to one, but more efficient, in *K. lactis*. The reverse situation is found in *Y. lipolytica*. The *XYL1* gene is absent but the *GRE3* gene encoding an aldose reductase appears to be the best match. Again, its functionality in the xylose pathway was confirmed by the fact that overexpression of this gene in *S. cerevisiae* contributes to its growth on D-xylose. The third enzymatic step, catalyzed by xylutol dehydrogenase encoded by the *XKS1* gene, is present in all three organisms. One can also note that 6-phoshogluconate dehydogenase and transaldolase activities are encoded by only one gene in place of two in *S. cerevisiae*.

No significant variations could be otherwise observed in the TCA cycle and related pathways (Table 3, Part D) with the exception of one additional gene in *K. lactis*, which may encode a pyruvate carboxylase activity and an expansion of the *Y. lipolytica* gene family coding for components of the alpha-ketoglutarate dehydrogenase complex. This last observation may be explained by the fact that this species, which has to degrade and oxidize fatty acids may need very active peroxisomes. A recent publication (Smaczynska-de Rooij et al. 2004) has evidenced the need of this complex for peroxisome functioning.

4 Conclusions

In this short review, we have examined and compared several yeast genomes, sequenced, and analyzed during the two Genolevures programs. We have only focused on some aspects, both for the structural and functional aspects. Concerning structural aspects, other parameters (not treated here) were assayed like tRNA genes, codon usage, GC content, sexuality, chromosomic segregation *etc*. Each of them can be a guide to reconstruct some history of evolution.

Many more analyses can be - and must be - done on this robust material. Nevertheless the main lesson of this work is that evolution, in the well-established framework of chance and necessity, superimposes stochastic discrete events, such as different kinds of duplications or genomic rearrangements, to the continuous mutation flow.

For functional aspects, the exploration of Genolevures I and II data clearly show that partial sequencing can reveal only very general trends due to the contribution of numerous genes in the defined functional categories. Most of the subtle differences that modulate and fine-tune gene expression and its adaptation to different environments are missed in partial sequence data analysis. In this short review, we have tried to go through the information contained in the genomes recently sequenced of two yeasts, *K. lactis* and *Y. lipolytica*, focusing mostly on what makes their more striking differences in carbon assimilation and anaerobiosis. Doing so, we have worked mainly on one of the two possible hypotheses for evolution of biological functions, namely, the number of genes in each gene family. We have illustrated that most of the genes necessary for growth in anaerobiosis are not present in *K. lactis* and *Y. lipolytica*. Scrutinizing genes involved in central metabolism also reveals that, in general, these two yeasts exhibit a reduction in redundancy as compared to *S. cerevisiae*. While it is tempting to explain such an observation with the metabolic flux reduction in the glycolytic pathway in these two respiratory yeasts and the decrease in carbohydrate storage, we now need experimental evidence to support such model. Fortunately, the complete knowledge of the genome sequences as well as the fact that these two "non-conventional yeasts" are easily amenable to experimentation by classical and molecular genetics opens the way to many functional studies.

The second (and not exclusive) hypothesis based on different regulatory pathways has not been explored here because of its complexity. Several requirements should be met before this possible aspect of functional evolution can be seriously studied. In particular, we need a more detailed knowledge of regulatory networks in *S. cerevisiae*, a careful comparative analysis of transcriptional regulators as well as DNA micro-arrays for different yeast species. With such information and tools at hand, it will be possible in the future to approach this very exciting question.

Acknowledgements

We thank P. Durrens and D. Sherman for their help and all the Genolevures staff for helpful discussions. This work was supported by the Consortium National de Recherche en Génomique (to Genoscope and Institut Pasteur Génopole), the CNRS (GDR2354, Genolevures sequencing consortium), the Ministère de la Jeunesse, de l'Education et de la Recherche (ACI IMPBio n° IMPB114 "Genolevures en ligne"), the Conseil régional d'Aquitaine ("Genotypage et génomique comparée"), the IFR 115 "Genomes" and INRA.

References

Abramova N, Sertil O, Mehta S, Lowry CV (2001) Reciprocal regulation of anaerobic and aerobic cell wall mannoprotein gene expression in *Saccharomyces cerevisiae*. J Bacteriol 183:2881-2887

Ares M Jr, Grate L, Pauling MH (1999) A handful of intron-containing genes produces the lion's share of yeast mRNA. RNA 5:1138-1139

Arikawa Y, Enomoto K, Muratsubaki H, Okazaki M (1998) Soluble fumarate reductase isoenzymes from *Saccharomyces cerevisiae* are required for anaerobic growth. FEMS Microbiol Lett 165:111-116

Barnett JA (1976) The utilization of sugars by yeasts. Adv Carbohydr Chem Biochem 32:125-234

Bianchi MM, Tizzani L, Destruelle M, Frontali L, Wesolowski-Louvel M (1996) The 'petite-negative' yeast *Kluyveromyces lactis* has a single gene expressing pyruvate decarboxylase activity. Mol Microbiol 19:27-36

Bon E, Casaregola S, Blandin G, Llorente B, Neuveglise C, Munsterkotter M, Guldener U, Mewes HW, Van Helden J, Dujon B, Gaillardin C (2003) Molecular evolution of eukaryotic genomes: hemiascomycetous yeast spliceosomal introns. Nucleic Acids Res 31:1121-1135

Bourgarel D, Nguyen CC, Bolotin-Fukuhara M (1999) HAP4, the glucose-repressed regulated subunit of the HAP transcriptional complex involved in the fermentation-respiration shift, has a functional homologue in the respiratory yeast *Kluyveromyces lactis*. Mol Microbiol 31:1205-1215

Breunig KD, Bolotin-Fukuhara M, Bianchi MM, Bourgarel D, Falcone C, Ferrero II, Frontali L, Goffrini P, Krijger JJ, Mazzoni C, Milkowski C, Steensma HY, Wesolowski-Louvel M, Zeeman AM (2000) Regulation of primary carbon metabolism in *Kluyveromyces lactis*. Enzyme Microb Technol 26:771-780

Bussereau F, Lafay JF, Bolotin-Fukuhara M (2004) Zinc finger transcriptional activators of yeasts. FEMS Yeast Res 4:445-458

Casaregola S, Neuveglise C, Bon E, Gaillardin C (2002) Ylli, a non-LTR retrotransposon L1 family in the dimorphic yeast *Yarrowia lipolytica*. Mol Biol Evol 19:664-677

Cliften P, Sudarsanam P, Desikan A, Fulton L, Fulton B, Majors J, Waterston R, Cohen BA, Johnston M (2003) Finding functional features in *Saccharomyces* genomes by phylogenetic footprinting. Science 301:71-76

Cohen BD, Sertil O, Abramova NE, Davies KJ, Lowry CV (2001) Induction and repression of DAN1 and the family of anaerobic mannoprotein genes in *Saccharomyces cerevisiae* occurs through a complex array of regulatory sites. Nucleic Acids Res 29:799-808

Dietrich FS, Voegeli S, Brachat S, Lerch A, Gates K, Steiner S, Mohr C, Pohlmann R, Luedi P, Choi S, Wing RA, Flavier A, Gaffney TD, Philippsen P (2004) The *Ashbya gossypii* genome as a tool for mapping the ancient *Saccharomyces cerevisiae* genome. Science 304:304-307

Dujon B, Sherman D, Fischer G, Durrens P, Casaregola S, Lafontaine I, De Montigny J, Marck C, Neuveglise C, Talla E, Goffard N, Frangeul L, Aigle M, Anthouard V, Babour A, Barbe V, Barnay S, Blanchin S, Beckerich JM, Beyne E, Bleykasten C, Boisrame A, Boyer J, Cattolico L, Confanioleri F, De Daruvar A, Despons L, Fabre E, Fairhead C, Ferry-Dumazet H, Groppi A, Hantraye F, Hennequin C, Jauniaux N, Joyet P, Kachouri R, Kerrest A, Koszul R, Lemaire M, Lesur I, Ma L, Muller H, Nicaud JM, Nikolski M, Oztas S, Ozier-Kalogeropoulos O, Pellenz S, Potier S, Richard GF, Straub ML, Suleau A, Swennen D, Tekaia F, Wesolowski-Louvel M, Westhof E, Wirth B, Zeniou-Meyer M, Zivanovic I, Bolotin-Fukuhara M, Thierry A, Bouchier C, Caudron B, Scarpelli C, Gaillardin C, Weissenbach J, Wincker P, Souciet JL (2004) Genome evolution in yeasts. Nature 430:35-44

Fink GR (1987) Pseudogenes in yeast? Cell 49:5-6

Flores CL, Rodriguez C, Petit T, Gancedo C (2000). Carbohydrate and energy-yielding metabolism in non-conventional yeasts. FEMS Microbiol Rev 24:507-529

Fraser CM, Gocayne JD, White O, Adams MD, Clayton RA, Fleischmann RD, Bult CJ, Kerlavage AR, Sutton G, Kelley JM (1995) The minimal gene complement of *Mycoplasma genitalium*. Science 270:397-403

Fukuhara H (2003) The Kluyver effect revisited. FEMS Yeast Res 3:327-331

Gaillardin C, Duchateau-Nguyen G, Tekaia F, Llorente B, Casaregola S, Toffano-Nioche C, Aigle M, Artiguenave F, Blandin G, Bolotin-Fukuhara M, Bon E, Brottier P, de Montigny J, Dujon B, Durrens P, Lepingle A, Malpertuy A, Neuveglise C, Ozier-Kalogeropoulos O, Potier S, Saurin W, Termier M, Wesolowski-Louvel M, Wincker P, Souciet J, Weissenbach J (2000) Genomic exploration of the hemiascomycetous yeasts: 21. Comparative functional classification of genes. FEBS Lett 487:134-149

Goffeau A, Barrell BG, Bussey H, Davis RW, Dujon B, Feldmann H, Galibert F, Hoheisel JD, Jacq C, Johnston M, Louis EJ, Mewes HW, Murakami Y, Philippsen P, Tettelin H, Oliver SG (1996) Life with 6,000 genes. Science 274:546, 563-547

Gojkovic Z, Knecht W, Zameitat E, Warneboldt J, Coutelis JB, Pynyaha Y, Neuveglise C, Moller K, Loffler M, Piskur J (2004) Horizontal gene transfer promoted evolution of the ability to propagate under anaerobic conditions in yeasts. Mol Genet Genomics 271:387-393

Gu Z, Rifkin SA, White KP, Li WH (2004) Duplicate genes increase gene expression diversity within and between species. Nat Genet 36:577-579

Kellis M, Birren BW, Lander ES (2004) Proof and evolutionary analysis of ancient genome duplication in the yeast *Saccharomyces cerevisiae*. Nature 428:617-624

Kellis M, Patterson N, Endrizzi M, Birren B, Lander ES (2003) Sequencing and comparison of yeast species to identify genes and regulatory elements. Nature 423:241-254

Kurtzman CP, Fell JW (1998) The Yeasts, a taxonomic study, 4th edition. Elsevier, Amsterdam

Langkjaer RB, Cliften PF, Johnston M, Piskur J (2003) Yeast genome duplication was followed by asynchronous differentiation of duplicated genes. Nature 421:848-852

Malpertuy A, Tekaia F, Casaregola S, Aigle M, Artiguenave F, Blandin G, Bolotin-Fukuhara M, Bon E, Brottier P, de Montigny J, Durrens P, Gaillardin C, Lepingle A, Llorente B, Neuveglise C, Ozier-Kalogeropoulos O, Potier S, Saurin W, Toffano-Nioche C, Wesolowski-Louvel M, Wincker P, Weissenbach J, Souciet J, Dujon B (2000) Genomic exploration of the hemiascomycetous yeasts: 19. Ascomycetes-specific genes. FEBS Lett 487:113-121

Moller K, Langkjaer RB, Nielsen J, Piskur J, Olsson L (2004) Pyruvate decarboxylases from the petite-negative yeast *Saccharomyces kluyveri*. Mol Genet Genomics 270:558-568

Neuvéglise C, Chalvet F, Wincker P, Gaillardin C, Casaregola S (submitted) A genome wide survey reveals an unusual *Mutator*-like element in the dimorphic yeast *Yarrowia lipolytica*.

Neuveglise C, Feldmann H, Bon E, Gaillardin C, Casaregola S (2002) Genomic evolution of the long terminal repeat retrotransposons in hemiascomycetous yeasts. Genome Res 12:930-943

Ohno S (1970) Evolution by gene duplication. Springer-Verlag, Heidelberg

Papp B, Pal C, Hurst LD (2003) Evolution of cis-regulatory elements in duplicated genes of yeast. Trends Genet 19:417-422

Passoth V, Zimmermann M, Klinner U (1996) Peculiarities of the regulation of fermentation and respiration in the crabtree-negative, xylose-fermenting yeast *Pichia stipitis*. Appl Biochem Biotechnol 57-58:201-212

Rodriguez-Navarro S, Strasser K, Hurt E (2002) An intron in the YRA1 gene is required to control Yra1 protein expression and mRNA export in yeast. EMBO Rep 3:438-442

Smaczynska-de Rooij I, Migdalski A, Rytka J (2004) Alpha-Ketoglutarate dehydrogenase and lipoic acid synthase are important for the functioning of peroxisomes of *Saccharomyces cerevisiae*. Cell Mol Biol Lett 9:271-286

Sybirna K, Guiard B, Li YF, Bao WG, Bolotin-Fukuhara M, Delahodde A (2005) A new *H. polymorpha HAP4* homologue which contains only the N-terminal conserved domain of the protein is fully functional in *S. cerevisiae*. Curr Genet 47:172-181

Ter Linde JJ, Liang H, Davis RW, Steensma HY, van Dijken JP, Pronk JT (1999) Genome-wide transcriptional analysis of aerobic and anaerobic chemostat cultures of *Saccharomyces cerevisiae*. J Bacteriol 181:7409-7413

Ter Linde JJ, Steensma HY (2002) A microarray-assisted screen for potential Hap1 and Rox1 target genes in *Saccharomyces cerevisiae*. Yeast 19:825-840

The Génolevures consortium (2000) Génolevures Special Issue. FEBS Lett 487

The yeast genome consortium (1997) The yeast genome. Nature 387

Thompson JD, Gibson TJ, Plewniak F, Jeanmougin F, Higgins DG (1997) The CLUSTAL_X windows interface: flexible strategies for multiple sequence alignment aided by quality analysis tools. Nucleic Acids Res 25:4876-4882

Tiuryn J, J.P. R, Slonimski PP (1999) Striking properties of duplicating DNA molecules. A Markov chain model demonstrates the convergence of amplified molecules to regular series of multiples of 2. CR Acad Sci Paris 322:455-459

Todd RB, Andrianopoulos A (1997) Evolution of a fungal regulatory gene family: the Zn(II)2Cys6 binuclear cluster DNA binding motif. Fungal Genet Biol 21:388-405

Toivari MH, Salusjarvi L, Ruohonen L, Penttila M (2004) Endogenous xylose pathway in *Saccharomyces cerevisiae*. Appl Environ Microbiol 70:3681-3686

Verho R, Richard P, Jonson PH, Sundqvist L, Londesborough J, Penttila M (2002) Identification of the first fungal NADP-GAPDH from *Kluyveromyces lactis*. Biochemistry 41:13833-13838

Visser W, Scheffers WA, Batenburg-van der Vegte WH, van Dijken JP (1990) Oxygen requirements of yeasts. Appl Environ Microbiol 56:3785-3792

Walfridsson M, Hallborn J, Penttila M, Keranen S, Hahn-Hagerdal B (1995) Xylose-metabolizing *Saccharomyces cerevisiae* strains overexpressing the TKL1 and TAL1 genes encoding the pentose phosphate pathway enzymes transketolase and transaldolase. Appl Environ Microbiol 61:4184-4190

Wilcox LJ, Balderes DA, Wharton B, Tinkelenberg AH, Rao G, Sturley SL (2002) Transcriptional profiling identifies two members of the ATP-binding cassette transporter superfamily required for sterol uptake in yeast. J Biol Chem 277:32466-32472

Wolfe KH, Shields DC (1997) Molecular evidence for an ancient duplication of the entire yeast genome. Nature 387:708-713

Wood V, Gwilliam R, Rajandream MA, Lyne M, Lyne R, Stewart A, Sgouros J, Peat N, Hayles J, Baker S, Basham D, Bowman S, Brooks K, Brown D, Brown S, Chillingworth T, Churcher C, Collins M, Connor R, Cronin A, Davis P, Feltwell T, Fraser A, Gentles S, Goble A, Hamlin N, Harris D, Hidalgo J, Hodgson G, Holroyd S, Hornsby T, Howarth S, Huckle EJ, Hunt S, Jagels K, James K, Jones L, Jones M, Leather S, McDonald S, McLean J, Mooney P, Moule S, Mungall K, Murphy L, Niblett D, Odell C, Oliver K, O'Neil S, Pearson D, Quail MA, Rabbinowitsch E, Rutherford K, Rutter S, Saunders D, Seeger K, Sharp S, Skelton J, Simmonds M, Squares R, Squares S, Stevens K, Taylor K, Taylor RG, Tivey A, Walsh S, Warren T, Whitehead S, Woodward J, Volckaert G, Aert R, Robben J, Grymonprez B, Weltjens I, Vanstreels E, Rieger M, Schafer M, Muller-Auer S, Gabel C, Fuchs M, Dusterhoft A, Fritzc C, Holzer E, Moestl D, Hilbert H, Borzym K, Langer I, Beck A, Lehrach H, Reinhardt R, Pohl TM, Eger P, Zimmermann W, Wedler H, Wambutt R, Purnelle B, Goffeau A, Cadieu E, Dreano S, Gloux S, et al. (2002) The genome sequence of *Schizosaccharomyces pombe*. Nature 415:871-880

Zhang Z, Kishino H (2004) Genomic background predicts the fate of duplicated genes evidence from the yeast genome. Genetics 166:1995-1999

Aigle, Michel
   IBGC. CNRS/Université Victor Segalen Bordeaux 2, rue Camille Saint Saens, 233077 Bordeaux Cedex, France

Bolotin-Fukuhara, Monique
Institut de Génétique et Microbiologie, Université Paris Sud/CNRS UMR 8621, 91405 Orsay Cedex, France
bolotin@igmors.u-psud.fr

Casaregola, Serge
Microbiologie et Génétique Moléculaire, INRA UMR 1238/ CNRS UMR 2585, INA-PG; 78850 Thiverval-Grignon, France

The genome of the filamentous fungus *Ashbya gossypii*: annotation and evolutionary implications

Sophie Brachat, Fred Dietrich, Sylvia Voegeli, Tom Gaffney, and Peter Philippsen

Abstract

The 9.2 Mb genome of the filamentous fungus *Ashbya gossypii* consists of seven chromosomes carrying 4718 protein coding genes, 194 tRNA genes, at least 60 small RNA genes, and 40-50 copies of rRNA genes. With respect to both, the size and the number of genes, this presently represents the smallest known genome of a free-living eukaryote. Over 95% of the *A. gossypii* open reading frames encode proteins with homology to *Saccharomyces cerevisiae* proteins. In addition, 90% of *A. gossypii* genes show both, homology and a particular pattern of synteny (conservation of gene order), with the genome of budding yeast. Gene orders in the two species are not strictly co-linear because individual clusters of *A. gossypii* genes are always syntenic to two distinct *S. cerevisiae* chromosomal regions but frequently homologous genes are missing in either of the two regions. These gene clusters of ancient synteny (CLAS) were found to cover over 90% of both genomes. Specifically, 95% of the *S. cerevisiae* genes can be paired in duplicate blocks that match the gene order of single *A. gossypii* gene groups. The almost complete coverage of both genomes by clusters of ancient synteny provides compelling evidence that both species originate from a common ancestor and that the evolution of *S. cerevisiae* included a whole genome duplication subsequently followed by random deletions of one gene copy in 90% of the duplicated genes. The alignment of both genomes revealed a complete list of the 10% still remaining duplicated genes (twin genes) in today's genome of *S. cerevisiae*. The analysis of this comprehensive set of ancient twin genes in *S. cerevisiae* suggests that their evolution is asynchronous. Finally, interpretation of the synteny pattern between the sixteen *S. cerevisiae* centromere regions and the homologous gene regions in *A. gossypii* suggests that the common ancestor of the two species most likely carried eight chromosomes. The postulated reduction to seven chromosomes in the *A. gossypii* lineage very likely marked a key event in the development of this filamentous yeast as a novel species.

Topics in Current Genetics, Vol. 15
P. Sunnerhagen, J. Piškur (Eds.): Comparative Genomics
DOI 10.1007/4735_114 / Published online: 18 November 2005
© Springer-Verlag Berlin Heidelberg 2005

(A) 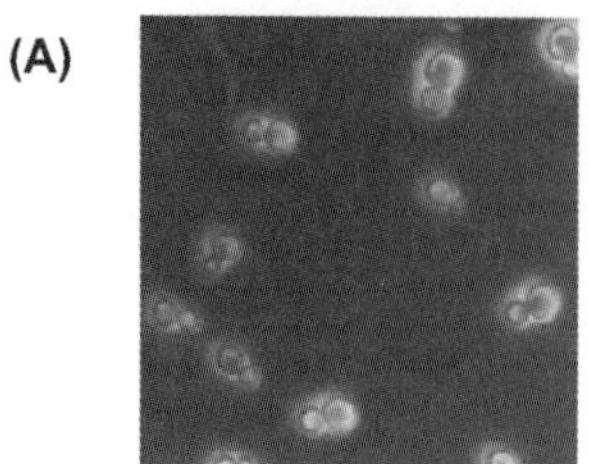    **(B)** 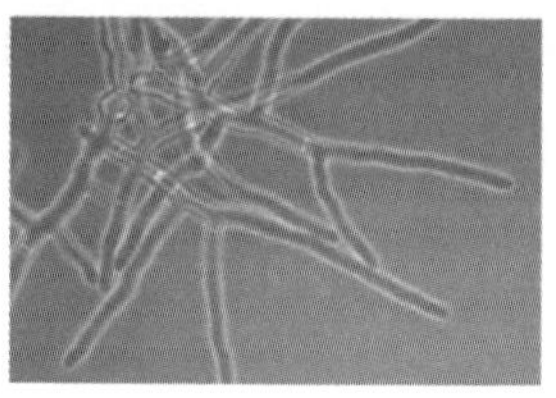

* Unicellular
* Mononucleated
* Switch between isotropic and polar growth
* Synchronous mitosis
* Short range nuclear migration
* No growth on lipids as sole carbon source
* Growth on galactose and maltose
* Sexual cycle

* Coenocytic
* Multinucleated
* Sustained polar growth
* Asynchronous mitosis
* Long range nuclear migration
* Growth on lipids as sole carbon source
* Overproducer of riboflavin
* No known sexual cycle

Fig. 1. Light microscopy images of *S. cerevisiae* (A) and *A. gossypii* (B). Each picture is provided with a non-exhaustive list of phenotypic differences between the two fungi.

1 Introduction

Eighteen years ago we initiated molecular genetics with *A. gossypii* in order to complement our studies with *S. cerevisiae*. We were intrigued by the fact that this hemiascomycete exclusively grows like a filamentous fungus (Fig. 1) though phylogenetic analyses placed *A. gossypii* next to *Saccharomyces* and *Kluyveromyces* species as well as the phytopathogenic fungi *Erymothecium*, *Holleya*, and *Nematospora* (Prillinger et al. 1997). Staining of nuclei in *A. gossypii* hyphae shows that the filamentous growth of this ascomycete has little in common with pseudohyphal growth of *S. cerevisiae* or *Candida albicans* (Whiteway and Oberholzer 2004), and very much resembles growth of multinucleated and branching hyphae of *Aspergillus* or *Neurospora* (Philippsen et al. 2005). However, unlike higher ascomycetes, *A. gossypii* lacks complex developmental programmes for conidiation.

During the first phase of our *A. gossypii* studies, we developed tools which allowed targeted deletions of genes and complementation studies with self-replicating plasmids (Wright and Philippsen 1991; Steiner et al. 1995; Wendland et al. 2000). In addition, we discovered that the gene order in one third of the genomic clones was syntenic to the gene order of their homologues in *S. cerevisiae*, and that the *A. gossypii* genome most likely lacked duplications, and that it was significantly smaller than the genome of *S. cerevisiae* (Steiner and Philippsen 1994; Mohr 1995; Altmann-Jöhl and Philippsen 1996; Pöhlmann 1996; Wendland et al. 1999). Based on these results and our several years experience with the *S. cerevisiae* genome project, we decided in 1996 to aim at the complete genome sequence of *A. gossypii*. We wanted to gain the necessary information basis for functional characterization of all genes involved in cellular pathways controlling growth, nuclear division, and long-range organelle migration. In addition, we envisaged that the *A. gossypii* gene order, once established, would serve as a Rosetta stone to gain essential insights into the evolution of the *S. cerevisiae* genome.

The complete sequence was finished early 2002. Completion of its annotation, and the synteny map of the *A. gossypii* and the *S. cerevisiae* genome took an additional two years. The majority of the work was performed by a small team of scientists working at the Biozentrum at the University of Basel and at Syngenta Biotechnology Inc. in Research Triangle Park, with essential help in the finishing phase from the Genome Center at Duke University in Durham. The genome was sequenced by combining three strategies: end-sequencing of 12000 chromosome-sorted plasmid and 1000 BAC clones, shotgun-sequencing of 32000 sheared genomic DNA fragments, and extensive gap filling by 12000 primer walking and PCR guided sequencing. The sequence assembly was performed in 8 individual batches (seven chromosomes and the mitochondrial DNA). This strategy led to the availability of close to 90% of the sequence three years after the launch of the project, and provided ample information for functional studies of *A. gossypii* genes (Ayad-Durieux et al. 2000; Wendland and Philippsen 2000, 2001; Alberti-Segui et al. 2001; Knechtle et al. 2003; Bauer et al. 2004). The remaining time was used to close sequence gaps, to annotate all genes, to complete the synteny map with *S. cerevisiae*, and to determine the endpoints of over 300 inversions and translocations in both genomes for the final publication (Dietrich et al. 2004).

This review describes first the sequence analysis and major results from its annotation. The second part focuses on evolutionary implication for both genomes as revealed from close inspection of the clusters of ancient synteny. This part also includes results of an attempt to reconstruct the *S. cerevisiae* genome from 126 duplication blocks. The third part briefly summarizes different fates of twin genes with respect to sequence conservation and divergence. Finally, in the fourth part, we try to address whether the common ancestor of both organisms carried seven chromosomes like *A. gossypii* and gained one chromosome in the *S. cerevisiae* lineage prior to the whole genome duplication or whether it carried eight chromosomes with subsequent loss of one chromosome in the *A. gossypii* lineage.

2 Sequence analysis and annotation of the *A. gossypii* genome

2.1 General features of the genome sequence

The *A. gossypii* genome was sequenced and assembled with a 4.2-fold sequence coverage (Phrap quality > 20) and an estimated accuracy of 99.8%. Three short regions from 300 bp to 1400 bp could not be sequenced most likely due to a very high GC-content. Genomic features of the seven chromosomes are summarized in Table 1. The nuclear genome has a total size of 9.2Mb including rDNA. This is much smaller than the estimated sizes for other filamentous fungal genomes that

Table 1. *A. gossypii* genome features

	GenBank Accession Number	Size (Kb)	ORFs	Intron containing genes [a]	Coding DNA (Kb)	Coding DNA (%)
ChrI	AE016814	692	381	17	545	78.75
ChrII	AE016815	868	462	19	693	79.85
ChrIII	AE016816	907	497	26	728	80.26
ChrIV	AE016817	1467	819	35	1150	78.38
ChrV	AE016818	1519	800	40	1209	79.57
ChrVI	AE016819	1813	982	55	1448	79.87
ChrVII [b]	AE016820	1476	777	25	1187	80.41
Genome		8741	4718	217	6970	79.52

[a] This number does not include tRNA genes containing introns, of which there are a total of 50. There are five protein-coding genes that encode two introns, so the total number of introns is 222.

[b] The ribosomal DNA Repeat is 8197bp with about 50 copies present on chromosome VII (Wendland, *et al.*, 1999). Thus the total size of chromosome VII is approximately 1900 kb - consistent with the observation that this chromosome corresponds to the slowest migrating band in pulsed field gels (Wendland et al. 1999).

are in the range of 20-50Mb (Osiewacz and Ridder 1991; Kupfer et al. 1997; Chavez et al. 2001; Galagan et al. 2003). Surprisingly, it is also 30% smaller than the *S. cerevisiae* genome (13Mb including rDNA). The small size of the *A. gossypii* genome is in part due to the almost complete lack of gene duplications (discussed below) and a very high genetic information density. On average, the *A. gossypii* genome carries one protein-coding gene per 1.86 kb. In *S. cerevisiae*, the average genetic density is one gene per 2.1 kb and in *Schizosaccharomyces pombe* one gene per 2.5 kb (Goffeau et al. 1996; Wood et al. 2002). The average inter-ORF region in *A. gossypii* is over 200 bp shorter than in *S. cerevisiae* indicating that sequences controlling transcription in *A. gossypii* are probably shorter than in *S. cerevisiae*. The average ORF size is only slightly shorter in *A. gossypii* compared with *S. cerevisiae*. The paucity of introns (see Table 1) and the lack of transposable elements also contribute to the small size of the *A. gossypii* genome. Another notable difference is the average GC content which is much higher than that of *S. cerevisiae* (52% vs. 38%), a fact that could reflect the different temperatures in their respective habitats (Ashby and Nowell 1926).

2.2 Annotation of the assembled DNA sequences

An essential component of any sequencing effort is the conversion of the DNA sequence data into useful information. The identification of genomic features such as protein or RNA coding genes, telomeres, and centromeres and their labelling on the genome sequence is the first step in this conversion and is referred to as structural annotation. Once structural annotation is available, functional information for the discovered genes can be inferred from sequence and domain database searches and finally from experimental results.

The annotation strategy was designed to make use of the high degree of sequence and gene order conservation between the *A. gossypii* and the *S. cerevisiae* genome. As automatic annotation remains prone to errors, our method implied both an automatic phase and an extensive manual revision. First, all ORFs sharing homology to *S. cerevisiae* proteins as available from SGD (Cherry et al. 1998; Saccharomyces Genome Database, http://www.yeastgenome.org) were automatically extracted and annotated. Based on homology and synteny, other genomic features, such as tRNA genes, snRNA genes, telomeres and centromeres, were manually identified and annotated. A careful manual recheck of all annotations was then conducted in order to disregard ORFs with very low homology to *S. cerevisiae*, which lacked synteny, to confirm low homology ORFs, which showed synteny, to annotate overlooked syntenic ORFs with low homology to *S. cerevisiae*, and to annotate introns. In a second step, we identified overlooked protein-coding genes in both genomes by comparing translated DNA. This resulted in the discovery of 46 novel ORFs and the identification of 72 putative annotation errors in the *S. cerevisiae* genome (Brachat et al. 2003). A sequence search conducted on available fungal databases including unfinished genomes and on GenBank, allowed us to identify *A. gossypii* ORFs that do not have a homologue in *S. cerevisiae* but in other species. For this annotation step, we did not apply any size restriction for *A. gossypii* ORFs. Finally, we annotated the remaining *A. gossypii* ORFs larger than 150 codons. Systematic *A. gossypii* gene names were given following the nomenclature procedure used for budding yeast: three letters describing the organism (A-Ashbya), the chromosome (A-G), and the chromosome arm (L, R), followed by the gene number counting outwards from the centromere and the DNA strand (W, C). In contrast to *S. cerevisiae*, all centromeres have the same orientation with respect to the conserved DNA elements CDEI, CDEII, and CDEIII (Hieter et al. 1985). For example, AAL001W is the first protein coding gene to the left of the centromere on chromosome I and is coded by the Watson strand. To facilitate functional comparisons we use for syntenic ORFs, in addition to the systematic name, the common yeast gene name with a prefix Ag, e.g., ScCDC14 and AgCDC14. The manual evaluation of start and end points of synteny was performed by two individuals; inconsistent results were rechecked. The final annotations of the seven chromosomes and the mitochondrial DNA were submitted to GenBank (Accession numbers: AE016814 to AE016821). These data and additional information can be accessed via the Ashbya Genome Database (Hermida et al. 2005; http://agd.unibas.ch). For a representative section of the annotation table for all *A. gossypii* genes see Table 2.

2.3 Protein coding genes

We have identified 4718 ORFs in the nuclear genome of *A. gossypii*. When compared with the 6000 ORFs found in *S. cerevisiae* (Goffeau et al. 1996) and the 4824 ORFs of the fission yeast *S. pombe* (Wood et al. 2002), *A. gossypii* contains

Table 2. Features of the *A. gossypii* genome annotation implemented as spreadsheet.

Feature	Start	End	inter ORF size	A.g. ORF name	S.c.1° homologue	S.c.2° homologue	Synteny for 1° homologue	Synteny for 2° homologue	S.c.1° homologue common name	S.c.2° homologue common name	Comment1
CDS	566187	566612	653	AEL036W	NOHBY502		NOHBY502				No homologue in S. cerevisiae
CDS	566819	568057	207	AEL035W	NOHBY501		NOHBY501				No homologue in S. cerevisiae
CDS	568246	570261	189	AEL034W	YOL023W		SH		IFM1		
CDS	570391	571677	130	AEL033C	YOL022C		SH				
tRNA	572004	572074	327	tRNA-Gly	tRNA-Gly						Anticodon-GCC
CDS	572196	574457	122	AEL032W	YFR009W		SH		GCN20		
CDS	574534	577527	77	AEL031C	YOL021C		SH		DIS3		
CDS	577332	579551	-195	AEL030W	YOL020W		SH		TAT2		
CDS	580062	581402	511	AEL029W	YFR010W		SH		UBP6		
CDS	581563	582261	161	AEL028C	YFR011C		SH				
CDS	582886	584916	625	AEL027W	YOL019W	YFR012W	SH	SH	TOS7		
CDS	585366	586415	450	AEL026C	YOL018C		SH		TLG2		
tRNA	586537	586621	122	tRNA-Tyr	tRNA-Tyr						Anticodon-GUA
CDS	586885	588495	264	AEL025W	YFR028C		SH		CDC14		
CDS	588771	589571	276	AEL024C	YFR027W		SH		ECO1		
CDS	589949	594325	378	AEL023C	YIR019C		NSH		MUC1		
CDS	594977	596011	652	AEL022W	YFR025C		SH		HIS2		
CDS	596131	598143	120	AEL021C	YHR020W		SH				
CDS	598366	600024	223	AEL020W	YHR019C		SH		DED81		
CDS	600410	601819	386	AEL019W	YHR018C		SH		ARG4		
CDS	601945	603096	126	AEL018C	YHR017W		SH		YSC83		
CDS	603303	604654	207	AEL017W	YFR024C-A	YHR016C	SH	SH	LSB3	YSC84	1-intron
CDS	605004	607040	350	AEL016C	YFR023W	YHR015W	SH	SH	PES4	MIP6	
tRNA	608101	608173	1061	tRNA-Ala	tRNA-Ala						Anticodon-AGC
CDS	608518	609885	345	AEL015W	YOL069W		SH		NUF2		
CDS	609930	610697	45	AEL014C	YDL043C		SH		PRP11		
CDS	611020	612699	323	AEL013C	YDL042C	YOL068C	SH	SH	SIR2	HST1	
CDS	612993	613559	294	AEL012C	YOL067C		SH		RTG1		

Footnote to Table 2 overleaf.
This section includes information for 25 protein-coding genes and three tRNA gene. Four *A. gossypii* genes are syntenic homologues to two *S. cerevisiae* genes (twins). Data is automatically extracted from the annotation files and processed manually in Excel®. The "Feature" row describes the nature of each genomic element: CDS stands for Coding DNA Strand and refers to protein-coding genes. "Start" and "End" rows describe the gene coordinates on the chromosomes. Inter-ORF sizes are provided to highlight short or long noncoding genomic regions and are useful to detect potential gene overlaps (yellow) or inter-ORF regions longer than the average 500 bp (purple). The annotated overlap of 195 bp for AEL031C and AEL030W was investigated by the 5' race method, and it was found that the fourth ATG of AEL030W is the start codon which generates an inter-ORF size of 276 bp. Systematic and common names for *S. cerevisiae* homologues are provided. These cells are filled using colours corresponding to each of the sixteen *S. cerevisiae* chromosomes and allowing a rapid estimation of ancient synteny . The "synteny" rows depict the *A. gossypii* gene categories: syntenic homologue (SH), non-syntenic homologue (NSH) or no homologue in baker's yeast (NOHBY).

the smallest set of protein coding genes presently described for a free-living eukaryote. This suggests that only slightly more than 4500 proteins are required for an eukaryotic life style which is similar to the 4300 protein coding genes reported for the prokaryotes *E. coli* and *B. subtilis* (Blattner et al. 1997; Kunst et al. 1997) and less than the 5000 genes predicted for Pseudomonas (Stover et al. 2000), Streptomyces (Bentley et al. 2002) and Mycobacterium (Cole et al. 1998). The *A. gossypii* ORF sizes range from 25 codons for ADR103C, which encodes the homologue of the *S. cerevisiae* ribosomal proteins RPL41A and RPL41B, to 4899 codons for AGR074C, which encodes the homologue of the *S. cerevisiae* protein Mdn1p. We annotated 217 (4.6%) intron-containing ORFs. Five genes were found to contain two introns (AFL082W, AEL145W, ADR193W, AFR579W, AEL262C-A). Similar to *S. cerevisiae*, introns are small (115bp on average, ranging from 25bp to 667bp). Most introns are located close to the 5' end of ORFs and 24 introns interrupt the start codon. Interestingly, intron positions within ORFs are often conserved in the two genomes. These findings were used to identify potentially overlooked introns in the *S. cerevisiae* genome (Brachat et al. 2003). We found that 66 *A. gossypii* ORFs carry an intron, which is not present in the *S. cerevisiae* homologue and that for 46 intron-containing *S. cerevisiae* ORFs the *A. gossypii* homologues are intron-free. Like in *S. cerevisiae*, intron-containing *A. gossypii* genes often belong to particular functional categories. For example, 47 encode ribosomal protein genes, 20 encode proteins involved in cellular trafficking and protein sorting, and 13 are homologues of *S. cerevisiae* genes encoding cytoskeleton proteins.

2.4 Sequence conservation of proteins

Our annotation procedure detected, for 95% of the *A. gossypii* ORFs, homologues in the budding yeast. The conservation, at the protein level, between these homologues is very variable ranging from 18.1% (homologues of Prm2) and 100% (homologues of histone H3) identical amino acids, with an average of 53.7%. In

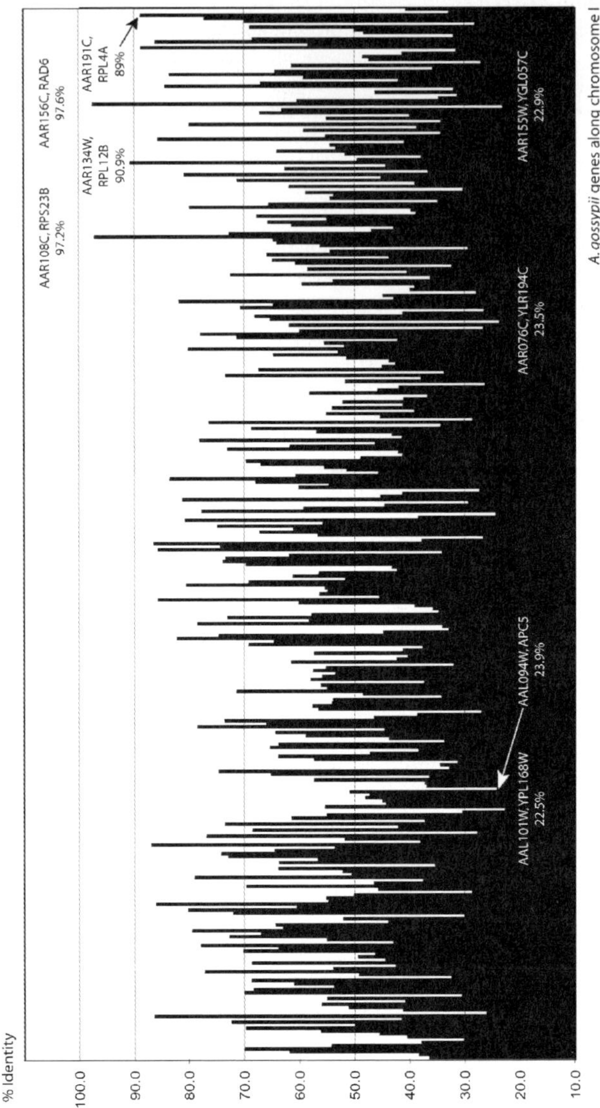

Fig. 2. Identity (at the protein level) between *A. gossypii* and *S. cerevisiae* syntenic homologues is independent from their genomic location. Percent identity between *A. gossypii* and *S. cerevisiae* syntenic homologues was evaluated by pairwise comparison of the sequences using Gap (GCG®) and plotted along the *A. gossypii* chromosome I. Each line represents a single *A. gossypii* gene. The identity level varies between genes independently from their relative location on the chromosome. The four genes with highest and lowest identity to their *S. cerevisiae* syntenic homologues are mentioned. Random distribution of homology levels along the chromosomes was also observed for the remaining six chromosomes.

order to estimate whether particular regions of the *A. gossypii* genome are more susceptible to sequence evolution than others, we plotted levels of identity between homologues along the *A. gossypii* chromosomes. Figure 2 shows this plot for chromosome I. Highly conserved protein coding genes alternate with less conserved genes in a seemingly random pattern. This indicates that the natural selection unit is a single gene rather than a chromosomal region. We used functional Gene Ontology categories (Ashburner et al. 2000) to classify the most similar (>90% identity) and least similar (< 30%) proteins between the two species. This analysis confirmed that, as generally observed in eukaryotes, ribosomal proteins, histones and proteasomal subunits are among the most conserved orthologues. It also revealed that meiotic proteins are very poorly conserved between the two fungi. Interestingly, five *S. cerevisiae* proteins among the 200 most conserved (>90% similarity) are so far of unknown function (YBR025C, YDR339C, YGR086C, YGR173W, YPL225W). This implies a critical cellular role for the encoded proteins. Consistent with this view, the proportion of essential genes, based on the yeast gene deletion project (Giaever et al. 2002), is nearly twice as large in the highly conserved gene group (45%) compared to the average of all *S. cerevisiae* genes having a homologue in *A. gossypii* (23%).

2.5 Species-specific proteins

The combination of homology and synteny was used to classify *A. gossypii* genes into three categories: Category 1 (syntenic homologue, SH) includes all genes that are homologous to and in synteny to an *S. cerevisiae* gene (90% of all *A. gossypii* genes). Category 2 (non-syntenic homologues, NSH) includes genes with homology to genes in *S. cerevisiae* but lack synteny (5% of all *A. gossypii* genes). Category 3 (NOHBYs: No Homologue in Baker's Yeast) includes the remaining genes, those with no homology in *S. cerevisiae* (5% of all *A. gossypii* genes). We mapped the three gene categories along *A. gossypii* chromosomes. The three gene types are, for the most part, evenly distributed across the genome. Gene category distribution is thus independent from the genomic location. Importantly, this demonstrates that neither NOHBYs nor non-syntenic homologues cluster at specific regions of the genome but that they are interspersed within syntenic regions, implying an independent evolution of individual genes within these two categories.

The annotation of NOHBYs that can be referred to as "*A. gossypii*-specific genes" was challenging and their authenticity can be questioned. However, 170 of the 262 annotated NOHBYs have homologues in other organisms or encode proteins with known functional domains, confirming that they do not represent annotation artifacts. According to this domain analysis, NOHBYs encode proteins of diverse functions from enzymes to membrane proteins. The remaining 92 NOHBYs encode putative proteins longer than 150 amino acids (34 longer than 250 amino acids). The annotated NOHBYs represent only 5% of all *A. gossypii* protein-coding genes, which points to a surprisingly small difference in gene sets between the two fungi. In contrast to the low number of NOHBYs, we found many more (1417) *S. cerevisiae* genes lacking a homologue in *A. gossypii*. This group contains both, real genes and also close to 700 short ORFs, which very likely do not encode proteins (Wood et al. 2001; Brachat et al. 2003).

2.6 RNA-encoding genes

We have identified a total of 194 tRNA-encoding genes in the *A. gossypii* nuclear genome, fewer than the 275 tRNA genes of *S. cerevisiae*, and 50 of the tRNA genes have introns. The relative abundance of iso-acceptor tRNAs in *S. cerevisiae* is very similar in *A. gossypii*. Similarly as found for protein coding genes, most tRNA genes map at syntenic locations compared with gene orders in *S. cerevisiae*. Interestingly, they often map at the ends of synteny regions, which mark break points of genome rearrangements (see below). In *A. gossypii*, tRNA genes represent the only type of interspersed repeated sequences and, thus, are most likely sites at which genome rearrangements are initiated (Dietrich et al. 2004). The ribosomal RNA genes are clustered at a single locus on chromosome VII and has been previously described (Wendland et al. 1999). It is composed of approximately 50 tandem gene copies (8197bp) coding for the precursor for 18S, 5.8S, and 26S ribosomal RNA and the same number of gene copies, on the opposite strand, coding for the 5S ribosomal RNA. The same arrangement of ribosomal RNA genes is present in *S. cerevisiae* chromosome XII with conserved proteins coding genes on both sides of the rDNA clusters. We have so far identified 68 genes encoding snRNAs and snoRNAs most of which at syntenic positions with *S. cerevisiae* small RNA genes including four snoRNA genes nested within introns of protein-coding genes (AgSNR18, AgSNR24, AgSNR54, and AgSNR39/SNR59) and four clusters of snoRNA genes arranged in a polycistronic manner like their *S. cerevisiae* homologues. The total number of small RNAs in *A. gossypii* is presently lower than in *S. cerevisiae* where 84 small RNA products are known (Saccharomyces Genome Database; A Database for Small Nucleolar RNAs (snoRNAs) from the Yeast *Saccharomyces cerevisiae*, http://www.bio.umass.edu/biochem/rna-sequence/Yeast_snoRNA_Database). The discrepancy might result from the small size of these small RNA products and from limited homology between species both rendering their detection difficult. We assume that additional small RNA genes remain to be discovered in *A. gossypii*.

2.7 Transposable elements

Strikingly, we could not detect any complete transposable element in the *A. gossypii* genome. However, one ORF (AGL178W) on chromosome VII has homology to the reverse transcriptase gene of the *S. cerevisiae* transposable element Ty3. Despite this transcriptase-like gene no remnants of Ty elements could be detected. The lack of bona fide transposable elements is however intriguing because to our knowledge *A. gossypii* is the first reported eukaryote lacking mobile elements.

2.8 Centromeres and telomeres

Centromeres were located based on homology and synteny to *S. cerevisiae*. The three elements CDEI, CDEII, and CDEIII (Centromere DNA Element) (Panzeri et al. 1985; Hieter et al. 1985) could be identified. However, the *A. gossypii* CDEII is approximately twice as long as the CDEII of *S. cerevisiae* and, therefore, resembles more the *K. lactis* centromeres (Heus et al. 1990). The telomeres of *A. gossypii* have also been identified and carry the tandemly repeated 24-mer CGCTGAGAGACCCATACACCACAC. A non-telomeric copy of this repeat is present in the putative AgTLC1 gene, the *A. gossypii* homologue of the *S. cerevisiae* RNA template component of the telomerase, which is found at a syntenic position on chromosome I. The *A. gossypii* telomeres differ to some extent from the *S. cerevisiae* telomeres as they have multiple perfect copies of this 24 base pair repeat. In contrast, the *S. cerevisiae* repeat unit is composed of the fairly similar 17 base pair template $(CACCACACCCACACACA)_n$ but the telomeres contain mostly degenerate copies of the template (Singer and Gottschling 1994; Cohn et al. 1998).

3 Evolutionary implications of the *A. gossypii* genome sequences

3.1 Possible origins of duplicated gene segments in *S. cerevisiae*

When the genome sequence of *S. cerevisiae* was completed two groups proposed a duplication of the ancestral genome during evolution. Philippsen et al. (1997) discussed such event as one possible explanation to account for the duplication of six long gene clusters, representing two thirds of chromosome XIV, with gene clusters on other chromosomes. The duplicated genes within these 55 kb to 130 kb long regions were interspersed with single copy genes but displayed in most cases conserved gene orientation and order, a genomic feature termed "relaxed synteny". Using the information from the yeast genome consortium Wolfe and Shields (1997) found that duplicate gene pairs represent 13% of all *S. cerevisiae* genes and that two or more of these gene pairs define so-called "duplicate blocks". They estimated that duplicate blocks cover 50% of the yeast genome and proposed that the *S. cerevisiae* genome is actually the result of a whole-genome duplication event followed by frequent loss of one copy of the duplicated gene pairs. According to these two proposals the today's *S. cerevisiae* genome represents a mosaic of the ancestral genome as schematically shown in Figure 3A. Explicitly, genes that were adjacent on a single chromosomal segment in the ancestral genome are now found alternating between two different chromosomal regions, and the original

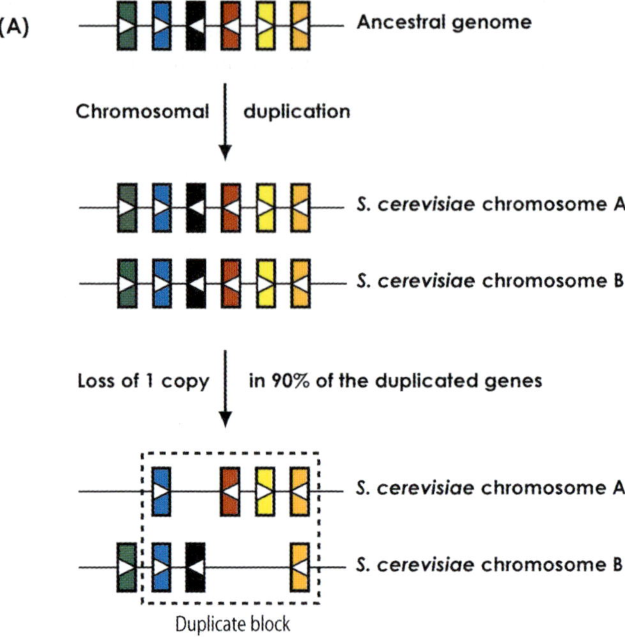

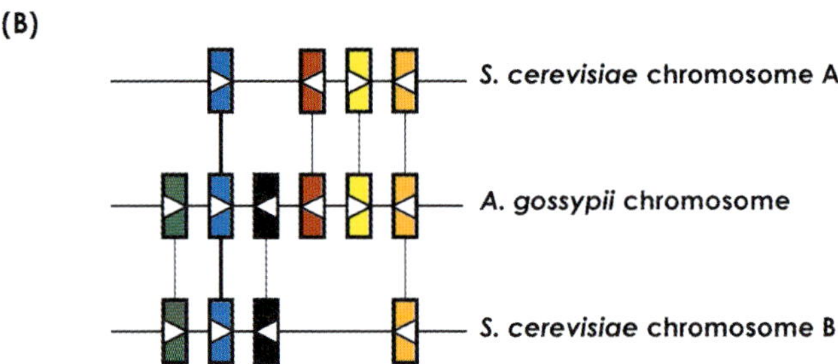

Fig. 3. Proposed fate of ancient duplications and reconstruction of ancient synteny. (A) Schematic representation of the origin of duplicate gene blocks in today's *S. cerevisae* genome by long-range duplications in the ancestral genome or by a single whole genome duplication event. The recognition of duplicate gene blocks is not possible in genome regions, which lack duplicate genes. In addition, reconstruction of the ancient gene order between duplicate genes is not possible. (B) Schematic illustration of a typical cluster of ancient synteny (CLAS) between *A. gossypii* and *S. cerevisiae* (CLAS). The gene order in *A. gossypii* is used as Rosetta stone to reconstruct the ancient gene order. In fact, identification of duplicate blocks in the *S. cerevisiae* genome no longer depends on the presence of duplicate genes. All *A. gossypii* gene regions aligning with homologues of two syntenic yeast chromosomes highlight duplicated *S. cerevisiae* gene regions and allow reconstruction of ancient gene orders.

order of genes between the still recognizable duplicated genes cannot be reconstructed without additional information. As an alternative hypothesis, Dujon and colleagues suggested that individual chromosomal segments were duplicated at different times (Llorente et al. 2000b; Fischer et al. 2001). This theory was considered less likely due to the absence of long triplicated regions in the *S. cerevisiae* genome, but based on the available *S. cerevisiae* genome information, it was a viable alternative explanation for the origin of duplicated segments (Piskur 2001). To find convincing evidence for one of the contradicting hypotheses and to be able to reconstruct the ancient gene order, it was necessary to align *S. cerevisiae* genes with homologous genes of a completely sequenced non-duplicated genome of a hemiascomycete as outlined in Figure 3B. Only a close to complete alignment of such genome with two gene segments of *S. cerevisiae* would provide compelling evidence in favour of the genome duplication hypothesis. The presence, in such synteny map, of long regions aligning only with single *S. cerevisiae* gene segments would favour the segmental duplication hypothesis.

3.2 Proof for an ancient whole-genome duplication in *S. cerevisiae*

A search for gene duplications in the complete *A. gossypii* gene set uncovered only a few tandemly duplicated genes but no gene cluster duplications. This search also revealed that 96% (see below) of the *A. gossypii* genome display synteny to the *S. cerevisiae* genome as indicated by several hundred clusters of ancient synteny like the one shown in Figure 3B with homology relations of single genes or groups of genes alternating between two *S. cerevisiae* regions. We did not find *A. gossypii* genomic segments, which were syntenic with more than two regions of the *S. cerevisiae* genome. This result can only be explained by a whole genome duplication in the *S. cerevisiae* lineage and does not support the segmental duplication hypothesis (Dietrich et al. 2004). The same conclusion was drawn from the analysis of genomic sequences of *Kluyveromyces waltii*, which carries a non-duplicated yeast genome (Kellis et al. 2004). We named the blocks of synteny between one *A. gossypii* and two *S. cerevisiae* gene groups "clusters of ancient synteny" (CLAS). The position and coding information of almost all gene deletions following the whole genome duplication can be inferred in each CLAS from the gaps in one chromosome segment and the gene still present in the other chromosome segment, respectively. On a whole genome scale this allows reconstruction of ancient gene orders prior to the genome duplication (Dietrich et al. 2004). Boundaries of a CLAS mark breaks of synteny either in one or in both *S. cerevisiae* chromosome segments. Such breaks define endpoints of viable translocations or inversions in the evolutionary past of both genomes, which is discussed in more detail in Chapter 3.6. The longest CLAS comprises 56 *A. gossypii* genes on chromosome III (ACR095 to ACR148, including two snRNA genes) syntenic with *S. cerevisiae* chromosomes XV (YOR212 to YOR245) and XVI (YPL127 to YPL158). This CLAS includes 10 duplicate gene pairs (twin genes). Neither of these three regions was rearranged by a translocation or inversion event since both

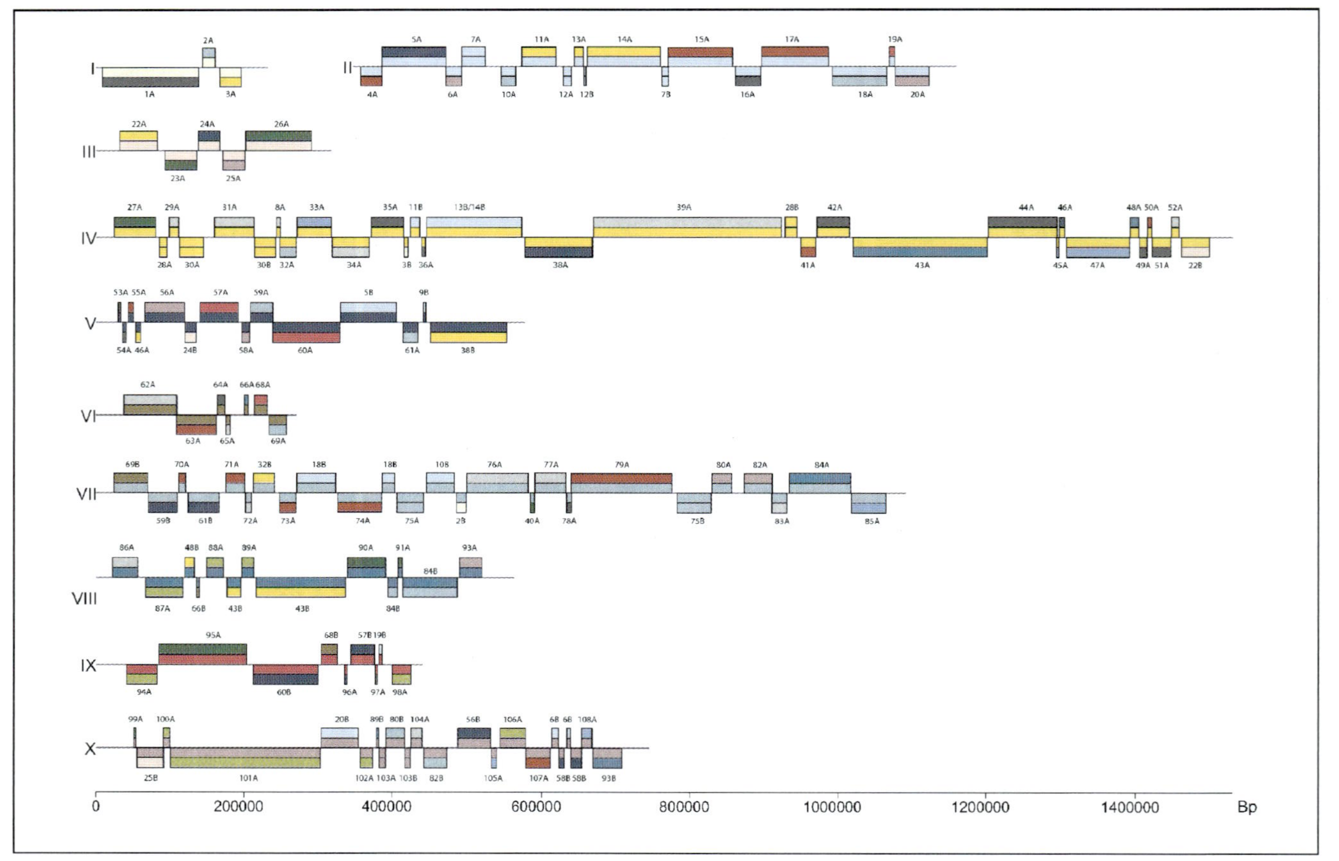

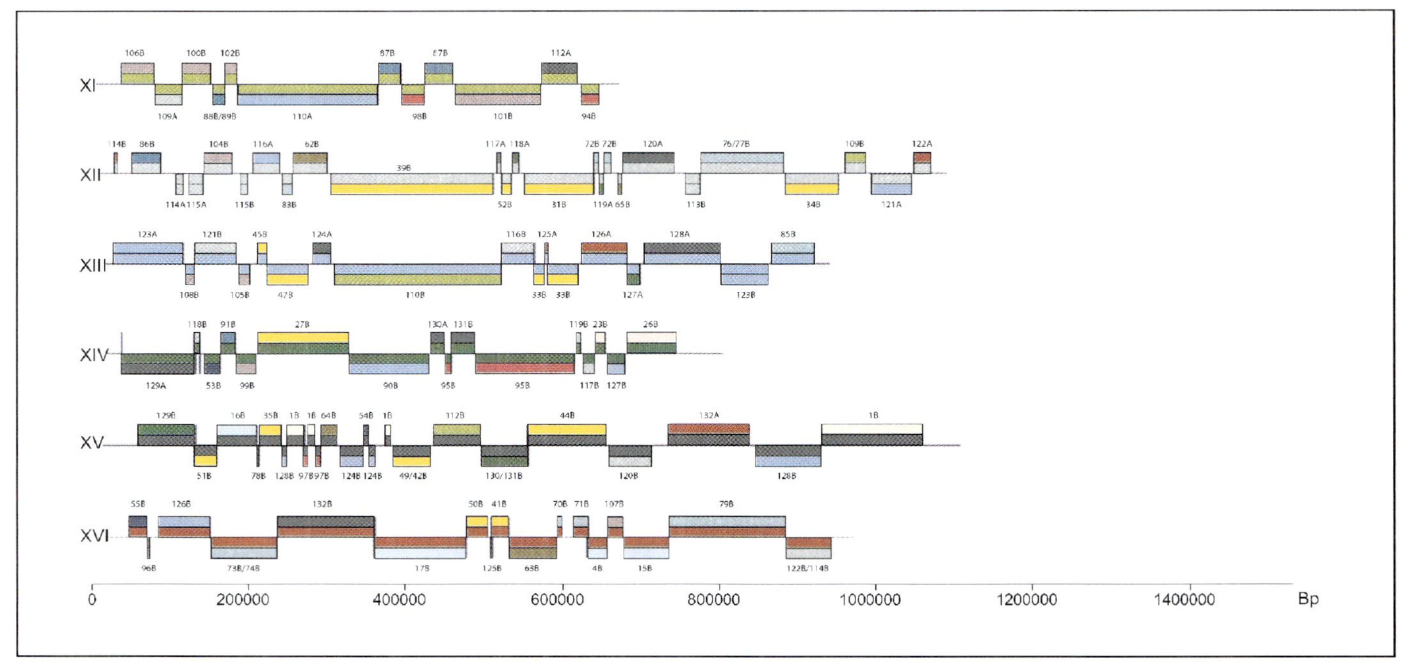

Fig. 4 (previous pages). Duplicate block map of the *S. cerevisiae* genome. Duplication blocks along the 16 *S. cerevisiae* chromosomes were reconstructed based on ancient synteny to the *A. gossypii* genome. Each block involves two chromosomes and each duplicate region is labelled as A and B. Blocks are depicted as boxes, divided in two coloured areas representing the two chromosomes. Blocks were drawn to scale and alternate above or below the chromosome lines for clarity. 95% of the *S. cerevisiae* protein-coding genes were found within anciently duplicated regions. Originally, 132 blocks were generated, but a visual inspection of boundaries led to the reassignment of six short blocks (21, 37, 67, 81, 92, 111) to adjacent blocks. A few short blocks appear more than two times (1B, 43B, 71B, 72B, 89B, 95B, 97B, 114B, 124B, 130B) most likely due to the mapping of gene family members. In this diagram only translocations are seen which occurred after the genome duplication; inversions remain undetected.

species diverged from a common ancester over hundred million years ago. There are longer regions in *A. gossypii* chromosomes which did not undergo a rearrangement whereas the syntenic *S. cerevisiae* regions rearranged after the duplication. The longest region comprises 182 genes (including six snRNA and two tRNA genes) in chromosome VI (AFL185 to AFL013).

3.3 NOHBY's and non-syntenic homologues in clusters of ancient synteny

In order to quantify the proportion of the *A. gossypii* genome that map within CLASs, we classified NOHBYs and non-syntenic homologues (NSHs) according to their location within or outside CLASs using the Ashbya Genome Database. For example, NOHBY AAR003W occurs within a CLAS whereas the NSH AAL120W maps at a break of synteny. We found that a total of 178 NOHBYs and 71 non-syntenic homologues map within CLASs. Thus, CLASs comprise 4276 syntenic homologues, 71 non-syntenic homologues and 178 NOHBYs of all 4718 *A. gossypii* protein-coding genes. In other words, 4525 of the *A. gossypii* protein-coding genes (96%) map in clusters of ancient synteny.

3.4 Update of duplicate gene blocks in *S. cerevisiae*

The almost complete coverage of the *A. gossypii* genome in clusters of ancient synteny raised the question which fraction of the *S. cerevisiae* genes belong to duplicated regions. We used ancient synteny information to reconstruct updated *S. cerevisiae* duplicate blocks and evaluated their coverage of the yeast genome. Technically, *S. cerevisiae* duplicate regions were first derived from each CLAS and the inferred duplicate loci were re-ordered along the *S. cerevisiae* chromosomes to generate a novel duplicate block map (Fig. 4).

We found that the *S. cerevisiae* genome can be re-constructed in 126 duplicate blocks. These blocks cover over 85% of the total genomic DNA (including the ribosomal DNA repeats and the centromeres, but excluding the ORF-free telomeric regions). We determined that 5969 *S. cerevisiae* ORFs are included within duplicated regions, representing over 95% of all protein coding genes. This is almost twice the number of genes previously estimated to participate in duplicated regions of the yeast genome. Interestingly, 13 duplicate blocks lost all "relics" of their ancient duplication since they lack duplicate genes (remaining twin genes). Furthermore, additional 28 blocks contain only a single twin pair. Thus, only the complete synteny map between *A. gossypii* and *S. cerevisiae* allowed the identification of most likely all duplicate blocks.

3.5 Loss of *S. cerevisiae* genes after the genome duplication

The synteny map implies that 90% of the gene duplicates lost one copy after the duplication. This corresponds to approximately 4000 viable gene losses since the genome duplication event, which happened over 100 million years ago (Piskur 2001). Different mechanisms of gene loss can be envisaged: progressive loss of function by accumulation of missense mutations or one-step loss of function by a nonsense mutation, a frame-shift mutation or a deletion event (involving one or more genes). One rare example for a relic of a duplicate gene was found in a CLAS involving genes from chromosome II and IV. The sequence between YBR060C and YBR061C has homology to the ORF YDR037W but the putative twin ORF on chromosome II is altered by six frame-shifts. The ancient synteny with *A. gossypii* suggests that YDR037W is a member of the duplication block 18, which involves YBR058C to YBR065C and YDR037W to YDR045C (Fig. 4). Therefore, the YBR061C/YBR060C inter-ORF region most likely still carries the highly mutated twin copy of YDR037W. We re-sequenced that region in the *S. cerevisiae* strain that was used for the yeast genome project and confirmed the published sequence. Most *S. cerevisiae* inter-ORF sequences are too short to contain highly degenerate copies for the majority of lost twins. Hence, loss of most genes of duplicate pairs most likely occurred via single or multiple gene deletions and is probably still ongoing.

3.6 Synteny breaks as marker of genome rearrangements

The large extent of synteny between *A. gossypii* and *S. cerevisiae* provided a unique opportunity to evaluate the degree of genomic rearrangements that shaped both genomes after the speciation event (Fig. 5A). As a matter of fact, breaks of synteny reflect chromosomal rearrangements that took place during evolution of the two genomes. As depicted in Figure 5B, a synteny break affecting only one of two *S. cerevisiae* chromosomes, is indicative of a translocation or inversion event that occurred in the *S. cerevisiae* lineage after the genome duplication (Period **c**, Fig. 5A). On the other hand, double breaks of ancient synteny can affect both *S. cerevisiae* duplicate segments, as shown in Figure 5C, and point to a translocation or inversion event that took place in the *A. gossypii* lineage or in the *S. cerevisiae* lineage prior to the doubling event (Periods **a** or **b**, Fig. 5A).

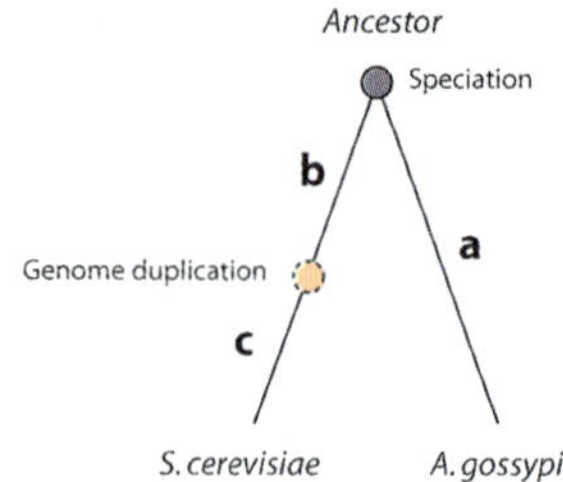

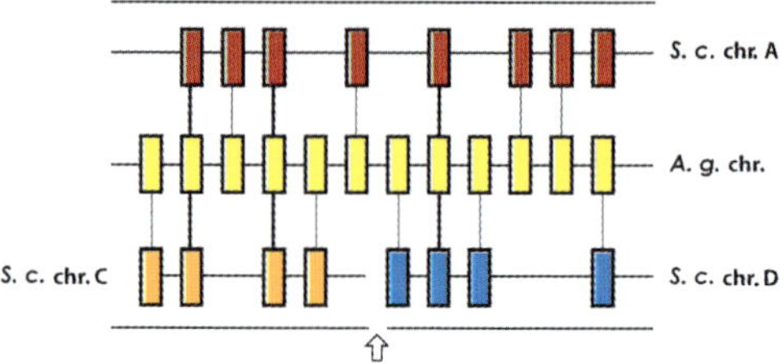

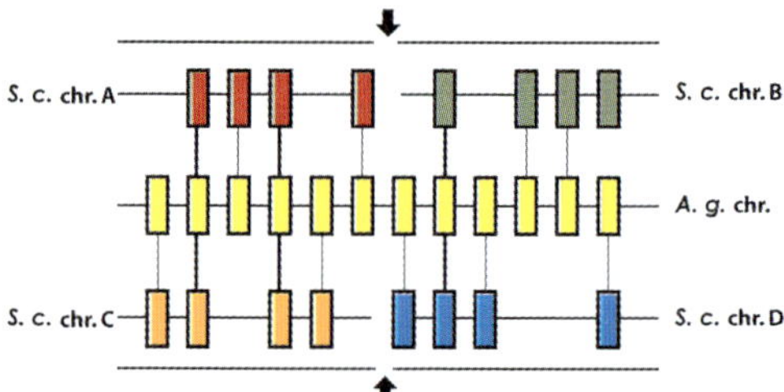

Fig. 5. Two types of synteny breaks define pre- and post-duplication rearrangements. (A) Evolution of *A. gossypii* and *S. cerevisiae*. Period **a** refers to the time between the speciation event and today. Periods **b** and **c** refer to pre- and post-duplication periods in the *S. cerevisiae* lineage. A dashed circle indicates the genome duplication event. (B) Single break of synteny (white arrow). Genomic rearrangements that occurred in phase **c** lead in two clusters of ancient synteny to reciprocal single synteny switches. One of these switches, from the blue to the orange chromosome, is shown. The other reciprocal switch can be searched for in the CLAS map (Dietrich et al. 2004). (C) Double break of synteny (black arrows). Genomic rearrangements in periods **a** and **b** lead in two clusters of ancient synteny to double synteny switches. One of these double synteny switches in the CLAS map is shown. The reciprocal double synteny switch is sometimes difficult to find in the CLAS map because subsequent rearrangements can involve previously used sites of rearrangements (Dietrich et al. 2004).

We determined the numbers of double and single synteny breaks on a whole genome scale (Dietrich et al. 2004). The map of all clusters of ancient synteny highlighted 328 double and 168 single breaks of synteny, suggesting 164 translocations/inversions during periods **a** and **b** and only 82 during period **c** (one genome rearrangement creates two break points). However, the number of genome rearrangements in period **c** is most likely higher because several single breaks of-synteny are sometimes masked by double breaks and because some sites of single breaks may represent an evolutionary hot spot for rearrangements, e.g., sites carrying a transposable element (Dietrich et al. 2004; Fischer et al. 2000). Indeed, we found that 58 synteny break points carry, in the *S. cerevisiae* genome, long terminal repeats of TY elements, which, in principle, can initiate genome rearrangements with over two hundred sites in *S. cerevisiae* chromosomes carrying these sequence repeats (Goffeau et al. 1996). A more realistic estimate taking into account repeated use of rearrangement sites in *S. cerevisiae* due to the presence of transposable elements predicts about 180 viable translocations or inversions in the *S. cerevisiae* lineage, 60 in period **b** and 120 in period **c** (Dietrich et al. 2004).

The map of reciprocal *S. cerevisiae* duplicate regions described in Chapter 3.4 revealed the presence of 126 duplication blocks (Fig. 4). Since inversions remain undetected in this figure, these blocks were presumably formed by successive reciprocal translocation events after the genome doubling and their number directly correlates with the number of these rearrangements. At the time of the genome duplication, the then sixteen chromosomes formed eight long duplicated regions. As each reciprocal translocation generates two additional duplicated regions, the 126 identified duplication blocks could have originated from some 60 independent translocations if the majority of the translocation endpoints were only used once. As the genome duplication took place approximately 100 millions years ago (Wolfe and Shields 1997; Piskur 2001), viable genome rearrangements occurred at very low frequency. This implies that those events are either rare or rarely maintained in the population due to detrimental effects for the organism. Our results corroborate an estimate of the number of post-duplication translocations by a simulation based on 55 blocks covering 50% of the yeast genome (Seoighe and Wolfe 1998). The authors estimated that between 70 and 100 events resulted in the 55 duplications blocks they had identified and that a total of 150 to 200 paired regions should be identifiable in the yeast genome from comparison to pre-duplication species.

4 Gene pairs (twins) originating from the genome duplication

4.1 Identification of twin ORFs

The ancient synteny map of both genomes allowed to identify 496 pairs of *S. cerevisiae* ORFs and their non-duplicated syntenic homologues in *A. gossypii* (see examples for the identification of four twin ORFs in Table 2). Each of these ORF pairs corresponds to duplicates that resulted from the whole genome duplication and which remained duplicated during evolution. This number of twin ORFs is significantly higher than the 406 pairs previously proposed (Wolfe and Shields 1997; Seoighe and Wolfe 1999). 367 pairs are shared between the two datasets. From the remaining 43 ORF pairs suggested by Wolfe and colleagues, 23 pairs correspond to ORFs lacking a homologue in *A. gossypii*. They are most likely real twin ORFs, which lost the syntenic *A. gossypii* homologue during evolution. Indeed, each of the 23 pairs can be assigned to syntenic positions in the yeast segments of clusters of ancient synteny, only that the *A. gossypii* gene segment lacks the homologue. In total, 129 novel pairs of *S. cerevisiae* twin ORFs were identified which represents a substantial amount of previously overlooked redundancy in the *S. cerevisiae* genome.

The earlier identification of genes duplicated in *S. cerevisiae* was based on an "all versus all" comparison of the *S. cerevisiae* protein-coding genes. To be counted as a duplicate pair, the respective sequences had to fulfil several criteria: Significant protein sequence homology (BLASTp E-value below 10^{-18}), less than 50 Kb distance between adjacent duplicate pairs of the same block, a minimum of three adjacent duplicate genes to confirm association of two chromosomal segments, and conserved gene orders and orientation in duplicate blocks. This understandably limited the identification of duplicate ORF pairs. Use of the ancient synteny map permitted detection of twins sharing homology below the previously used threshold. This is in agreement with the average amino acid identity between the 129 newly discovered twins, which is 10% lower compared to the original set of twins.

4.2 Genetic complexity caused by twin genes

Due to the absence of twin genes in *A. gossypii*, gene redundancy in this organism is restricted to a small number of tandem repeats and gene families. This simplifies for example the functional characterization of novel ORFs by reverse genetics because single gene knockouts show more often a phenotype compared to *S. cerevisiae*, making *A. gossypii* a promising fungal model organism. For example, the *A. gossypii* genome encodes only five homologues of CDK-regulating cyclins compared to nine in *S. cerevisiae*. Early cell cycle studies in budding yeast have shown that pairs of cyclins are functionally redundant (Futcher 1996) suggesting that the *A. gossypii* cell cycle may be regulated by simpler control system and that single deletions of cyclin genes may already affect this control system. Indeed, deletion of single B-type cyclin genes in *A. gossypii* results in strong phenotypes (K Hungerbühler and A Gladfelter, personal communication), whereas the deletion of three B-type cyclins is needed to lead to a defective cell cycle in budding yeast (Richardson et al. 1989; Futcher 1996). Additional examples of lack of functional redundancy in *A. gossypii* include ribosomal proteins, myosins, RAS GTPases, and guanine nucleotide exchange factors (GEFs) for small GTPases.

Table 3. List of novel *S. cerevisiae* twin gene pairs involving one functionally characterized gene and one gene of unknown function

| | *S. cerevisiae* twin 1 | | | | *S. cerevisiae* twin 2 | | |
Systematic name	Systematic name	Common name	SGD description line		Systematic name	common name	SGD description line
pAAL001W	YNL001W	DOM34	Involved in protein translation		YCL001W-A		Hypothetical ORF
pAAL061C	YDR300C	PRO1	gamma-glutamyl kinase		YHR033W		Hypothetical ORF
pAAL092C	YPL163C	SVS1	Serine and threonine rich protein.		YOR247W	SRL1	Suppressor of rad53 lethality
pACL072C	YLL048C	YBT1	bile acid transporter of ABC family		YHL035C		Hypothetical ORF
pACR052W	YKL148C	SDH1	flavoprotein subunit of succinate dehydrogenase		YJL045W		Hypothetical ORF
pACR165W	YKL020C	SPT23	suppressor protein		YIR033W	MGA2	Product of gene unknown
pADR233W	YOR083W	WHI5	Protein of unknown function		YKR091W	SRL3	Suppressor of rad53 lethality
pAEL159W	YJL065C		Hypothetical ORF		YBR278W	DPB3	C and C' subunits of DNA polymerase II
pAEL266W	YLR457C	NBP1	Nap1p binding protein		YPR174C		Hypothetical ORF
pAER122C	YLR437C		Hypothetical ORF		YML058W	SML1	Suppressor of mec lethality
pAER164C	YJR130C	STR2	Cystathionine gamma-synthase		YML082W		Hypothetical ORF
pAER190W	YCR011C	ADP1	Active transport ATPase		YOL075C		Hypothetical ORF
pAER232C	YHR030C	SLT2	serine/threonine MAP kinase		YKL161C		Hypothetical ORF
pAER279W	YMR264W	CUE1	Ubc7p binding and recruitment protein		YML101C	CUE4	Hypothetical ORF
pAFR325W	YGR142W	BTN2	Gene/protein whose expression is elevated in a btn1 minus/Btn1p lacking yeast strain.		YPR158W		Hypothetical ORF
pAFR606C	YFR013W	IOC3	ISW1 One Complex		YOL017W	ESC8	Hypothetical ORF
pAFR617C	YJL047C	RTT101	Regulator of Ty1 Transposition		YBR259W		Hypothetical ORF
pAFR684C	YNL278W	CAF120	CCR4 Associated Factor 120 kDa		YLR187W		Hypothetical ORF
pAGL082W	YKL035W	UGP1	Uridinephosphoglucose pyrophosphorylase		YHL012W		Hypothetical ORF
pAGL126C	YLR168C		possibly involved in intramitochondrial sorting		YDR185C		Hypothetical ORF
pAGL137W	YLR164W		Hypothetical ORF		YDR178W	SDH4	succinate dehydrogenase membrane anchor subunit

The majority of twin gene products have at least partially overlapping activities and functional analysis of twin genes will be complicated by this redundancy. The Yeast Gene Deletion Project provided phenotypic analysis for 451 pairs of twins (Giaever et al. 2002). For 413 pairs both twins are dispensable for cell survival and for 38 pairs one of the twins was found to be essential, the other not. This indicates that only 4% of the twins, compared to 15% for the whole genome, are actually essential, confirming a very high degree of functional redundancy of the duplicates. Furthermore, for none of the pairs both twin copies are essential for cell viability. The 38 pairs, for which only one of the twins is essential, might represent cases in which an essential gene duplicated and one copy functionally diverged due to mutations or due to significant decrease in the expression level, which prevents functional complementation. One example is the novel twin pair YDL239C and YNL225C with only 21% identical amino acids. YDL239Cp has been recently described as a component of the spindle pole body critical during meiosis (Nickas and Neiman 2002) but non-essential for mitosis. YNL225Cp has an essential function at the spindle pole body during mitosis (Brachat et al. 1998) and is also essential for meiosis. Table 3 lists 21 twin pairs for which only one of the two copies was functionally characterized and this knowledge is important to uncover the function of the other twin copy. Remarkably, for 40 novel twin pairs both copies are functionally uncharacterized. Only pairwise deletions of these twin genes might result in a detectable phenotype and may thus lead to the elucidation of previously unknown gene functions. Examples for such twin pairs were discovered when we screened the 497 twin genes for increased expression during meiosis taking advantage on microarray-based genome transcription profiling data (Primig et al. 2000). We found eight twin pairs for which both twin genes show increased expression levels during meiosis suggesting a redundant role in this process. For some of these twins, single deletion strain show no meiotic phenotype and we expect that double knockouts will affect meiosis.

4.3 Sequence divergence of twin genes

The complete inventory of *S. cerevisiae* twin genes is not only an important source of information on functional redundancy but also shows the whole spectrum of sequence evolution in twin genes. Gene duplication has long been recognized as an important mechanism for the creation of new gene functions. However, not all gene duplications result in the appearance of new functions as the new duplicates might remain redundant or one of the duplicates might be inactivated either by deletion or by accumulation of point mutations. Based on the extended inventory of *S. cerevisiae* twin genes, we re-investigated the evolution of their sequences. Importantly, the known syntenic *A. gossypii* homologues for each twin pair provided an essential reference for the degree of sequence divergence in the absence of a second gene copy. We performed a systematic sequence comparison of the twin gene products and their unique *A. gossypii* counterparts, creating a similarity tree for each of the "gene product sets". The comparison of these trees to the species tree allows the distinction of three types of protein set phylogenies: i) all the proteins in the set are equally distant to each other; ii) *S. cerevisiae* duplicates are closer to each other than to their *A. gossypii* counterparts; iii) the *A. gossypii* protein is more related to at least one of the twins than the twins are to each other. We defined the twin gene with the highest similarity to the *A. gossypii* homologue as twin 1. Figure 6 summarizes the results of the triple sequence comparisons. In 151

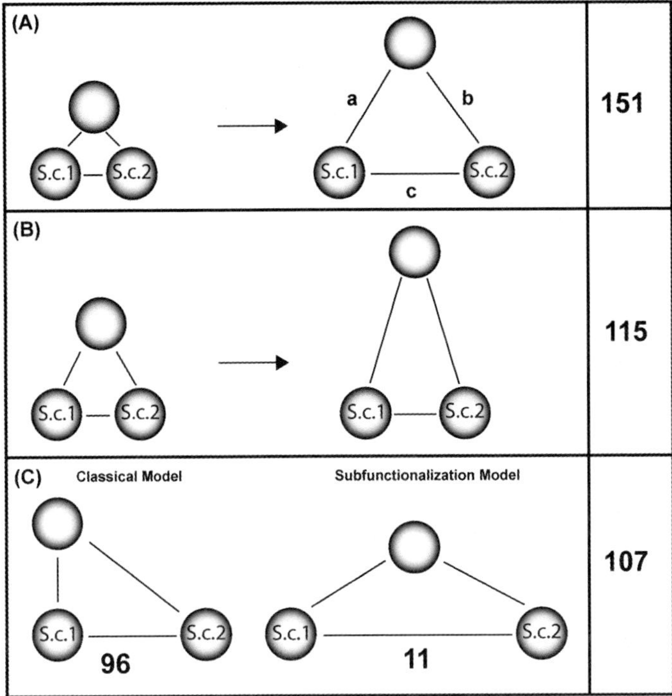

Fig. 6. Sequence divergence among *S. cerevisiae* twins and their *A. gossypii* homologues. (A) Triple sequence comparisons showing very similar levels of sequence divergence. 151 groups of twins plus the syntenic *A. gossypii* showed less than 5% similarity difference (344 when the threshold was relaxed to 10%). (B) Groups of proteins in which the twins were more diverged from the *A. gossypii* reference protein than from each other. 115 belong to this class when the threshold was set to 5%. (C) Groups of proteins in which twin 1 was more similar to *A. gossypii* reference protein than to its duplicate copy. These total 107 cases can be divided in two groups: 96 cases for which twin 2 is equally or more distant to the *A. gossypii* protein than to twin 1 suggesting a divergence as described by the classical model (Fig. 7). In the remaining 11 cases, both twins are closer to their *A. gossypii* homologue than to each other supporting the subfunctionalization model (Fig. 7).

cases the pairwise similarities are very close to each other (at the most 5% similarity difference) indicating similar selection pressures for the three genes. For 115 triple comparisons the twins are more closely related to each other than either of them to the *A. gossypii* homologue (5% similarity difference threshold). In these cases the *A. gossypii* gene may have been exposed to less selection pressure than the yeast counterparts resulting in a faster sequence divergence between the *A. gossypii* gene and both yeast twin genes than between the twin genes. Interestingly, 107 twin 1 proteins are more related to the *A. gossypii* homologue than to the respective twin 2 (5% similarity difference threshold) indicating a strong sequence divergence among the twins.

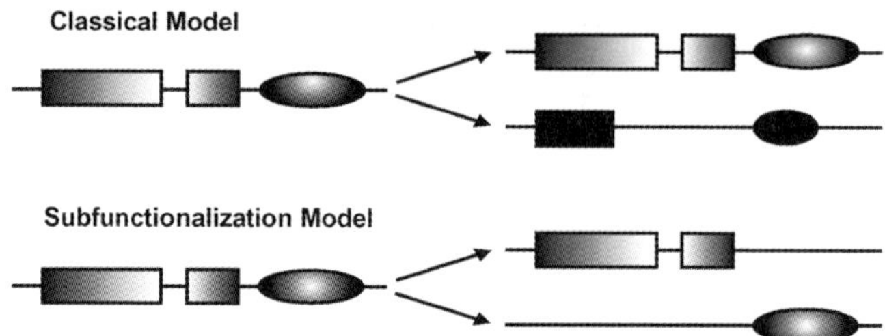

Fig. 7. Current models for the evolutionary divergence of duplicate genes. (A) The classical model suggests that one duplicate retains the ancestral characteristics whereas the second copy is allowed to freely diverge to acquire new properties. (B) In the subfunctionalization model, the ancestral functions are split among the two duplicates that can specialize.

It is generally believed that the presence of duplicates allows for diversification and adoption of new functions. The classical model proposes that one copy retains the original sequence and function while the second copy is allowed to diverge (Ohno 1970). An interesting alternative was put forward as the "subfunctionalization" model. In this model, both duplicates evolve and the ancestral function is-subdivided between the two duplicates (Force et al. 1999). Both models are summarized in Figure 7. The 107 divergent pairs of twin genes shown in Figure 6 provide important data for discussion of these models. For only eleven triple comparisons, we found that both twins are more closely related to the *A. gossypii* reference protein than the twins to each other (5% similarity difference threshold) as would be predicted by the sub-functionalization model. For the large majority of cases (96 out of 107), the *A. gossypii* reference protein is at least as distantly related to twin 2 than twin 1 is to twin 2. These results clearly favour the classical Ohno model for duplicate divergence.

This genome-scale analysis of gene duplicates together with pre-duplication reference genes provides strong evidence that different *S. cerevisiae* twin pairs, which arose at the same time, experienced diverse evolutionary fates and frequently evolved in an asynchronous manner. A similar analysis, based on the comparison of 38 duplicate *S. cerevisiae* gene pairs with homologues in other ascomycetes, already suggested this asynchrony (Langkjar et al. 2003).

5 Evolution of chromosome number in *A. gossypii*

Another evolutionary aspect that can be analyzed by the comparison of the *A. gossypii* and *S. cerevisiae* genomes concerns the evolution of the chromosome numbers in the two species. *S. cerevisiae* and closely related species bear sixteen chromosomes, implying that prior to the duplication event the ancestral genome carried eight chromosomes. The presence of only seven chromosomes in *A. gossypii* raises the question of how many chromosomes were present in the common ancestor of *S. cerevisiae* and *A. gossypii*. To address this question, we aligned all

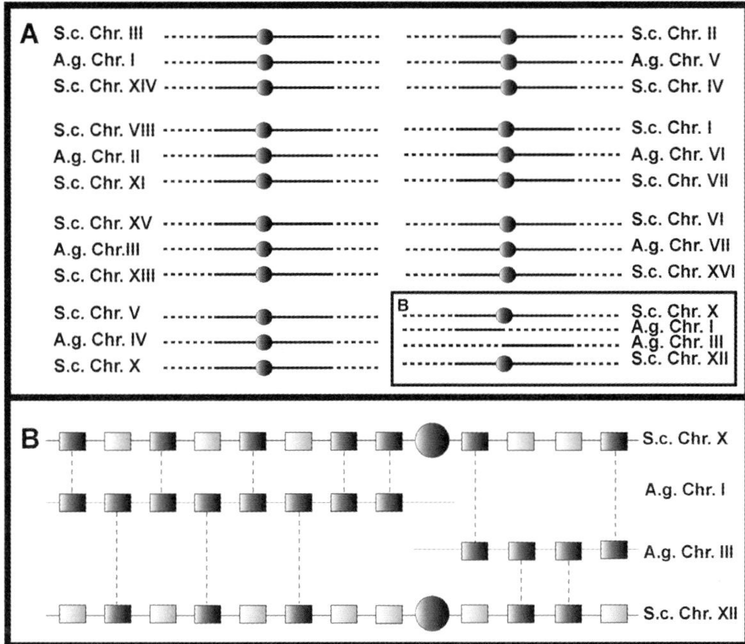

Fig. 8. Comparison of centromeric regions between *S. cerevisiae* and *A. gossypii*. (A) Evolutionary twin centromeres of *S. cerevisiae* were identified for fourteen out of sixteen chromosomes using synteny to *A. gossypii*. Syntenic sets of chromosomes are depicted in groups of three. The insert (B-1) schematically shows the alignment of the remaining two yeast chromosomes with two syntenic *A. gossypii* regions both lacking a centromere (pseudo-centromeric regions). (B-2) Pseudo-centromeric regions depicted in more detail, schematically showing the gene organization. Circles represent centromeres, genes are depicted by boxes and homologous genes are connected by dashed lines.

centromeric regions from *S. cerevisiae* to the *A. gossypii* genome. The seven *A. gossypii* centromere region align pairwise with fourteen centromere regions in *S. cerevisiae* as depicted in Figure 8A. The centromere regions of *S. cerevisiae* chromosomes X and XII did not align with any centromeric region in *A. gossypii*. But centromere-adjacent genes from the left arms of yeast chromosomes X and XII perfectly align with *A. gossypii* genes on chromosome I, and centromere-adjacent genes from the right arms of chromosomes X and XII align with *A. gossypii* genes on chromosome III (Fig. 8 B1 and B2). The pattern of ancient synteny of these alignments is reminiscent of a reciprocal translocation in the *A. gossypii* lineage. However, no relics of centromere sequences (pseudo-centromere) are present at the presumptive translocation sites in *A. gossypii*. Either, centromere sequences were never present in these regions, or they diverged beyond recognition.

The observed gene order and centromere alignment can be explained by two models depicted in Figures 9 and 10, respectively. As a first possibility, eight

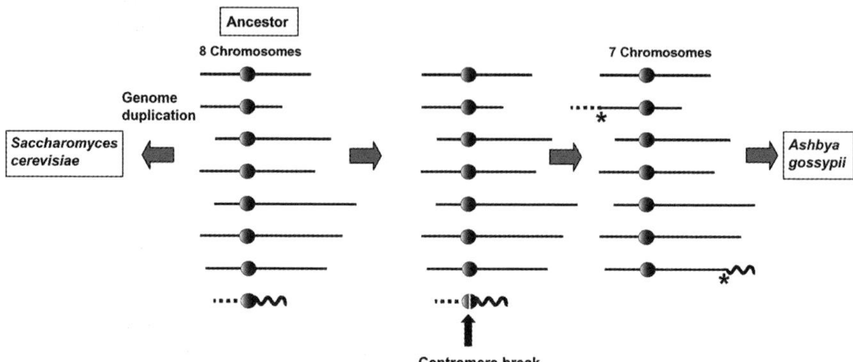

Fig. 9. Evolutionary model for an ancestral genome with eight chromosomes. One of the eight chromosomes was lost in the *A. gossypii* lineage by a centromere break, followed by the fusion of the two chromosome arms to the telomeric ends of remaining, intact chromosomes. The resulting genome consists of only seven chromosomes two of which with increased length. Asterisks label the positions of the two *A. gossypii* pseudo-centromeric regions.

chromosomes could have been present in the common ancestor and one chromosome was lost in the *A. gossypii* lineage. The breaking of a chromosome at its centromere would have resulted in the formation of two centromere-free chromosomal pieces. These could have been fused in a non-homologous manner to the ends of intact chromosomes (Fig. 9). As a second possibility, the common ancestor may have carried only seven chromosomes. As outlined in Figure 10, a centromere duplication event could have generated a dicentric chromosome in the *S. cerevisiae* lineage, causing chromosome breaks between the centromeres. Such breaks can be healed by addition of telomere sequences, thus, creating two smaller chromosomes as previously shown with artificially generated dicentric yeast chromosomes (Haber et al. 1984; Jäger and Philippsen 1989a). However, the current pattern of ancient synteny across the *A. gossypii* pseudo-centromere would only be observed if the position of centromere appearance in the *S. cerevisiae* lineage coincided with the position of a perfect reciprocal translocation in the *A. gossypii* lineage. This model is less likely as it implies two independent and different events at two syntenic loci in both genomes.

The analysis of the karyotype of other ascomycetes revealed a chromosome number ranging from six to sixteen (Jäger and Philippsen 1989b; Lankjaer et al. 2000). Species that diverged prior to genome duplication tend to have six or seven chromosomes. This suggests that alteration of the chromosome number was recurrent in the history of ascomycetes. The fact that many species that diverged earlier than *S. cerevisiae* and *A. gossypii* carry less than eight chromosomes could argue for a seven-chromosome ancestor. Although our comparative analysis of the two genomes favours the eight-chromosome hypothesis, further evidence will be needed to distinguish between the two possibilities, e.g., by investigating syntenic centromere and pseudo-centromere regions in other hemiascomycetes.

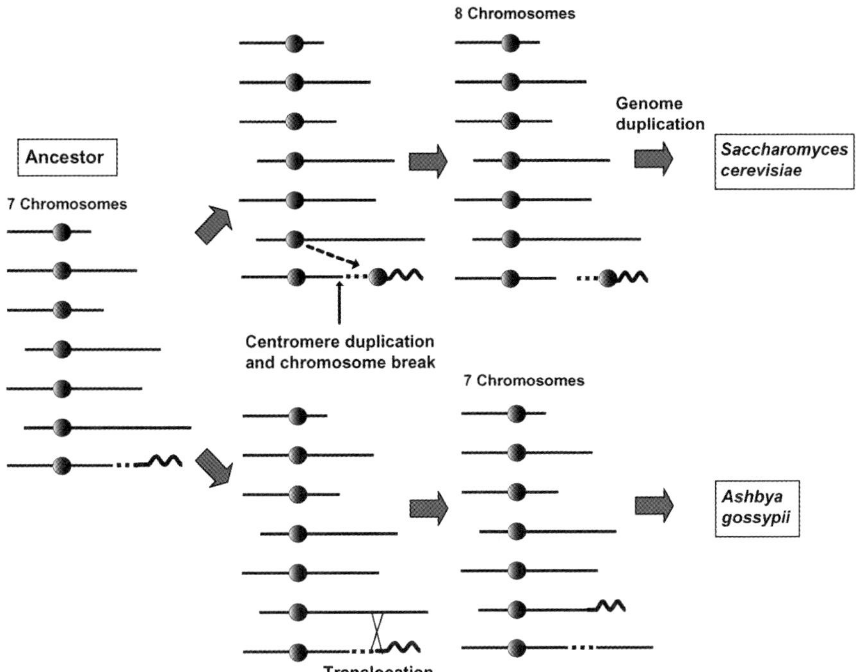

Fig. 10. Evolutionary model for an ancestral genome with seven chromosomes. Two independent events lead to the generation of the *A. gossypii* and *S. cerevisiae* ancestors respectively. In the *S. cerevisiae* lineage, a centromere duplication event, followed by a chromosome break gave rise to an organism with eight chromosomes that subsequently underwent a whole-genome duplication. In the *A. gossypii* lineage, a reciprocal translocation occurred (crossing lines) very close to the location of the centromere acquisition in the *S. cerevisiae* lineage.

6 Material and methods

6.1 Sequence and assembly quality

The quality of individual sequence reads was evaluated using Phred quality scores and the overall quality of the sequence assembly was evaluated using Phrap quality scores (Ewing and Green 1998; Ewing et al. 1998). When potential frame shift or stop codons were detected in protein coding genes, these sequences were inspected and edited or re-sequenced. Based on the Phrap scores >20 (Ewing and Green 1998; Ewing et al. 1998) and on our examination of the genomic sequence data, we estimate that the overall sequence coverage is approximately 4.2-fold and the average accuracy of the sequence data is 99.8%, with less than 4.5% of the genome covered by a single sequence read.

6.2 Annotation

Open reading frames (ORFs) larger than 150bp were extracted and searched against the *S. cerevisiae* protein dataset as available from SGD (Cherry et al. 1998; Saccharomyces Genome Database) using tFASTA (Pearson and Lipman 1988; Pearson 1994). ORFs with identity to *S. cerevisiae* proteins ranging from less than 20% to 100% were retained and automatically annotated. ORFs without similarity but larger than 450bp were also automatically annotated. Originally, the most upstream ATG was annotated as start codon but we later reported several adjustments to NCBI after experimental determination of several transcription start sites (S. Voegeli, unpublished). Initial analysis revealed that the intron splice rules for *A. gossypii* appear to be identical to those of *S. cerevisiae* (Rymond and Rosbash 1992; Davis et al. 2000). These splice rules where then used to identify and annotate putative intron containing genes. Annotations were subsequently manually checked. tRNAs, snRNAs, centromeres, and telomeres were annotated by homology to *S. cerevisiae*. The tRNA annotation was compared to tRNA-scan-SE results (Lowe 1997) but this did not lead to the identification of additional tRNA genes. As snRNA homology is in general weaker than that observed for tRNA, all *A. gossypii* genomic regions syntenic to the 72 snRNA positions in the *S. cerevisiae* genome were manually inspected for missed syntenic snRNAs. Inter-ORF regions were screened against a *S. cerevisiae* translated genome dataset using TBlastx and led us to the identification of overlooked protein-coding genes in both *A. gossypii* and *S. cerevisiae* (see Chapter 2). Resulting inter-ORF regions were then searched against other databases including *C. albicans* (Stanford Genome Technology Center), Génolevure (Llorente et al. 2000b), *N. crassa* (Neurospora Sequencing Project; Whitehead Institute/MIT Center for Genome Research; Galagan et al. 2003) and *S. pombe* (Wood et al. 2002). This allowed the annotation of *A. gossypii* genes without homologue in *S. cerevisiae* (NOHBYs) but having homologues in other species. Finally, remaining ORFs longer than 150 codons and non-overlapping other features were annotated as NOHBYs. Gene names were given following the *S. cerevisiae* nomenclature from the yeast genome consortium. Chromosome sequences were prepared for submission to GenBank using sequin (Benson et al. 2003). A minimal set of BAC and plasmid clones covering the genome sequence was extracted from the assembly information and clone ends were automatically annotated. Sequence quality information was derived from the assembly and used to extract regions of the genomes with less than 90% sequence confidence. These "low quality regions" were also automatically annotated. Genes having homologues in *S. cerevisiae* were manually assessed for synteny by two individuals and the label "syntenic homologue to *S. cerevisiae*" or "non-syntenic homologue to *S. cerevisiae*" was automatically added to the note descriptor of each gene. When more than one *S. cerevisiae* homologue was found for a single *A. gossypii* gene, the gene with the highest homology is always mentioned first in the description.

6.3 Data analysis

The data were converted from the GenBank format to the *A. gossypii* genome spreadsheet (Table 2) using the Readseq program (Readseq, http://iubio.bio.indiana.edu/soft/molbio/readseq/), which produces GFF (general feature format) tables, and manually edited in Excel®. Additional analyses of the data were done using BLAST (Altschul et al. 1990), Fasta (Pearson 1994), and the GCG® Wisconsin Package® (Accelerys). Gene conservation levels were obtained from systematic pairwise sequence alignments of homologous genes using Gap (GCG® Wisconsin Package® (Accelerys)). The map of the distribution of *A. gossypii* gene categories was created using GeneSpring®. Ancient synteny information was directly derived from the *A. gossypii* genome spreadsheet and manually drawn to produce the CLAS map. Gene viability information was taken from the Yeast Deletion Project data (Giaever et al. 2002) and functional information for the yeast genes were obtained from SGD (Cherry et al. 1998; Saccharomyces Genome Database) and the Gene Ontology Consortium (Ashburner et al. 2000). Sequence data was organized and maintained within local GCG® sequence databases. Annotation information and all data mining results were organized and maintained within a local FileMakerPro® database.

6.4 Creation of the map of Duplicate Blocks

The synteny information from the *A. gossypii* genome spreadsheet was used to associate each pair of *S. cerevisiae* chromosomal regions homologous and syntenic to single *A. gossypii* loci. *S. cerevisiae* genes were then reordered along the *S. cerevisiae* chromosomes. Duplication blocks were identified as stretches of the genome with paired regions involving similar chromosomes. Blocks were numbered starting from the left arm of chromosome I. First and last gene members of a block were used to label the block edges. Block edges were then mapped on the complete set of *S. cerevisiae* features to recover genes absent in *A. gossypii*. Duplicate block break points were manually inspected for the presence of transposable elements and LTRs. The duplicate block map was partially generated using GeneSpring®.

Acknowledgments

We thank A Lerch, K Gates, A Flavier, S Choi, R Wing, S Steiner, C Mohr, R Pöhlmann, Ph Luedi, Y Bauer, A Binder, K Gaudenz, S Goff, J Hoheisel, M Jacquot, P Knechtle, M Primig, C Rebischung, H-P Schmitz, J Wendland, and the Syngenta (formerly Novartis) sequencing facility at Research Triangle Park for their assistance with this project and for valuable discussions. We also acknowledge the help of Iza Kaminski in drawing ancient synteny maps. This work was supported by major funding obtained from the University of Basel and Syngenta

Biotechnology Inc. (formerly Novartis Agribusiness Biotechnology Research Inc). We also acknowledge support from the Duke University's Young Investigator Start-Up Fund.

References

Alberti-Segui C, Dietrich F, Altmann-Johl R, Hoepfner D, Philippsen P (2001) Cytoplasmic dynein is required to oppose the force that moves nuclei towards the hyphal tip in the filamentous ascomycete *Ashbya gossypii*. J Cell Sci 114:975-986

Altmann-Jöhl R, Philippsen P (1996) AgTHR4: a new selection marker for transformation of the filamentous fungus *Ashbya gossypii*, maps in a four-gene cluster that is conserved between *A. gossypii* and *Saccharomyces cerevisiae*. Mol Gen Genet 250:69-80

Altschul SF, Gish W, Miller W, Myers EW, Lipman DJ (1990) Basic local alignment search tool. J Mol Biol 215:403-410

Ashburner M, Ball CA, Blake JA, Botstein D, Butler H, Cherry JM, Davis AP, Dolinski K, Dwight SS, Eppig JT, Harris MA, Hill DP, Issel-Tarver L, Kasarskis A, Lewis S, Matese JC, Richardson JE, Ringwald M, Rubin GM, Sherlock G (2000) Gene ontology: tool for the unification of biology. The Gene Ontology Consortium. Nat Genet 25:25-29

Ashby SF, Nowell W (1926) The fungi of stigmatomycosis. Ann Botany 40:69-84

Ayad-Durieux Y, Knechtle P, Goff S, Dietrich F, Philippsen P (2000) A PAK-like protein kinase is required for maturation of young hyphae and septation in the filamentous ascomycete *Ashbya gossypii*. J Cell Sci 113(Pt 24):4563-4575

Benson DA, Karsch-Mizrachi I, Lipman DJ, Ostell J, Wheeler DL (2003) GenBank. Nucleic Acids Res 31:23-27

Bentley SD, Chater KF, Cerdeno-Tarraga AM, Challis GL, Thomson NR, James KD, Harris DE, Quail MA, Kieser H, Harper D, Bateman A, Brown S, Chandra G, Chen CW, Collins M, Cronin A, Fraser A, Goble A, Hidalgo J, Hornsby T, Howarth S, Huang CH, Kieser T, Larke L, Murphy L, Oliver K, O'Neil S, Rabbinowitsch E, Rajandream MA, Rutherford K, Rutter S, Seeger K, Saunders D, Sharp S, Squares R, Squares S, Taylor K, Warren T, Wietzorrek A, Woodward J, Barrell BG, Parkhill J, Hopwood DA (2002) Complete genome sequence of the model actinomycete *Streptomyces coelicolor* A3(2) Nature 417:141-147

Blandin G, Durrens P, Tekaia F, Aigle M, Bolotin-Fukuhara M, Bon E, Casaregola S, de Montigny J, Gaillardin C, Lepingle A (2000a) Genomic exploration of the hemiascomycetous yeasts: 4. The genome of *Saccharomyces cerevisiae* revisited. FEBS Lett 487:31-36

Blattner FR, Plunkett G 3rd, Bloch CA, Perna NT, Burland V, Riley M, Collado-Vides J, Glasner JD, Rode CK, Mayhew GF, Gregor J, Davis NW, Kirkpatrick HA, Goeden MA, Rose DJ, Mau B, Shao Y (1997) The complete genome sequence of *Escherichia coli* K-12. Science 277:1453-1474

Bosco G, Haber JE (1998) Chromosome break-induced DNA replication leads to non-reciprocal translocations and telomere capture. Genetics 150:1037-1047

Brachat A, Kilmartin JV, Wach A, Philippsen P (1998) *Saccharomyces cerevisiae* cells with defective spindle pole body outer plaques accomplish nuclear migration via half-bridge-organized microtubules. Mol Biol Cell 9:977-991

Brachat S, Dietrich FS, Voegeli S, Zhang Z, Stuart L, Lerch A, Gates K, Gaffney T, Philippsen P (2003) Reinvestigation of the *Saccharomyces cerevisiae* genome annotation by comparison to the genome of a related fungus: *Ashbya gossypii*. Genome Biol 4:R45

Bauer Y, Knechtle P, Wendland J, Helfer H, Philippsen P, (2004) A Ras-like GTPase is involved in hyphal growth guidance in the filamentous fungus *Ashbya gossypii*. Mol Biol Cell 15:4622-4632

Chavez R, Fierro F, Gordillo F, Francisco Martin J, Eyzaguirre J (2001) Electrophoretic karyotype of the filamentous fungus *Penicillium purpurogenum* and chromosomal location of several xylanolytic genes. FEMS Microbiol Lett 205:379-383

Cherry JM, Adler C, Ball C, Chervitz SA, Dwight SS, Hester ET, Jia Y, Juvik G, Roe T, Schroeder M, Weng S, Botstein D (1998) SGD: *Saccharomyces* Genome Database. Nucleic Acids Res 26:73-79

Cliften P, Sudarsanam P, Desikan A, Fulton L, Fulton B, Majors J, Waterston R, Cohen BA, Johnston M (2003) Finding functional features in *Saccharomyces* genomes by phylogenetic footprinting. Science 301:71-76

Cohn M, McEachern MJ, Blackburn EH (1998) Telomeric sequence diversity within the genus *Saccharomyces*. Curr Genet 33:83-91

Cole ST, Brosch R, Parkhill J, Garnier T, Churcher C, Harris D, Gordon SV, Eiglmeier K, Gas S, Barry CE 3rd, Tekaia F, Badcock K, Basham D, Brown D, Chillingworth T, Connor R, Davies R, Devlin K, Feltwell T, Gentles S, Hamlin N, Holroyd S, Hornsby T, Jagels K, Barrell BG (1998) Deciphering the biology of *Mycobacterium* tuberculosis from the complete genome sequence. Nature 393:537-544

Demain AL (1972) Riboflavin oversynthesis. Annu Rev Microbiol 26:369-388

Dietrich FS, Voegeli S, Brachat S, Lerch A, Gates K, Mohr C, Steiner S, Luedi P, Pöhlmann R, Flavier A, Choi S, Wing RA, Goff SA, Hoheisel JD, Gaffney T, Philippsen P (2004) *Ashbya gossypii* - experimental evidence of a whole genome duplication in *Saccharomyces cerevisiae*. Science 304:304-307

Ewing B, Green P (1998) Base-calling of automated sequencer traces using phred. II. Error probabilities. Genome Res 8:186-194

Ewing B, Hillier L, Wendl MC, Green P (1998) Base-calling of automated sequencer traces using phred. I. Accuracy assessment. Genome Res 8:175-185

Fischer G, James SA, Roberts IN, Oliver SG, Louis EJ (2000) Chromosomal evolution in *Saccharomyces*. Nature 405:451-454

Fischer G, Neuveglise C, Durrens P, Gaillardin C, Dujon B (2001) Evolution of gene order in the genomes of two related yeast species. Genome Res 11:2009-2019

Force A, Lynch M, Pickett FB, Amores A, Yan Y, Postlehwait J (1999) Preservation of duplicate genes by complementary, degenerative mutations. Genetics 151:1531-1545

Futcher B (1996) Cyclins and the wiring of the yeast cell cycle. Yeast 12:1635-1646

Galagan JE, Calvo SE, Borkovich KA, Selker EU, Read ND, Jaffe D, FitzHugh W, Ma LJ, Smirnov S, Purcell S, Rehman B, Elkins T, Engels R, Wang S, Nielsen CB, Butler J, Endrizzi M, Qui D, Ianakiev P, Bell-Pedersen D, Nelson MA, Werner-Washburne M, Selitrennikoff CP, Kinsey JA, Braun EL, Zelter A, Schulte U, Kothe GO, Jedd G, Mewes W, Staben C, Marcotte E, Greenberg D, Roy A, Foley K, Naylor J, Stange-Thomann N, Barrett R, Gnerre S, Kamal M, Kamvysselis M, Mauceli E, Bielke C, Rudd S, Frishman D, Krystofova S, Rasmussen C, Metzenberg RL, Perkins DD, Kroken S, Cogoni C, Macino G, Catcheside D, Li W, Pratt RJ, Osmani SA, DeSouza CP, Glass L, Orbach MJ, Berglund JA, Voelker R, Yarden O, Plamann M, Seiler S, Dunlap

J, Radford A, Aramayo R, Natvig DO, Alex LA, Mannhaupt G, Ebbole DJ, Freitag M, Paulsen I, Sachs MS, Lander ES, Nusbaum C, Birren B (2003) The genome sequence of the filamentous fungus *Neurospora crassa*. Nature 422:859-868

Giaever G, Chu AM, Ni L, Connelly C, Riles L, Veronneau S, Dow S, Lucau-Danila A, Anderson K, Andre B, Arkin AP, Astromoff A, El-Bakkoury M, Bangham R, Benito R, Brachat S, Campanaro S, Curtiss M, Davis K, Deutschbauer A, Entian KD, Flaherty P, Foury F, Garfinkel DJ, Gerstein M, Gotte D, Guldener U, Hegemann JH, Hempel S, Herman Z, Jaramillo DF, Kelly DE, Kelly SL, Kotter P, LaBonte D, Lamb DC, Lan N, Liang H, Liao H, Liu L, Luo C, Lussier M, Mao R, Menard P, Ooi SL, Revuelta JL, Roberts CJ, Rose M, Ross-Macdonald P, Scherens B, Schimmack G, Shafer B, Shoemaker DD, Sookhai-Mahadeo S, Storms RK, Strathern JN, Valle G, Voet M, Volckaert G, Wang CY, Ward TR, Wilhelmy J, Winzeler EA, Yang Y, Yen G, Youngman E, Yu K, Bussey H, Boeke JD, Snyder M, Philippsen P, Davis RW, Johnston M (2002) Functional profiling of the *Saccharomyces cerevisiae* genome. Nature 418:387-391

Goffeau A, Barrell BG, Bussey H, Davis RW, Dujon B, Feldmann H, Galibert F, Hoheisel JD, Jacq C, Johnston M, Louis EJ, Mewes HW, Murakami Y, Philippsen P, Tettelin H, Oliver SG (1996) Life with 6000 genes. Science 274:546, 563-567

Haber JE, Thorburn PC, Rogers D (1984) Meiotic and mitotic behavior of dicentric chromosomes in *Saccharomyces cerevisiae*. Genetics 106:185-205

Hieter P, Pridmore D, Hegemann J, Thomas M, Davis R, Philippsen P (1985) Functional selection and analysis of yeast centromeric DNA. Cell 42:913-921

Heus JJ, Zonneveld BJ, Steensma HY, Van den Berg JA (1990) Centromeric DNA of *Kluyveromyces lactis*. Curr Genet 18:517-522

Hermida L, Brachat S, Voegeli S, Philippsen P, Primig M (2005) The Ashbya Genome Database (AGD)-a tool for the yeast community and genome biologists. Database issue doi:10.1093/nar/gki009 33:D348-D352

Jäger D, Philippsen P (1989a) Stabilization of dicentric chromosomes in *Saccharomyces cerevisiae* by telomere addition to broken ends or by centromere deletion. EMBO J 8:247-254

Jäger D, Philippsen P (1989b) Many yeast chromosomes lack the telomere-specific Y' sequence. Mol Cell Biol 89:5754-5757

Kellis M, Patterson N, Endrizzi M, Birren BW, Lander ES (2003) Sequencing and comparison of yeast species to identify genes and regulatory elements. Nature 423:241-254

Kellis M, Birren BW, Lander E (2004) Proof and evolutionary analysis of ancient genome duplication in the yeast *Saccharomyces cerevisiae*. Nature 428:617-624

Kunst F, Ogasawara N, Moszer I, Albertini AM, Alloni G, Azevedo V, Bertero MG, Bessieres P, Bolotin A, Borchert S, Borriss R, Boursier L, Brans A, Braun M, Brignell SC, Bron S, Brouillet S, Bruschi CV, Caldwell B, Capuano V, Carter NM, Choi SK, Codani JJ, Connerton IF, Danchin A, et al. (1997) The complete genome sequence of the gram-positive bacterium *Bacillus subtilis*. Nature 390:249-256

Kupfer DM, Reece CA, Clifton SW, Roe BA, Prade RA (1997) Multicellular ascomycetous fungal genomes contain more than 8000 genes. Fungal Genet Biol 21:364-372

Kurtzman CP, Robnett C (2003) Phylogenetic relationships among yeasts of the "Saccharomyces complex" determined from multigene sequence analyses. FEMS Yeast Research 3:417-432

Langkjar RB, Nielsen ML, Daugaard PR, Liu W, Piskur J (2000) Yeast chromosomes have been significantly reshaped during their evolutionary history. J Mol Biol 304:271-288

Langkjar RB, Cliften PF, Johnston M, Piskur J (2003) Yeast genome duplication was followed by asynchronous differentiation of duplicated genes. Nature 421:848-852

Llorente B, Malpertuy A, Blandin G, Artiguenave F, Wincker P, Dujon B (2000a) Genomic exploration of the hemiascomycetous yeasts: 12. *Kluyveromyces marxianus var. marxianus*. FEBS Lett 487:71-75

Llorente B, Malpertuy A, Neuveglise C, de Montigny J, Aigle M, Artiguenave F, Blandin G, Bolotin-Fukuhara M, Bon E, Brottier P, Casaregola S, Durrens P, Gaillardin C, Lepingle A, Ozier-Kalogeropoulos O, Potier S, Saurin W, Tekaia F, Toffano-Nioche C, Wesolowski-Louvel M, Wincker P, Weissenbach J, Souciet J, Dujon B (2000b) Genomic exploration of the hemiascomycetous yeasts: 18. Comparative analysis of chromosome maps and synteny with *Saccharomyces cerevisiae*. FEBS Lett 487:101-112

Lowe TM, Eddy SR (1997) tRNAscan-SE: a program for improved detection of transfer RNA genes in genomic sequence. Nucleic Acids Res 25:955-964

Melnick L, Sherman F (1993) The gene clusters ARC and COR on chromosomes 5 and 10: respectively, of *Saccharomyces cerevisiae* share a common ancestry. J Mol Biol 233:372-388

Mohr C (1995) Genetic engineering of the filamentous fungus *Ashbya gossypii*: construction of a genomic library, isolation of genes for beta-isopropylmalate-dehydrogenase (LEU2) and a protein kinase (APK1) by heterologous complementation, and characterization of non-reverting mutants. PhD thesis. Applied microbiology. Universitaet Basel

National Center for Biotechnology Information. http://www.ncbi.nih.nlm.gov

Ness F, Aigle M (1995) RTM1: a member of a new family of telomeric repeated genes in yeast. Genetics 140:945-956

Neurospora Sequencing Project. http://www-genome.wi.mit.edu

Nickas ME, Neiman AM (2002) Ady3p links spindle pole body function to spore wall synthesis in *Saccharomyces cerevisiae*. Genetics 160:1439-1450

Ohno S (1970) Evolution by gene duplication. Springer-Verlag Heidelberg

Osiewacz HD, Ridder R (1991) Genome analysis of imperfect fungi: electrophoretic karyotyping and characterization of the nuclear gene coding for glyceraldehyde-3-phosphate dehydrogenase (gpd) of *Curvularia lunata*. Curr Genet 20:151-155

Panzeri L, Landonio L, Stotz A, Philippsen P (1985) Role of conserved sequence elements in yeast centromere DNA. EMBO J 4:1867-1874

Pearson WR (1994) Using the FASTA program to search protein and DNA sequence databases. Methods Mol Biol 25:365-389

Pearson WR, Lipman DJ (1988) Improved tools for biological sequence comparison. Proc Natl Acad Sci USA 85:2444-2448

Philippsen P, Kleine K, Pohlmann R, Dusterhoft A, Hamberg K, Hegemann JH, Obermaier B, Urrestarazu LA, Aert R, Albermann K, Altmann R, Andre B, Baladron V, Ballesta JP, Becam AM, Beinhauer J, Boskovic J, Buitrago MJ, Bussereau F, Coster F, Crouzet M, D'Angelo M, Dal Pero F, De Antoni A, Hani J et al. (1997) The nucleotide sequence of *Saccharomyces cerevisiae* chromosome XIV and its evolutionary implications. Nature 387:93-98

Philippsen P, Kaufmann A, Schmitz H-P (2005) Homologues of yeast polarity genes control the development of multinucleated hyphae in *Ashbya gossypii*. Curr Opin Microbiol 8:370-377

Piskur J (2001) Origin of the duplicated regions in the yeast genomes. Trends Genet 17:302-303

Pöhlmann R (1996) Computational evalution of systematic and random genomic sequence data of the yeast *Saccharomyces cerevisiae* and the filamentous fungus *Ashbya gossypii*. PhD thesis. Applied Microbiology. Universitaet Basel

Prillinger H, Schweigkofler W, Breitenbach M, Briza P, Staudacher E, Lopandic K, Molnar O, Weigang F, Ibl M, Ellinger A (1997) Phytopathogenic filamentous (*Ashbya, Eremothecium*) and dimorphic fungi (Holleya, Nematospora) with needle-shaped ascospores as new members within the *Saccharomycetaceae*. Yeast 13:945-960

Primig M, Williams RM, Winzeler EA, Tevzadze GG, Conway AR, Hwang SY, Davis RW, Esposito RE (2000) The core meiotic transcriptome in budding yeasts. Nat Genet 26:415-423

Richardson HE, Wittenberg C, Cross F, Reed SI (1989) An essential G1 function for cyclin-like proteins in yeast. Cell 59:1127-1133

Rymond BC, Rosbash M (1992) The molecular and cellular biology of the yeast *Saccharomyces*; volume 2; gene expression. Cold Spring Harbor Laboratory Press

Saccharomyces Genome Database http://genome-www.stanford.edu/Saccharomyces/

Seoighe C, Wolfe KH (1998) Extent of genomic rearrangement after genome duplication in yeast. Proc Natl Acad Sci USA 95:4447-4452

Seoighe C, Wolfe KH (1999) Updated map of duplicated regions in the yeast genome. Gene 238:253-261

Singer MS, Gottschling DE (1994) TLC1: template RNA component of *Saccharomyces cerevisiae* telomerase. Science 266:404-409

Steiner S, Philippsen P (1994) Sequence and promoter analysis of the highly expressed TEF gene of the filamentous fungus *Ashbya gossypii*. Mol Gen Genet 242:263-271

Steiner S, Wendland J, Wright MC, Philippsen P (1995) Homologous recombination as the main mechanism for DNA integration and cause of rearrangements in the filamentous ascomycete *Ashbya gossypii*. Genetics 140:973-987

Stover CK, Pham XQ, Erwin AL, Mizoguchi SD, Warrener P, Hickey MJ, Brinkman FS, Hufnagle WO, Kowalik DJ, Lagrou M, Garber RL, Goltry L, Tolentino E, Westbrock-Wadman S, Yuan Y, Brody LL, Coulter SN, Folger KR, Kas A, Larbig K, Lim R, Smith K, Spencer D, Wong GK, Wu Z, Paulsen IT, Reizer J, Saier MH, Hancock RE, Lory S, Olson MV (2000) Complete genome sequence of *Pseudomonas aeruginosa* PA01: an opportunistic pathogen. Nature 406:959-964

Wendland J, Ayad-Durieux Y, Knechtle P, Rebischung C, Philippsen P (2000) PCR-based gene targeting in the filamentous fungus *Ashbya gossypii*. Gene 242:381-391

Wendland J, Philippsen P (2000) Determination of cell polarity in germinated spores and hyphal tips of the filamentous ascomycete *Ashbya gossypii* requires a rhoGAP homolog. J Cell Sci 113:1611-1621

Wendland J, Philippsen P (2001) Cell polarity and hyphal morphogenesis are controlled by multiple rho-protein modules in the filamentous ascomycete *Ashbya gossypii*. Genetics 157:601-610

Wendland J, Philippsen P (2002) An IQGAP-related protein, encoded by AgCYK1: is required for septation in the filamentous fungus *Ashbya gossypii*. Fungal Genet Biol 37:81

Wendland J, Pöhlmann R, Dietrich F, Steiner S, Mohr C, Philippsen P (1999) Compact organization of rRNA genes in the filamentous fungus *Ashbya gossypii*. Curr Genet 35:618-625

Wolfe KH, Shields DC (1997) Molecular evidence for an ancient duplication of the entire yeast genome. Nature 387:708-713

Wolfe K (2004) Evolutionary genomics: yeasts accelerate beyond BLAST. Current Biol 14:R392-R394

Wood V, Gwilliam R, Rajandream MA, Lyne M, Lyne R, Stewart A, Sgouros J, Peat N, Hayles J, Baker S, Basham D, Bowman S, Brooks K, Brown D, Brown S, Chillingworth T, Churcher C, Collins M, Connor R, Cronin A, Davis P, Feltwell T, Fraser A, Gentles S, Goble A, Hamlin N, Harris D, Hidalgo J, Hodgson G, Holroyd S, Hornsby T, Howarth S, Huckle EJ, Hunt S, Jagels K, James K, Jones L, Jones M, Leather S, McDonald S, McLean J, Mooney P, Moule S, Mungall K, Murphy L, Niblett D, Odell C, Oliver K, O'Neil S, Pearson D, Quail MA, Rabbinowitsch E, Rutherford K, Rutter S, Saunders D, Seeger K, Sharp S, Skelton J, Simmonds M, Squares R, Squares S, Stevens K, Taylor K, Taylor RG, Tivey A, Walsh S, Warren T, Whitehead S, Woodward J, Volckaert G, Aert R, Robben J, Grymonprez B, Weltjens I, Vanstreels E, Rieger M, Schafer M, Muller-Auer S, Gabel C, Fuchs M, Fritzc C, Holzer E, Moestl D, Hilbert H, Borzym K, Langer I, Beck A, Lehrach H, Reinhardt R, Pohl TM, Eger P, Zimmermann W, Wedler H, Wambutt R, Purnelle B, Goffeau A, Cadieu E, Dreano S, Gloux S, Lelaure V, Mottier S, Galibert F, Aves SJ, Xiang Z, Hunt C, Moore K, Hurst SM, Lucas M, Rochet M, Gaillardin C, Tallada VA, Garzon A, Thode G, Daga RR, Cruzado L, Jimenez J, Sanchez M, del Rey F, Benito J, Dominguez A, Revuelta JL, Moreno S, Armstrong J, Forsburg SL, Cerrutti L, Lowe T, McCombie WR, Paulsen I, Potashkin J, Shpakovski GV, Ussery D, Barrell BG, Nurse P (2002) The genome sequence of *Schizosaccharomyces pombe*. Nature 415:871-880

Wood V, Rutherford KM, Ivens A, Rajandream MA, Barrell B (2001) A reannotation of the *Saccharomyces cerevisiae* genome. Comp Func Genomics 2:143-154

Wright MC, Philippsen P (1991) Replicative transformation of the filamentous fungus *Ashbya gossypii* with plasmids containing *Saccharomyces cerevisiae* ARS elements. Gene 109:99-105

Whiteway M, Oberholzer U (2004) Candida morphogenesis and host-pathogen interactions. Curr Opin Microbiol 7:350-357

Brachat, Sophie
Biozentrum Universität Basel, Klingelbergstrasse 50, CH-4056 Basel, Switzerland and Novartis-Pharma (Basel)

Dietrich, Fred
Biozentrum Universität Basel, Klingelbergstrasse 50, CH-4056 Basel, Switzerland and Department of Molecular Genetics and Microbiology, Duke University, Medical Center, Research Drive, Durham, NC 27710, USA

Gaffney, Tom
Syngenta Biotechnology Inc., 3054 Cornwallis Road, Research Triangle Park, NC 27709, USA

Philippsen, Peter
Biozentrum Universität Basel, Klingelbergstrasse 50, CH-4056 Basel, Switzerland
Peter.Philippsen@unibas.ch

Voegeli, Sylvia
Biozentrum Universität Basel, Klingelbergstrasse 50, CH-4056 Basel, Switzerland

Schizosaccharomyces pombe comparative genomics; from sequence to systems

Valerie Wood

Abstract

The fission yeast *Schizosaccharomyces pombe* is becoming increasingly important as a model for the characterization and study of many globally conserved genes, second only in importance to the budding yeast *Saccharomyces cerevisiae*. This chapter provides an updated inventory of gene number and genome contents for fission yeast compared to budding yeast. Functional and comparative genomics studies, and the insights these have provided into how the different genome contents of these two yeasts are manifested in their individual biologies are reviewed. Phylogenetic analysis, comparative genomics and experimental research support the choice of *S. pombe* as a model for the dissection of many biological processes, which are often more similar to the analogous processes in higher eukaryotes than those of the *Saccharomytina*. The review underlines the advantages of exploiting this organism through the integration of bench science, functional genomics, phylogenomics and systems biology in order to identify and interpret the minimal requirements for a eukaryotic cell.

1 Introduction

Schizosaccharomyces pombe, or fission yeast, is a simple unicellular archiascomycete fungus. It was established as a model organism by the influential work which culminated in a universal model for control of the cell cycle (reviewed in Nurse 2000). The fission yeast and its distant relative, budding yeast (*Saccharomyces cerevisiae*), are estimated to have diverged 330-420 million years ago; in comparison to the metazoan split which is estimated to have occurred 1000-1200 million years ago (Berbee and Taylor 1993; Lum et al. 1996). Other estimates propose a radical adjustment of these figures to 1,144 and 1600 million years ago respectively (Heckman et al. 2001). Despite the variation in the predicted time of divergence, phylogenetic analyses and anecdotal evidence indicate that *S. pombe* gene sequences are often more similar to their mammalian counterparts than the equivalent *S. cerevisiae* genes (reviewed in Sipiczki 2001).

Completion of the *S. cerevisiae* genome sequence in 1996, was a landmark that changed the nature of experimental biology for this organism (Goffeau et al. 1996). The availability of the genome sequence of *S. pombe* has similarly revolutionised research for the expanding fission yeast community and made possible

Topics in Current Genetics, Vol. 15
P. Sunnerhagen, J. Piškur (Eds.): Comparative Genomics
DOI 10.1007/4735_97 / Published online: 2 August 2005
© Springer-Verlag Berlin Heidelberg 2005

tionised research for the expanding fission yeast community and made possible the first global comparative genomics of two free living fungal species (Wood et al. 2002). The completed *S. pombe* genome, coupled with the features which have made it a popular experimental model (sophisticated technologies for molecular and cell biology and well developed genetic techniques), also make it an attractive target organism for functional genomics and global systems approaches.

The evolutionary distance between these two yeasts allows their differing genome contents to be usefully compared and evaluated, not only to interpret their individual evolutionary histories in terms of functionality, but often to extrapolate these findings to higher eukaryotic systems. For, despite the length of time since fission yeast and budding yeast shared a common ancestor with humans, both organisms provide excellent experimental models for many essential eukaryotic processes because the majority of genes from both yeasts have predicted homologs in multicellular eukaryotes (Wood et al. 2002)[1]. A previous comparison between *S. cerevisiae* and *Caenorhabditis elegans,* using different thresholds, predicted that a minimum of 40% of budding yeast genes had a homolog in multicellular eukaryotes (Chervitz et al. 1998)[2]. Chervitz and colleagues also proposed that most core biological functions are carried out by orthologous pairs of conserved genes. Furthermore, they demonstrated that orthologs could usually be reliably identified on a genome-wide basis by simple sequence comparisons, even within families of highly similar proteins with many members. Initial comparisons using *S. pombe* also showed that genes which were highly conserved between the animal and plant kingdoms were also almost always conserved in both yeasts (Wood et al. 2002). These observations continue to be supported by the characterisation of many conserved genes involved in processes fundamental to the maintenance of all eukaryotic cells. Significantly, but not surprisingly, many universally conserved genes are required for genome stability, and their mutated forms are often implicated in human cancers.

Perhaps unexpectedly, considering its smaller proteome, substantial numbers of broadly conserved proteins are completely absent from *S. cerevisiae* but are present in *S. pombe* (Aravind et al. 2000). Consequently, when gene products conserved in higher eukaryotes are absent from the budding yeast but present in the fission yeast, the fission yeast processes display closer functional correspondence to those of more complex organisms. These processes include centromere structure and function (Kniola et al. 2001; Appelgren et al. 2003), RNA interference and heterochromatin formation (Volpe et al. 2002; Hall et al. 2002), nuclear mRNA splicing (Käufer and Potashkin 2000; Kuhn and Käufer 2003; Webb and Wise 2004), certain aspects of cell cycle progression (Mundt et al. 1999), and telomere function (Kanoh and Ishikawa 2003). However, because of the subtle nature of the variations in many of the regulatory circuits controlling these processes, ultimately both the similarities and differences between these two yeasts will continue to be informative for the understanding of basic biological phenomena (Forsburg 1999).

---

[1] using BLASTP with a cut-off E-value of 0.001
[2] using BLASTP with a cut-off P-value of 10^{-10}

S. pombe has lower protein redundancy than *S. cerevisiae* (inferred by fewer duplicate genes). This partially explains the apparent closer similarity of *S. pombe* to higher eukaryotes, because duplication is often accompanied by divergence (Langkjaer 2003; Kellis et al. 2004). Significantly, the evolution of some duplicated *S. cerevisiae* genes appears to have played a direct role in the transition to a fermentative lifestyle (Piskur 2001). Although *S. cerevisiae* will continue to be the most intensively studied because of its enormous industrial importance; *S. pombe* is more likely to resemble the cellular content of the common ancestor and may prove to be more suitable for the functional analysis of certain genes.

This chapter provides an updated inventory of the gene number and genome content of the fission yeast, *S. pombe,* as compared to the budding yeast, *S. cerevisiae,* and emphasises the importance of continual sequence analysis for the refinement of the primary data. Genome features and contents are interpreted in the context of published experimental research. The available functional and comparative genomics studies, and the associated insights into how the differing genome contents of these two yeasts are manifested in their individual biologies are reviewed. These include studies of transposon content, gene organization and regulation, microarray expression studies, proteome comparisons and orthology mapping. Finally, an overview of the current status of genome annotation and literature curation using Gene Ontology (GO) descriptors and a summary of the global similarities and differences between the 'high level' biological processes of these two important model yeasts are presented.

Phylogenetic analyses, comparative genomics and experimental research into chromosome structure and organization support the choice of *S. pombe* for the dissection of many processes which appear to be more similar to analogous processes in higher eukaryotes than to those of the *Saccharomycotina.* Drawing the cumulative body of research to date into a single unified review emphasises the advantages of exploiting this organism by the integration of bench science, functional genomics, phylogenomics and systems biology approaches in order to identify and interpret the minimal requirements for a single eukaryotic cell.

2 Genome features

2.1 Genome size and sequencing status

The *S. pombe* genome size was estimated to be 13.8 Mb by restriction mapping, compared to the 13.0 Mb genome of *S. cerevisiae* (Fan et al. 1988; Smith et al. 1987). Although the genome sizes are similar, *S. pombe* has only three chromosomes compared to *S. cerevisiae's* 16; their sizes being 5.7, 4.6 and 3.5 Mb for chromosomes I, II and III respectively. The smallest *S. pombe* chromosome is therefore over twice the length of the longest *S. cerevisiae* chromosome (1.5 Mb). For *S. cerevisiae* the increased chromosome number and smaller size is a consequence of the proposed whole genome duplication events in some yeast lineages. Most of the species which lie on the deeper branches of the ascomycete phylogeny have haploid chromosome numbers between six and eight. This implies an approximate doubling in the *Saccharomyces (sensu stricto)* group (Wolfe and Shields 1997; Keogh et al. 1998). Duplication appears to be accompanied by downsizing through deletion, because although chromosome number is often increased, total genome size is broadly similar. For *S. pombe* the lower chromosome number and larger size may indicate an absence of whole genome duplication events.

The contiguated fission yeast sequence is 12 571 419 bases, arranged in seven contigs with four sequence gaps (two centromeric and two telomeric). The published genome sequence excludes the ribosomal DNA (rDNA) repeats which are present in two tandem arrays on chromosome III. These arrays are estimated to be 1225 kb and 240 kb in size for the sequenced strain (972 h-), although dramatic length polymorphisms between closely related strains are reported for these regions (Pasero and Marilley 1993). The unsequenced subtelomeric regions for chromosomes I and II are approximately 80 kb +/- 20 kb (R. Hyppa and G. Smith, personal communication). The centromeric gaps are estimated to be less than 36 kb and are restricted to known repeats.

The sequenced genome size, together with estimated sizes of the unsequenced elements is 14.1 Mb, and compares well with the 13.8 Mb calculated earlier from *Not* I fragment sizes (see above). The estimated sizes of chromosome I and II are almost identical to earlier approximations. The majority of the observed size difference is between the chromosome III totals and may be due to the variable nature of the rDNA repeats.

The composite sequence is expected to be missing only repetitive regions, there should therefore be little, if any, unique sequence excluded from the present data[3]. Efforts are continuing to sequence the remaining centromeric and telomeric gaps, and the sequence status is continually updated at http://www.sanger.ac.uk/Projects/ S_pombe/status.shtml

2.2 Centromeres

The basic structure of the centromeres and their approximate sizes were determined prior to complete genome sequencing by Southern blotting and partial sequencing (Chikashige et al. 1989; Clarke and Baum 1990; Murakami et al. 1991). Centromeres 1, 2 and 3 were estimated at 40, 69 and 110 kb, respectively. These sizes are inversely proportional to the length of the chromosomes at 5.7, 4.6 and 3.5 Mb, and their structure was verified by the genome sequence. The centromere structure comprises a non-conserved central core sequence (*cnt*) flanked by inverted repeats (*Imr*L and *Imr*R) that display sequence identity with each other (Takahashi et al. 1992). These central elements are flanked by variable numbers of outer repeats (*otr*L and *otr*R). Initial studies showed that the central core is essential, but not sufficient for centromeric function, and at least a portion of the outer

---

[3] Based on the assumption that no unique protein coding genes exist at the telomeres of chromosome III proximal to the rDNA repeats.

repeat is required (Takahashi et al. 1992). These repeats contain a highly conserved region, *dg* or K, which was found to be critical, and additional repeats were
shown to have a positive effect on minichromosome stability (Baum et al. 1994).
The complex and diffuse centromere structure of fission yeast is more reminiscent
of higher organisms than of the 125 base pair structurally conserved element is
sufficient for centromere function in *S. cerevisiae* (Fitzgerald-Hayes et al. 1982).

Work on centromere and kinetochore function in fission yeast is beginning to
dissect the biological basis for these structural differences. Several centromeric
proteins have been identified in *S. pombe* which are conserved in mammals but are
absent from *S. cerevisiae*, including Swi6 and Chp1 (Lorentz et al. 1994; Ekwall
et al. 1995; Doe et al. 1998). These proteins have, like their mammalian counterparts, been linked with distinct structural and functional domains. Recently, an
important link has been made between the formation of silent centromeric heterochromatin and the RNA interference (RNAi) machinery (which is absent from *S.
cerevisiae* but conserved in plants, insects and mammals; Volpe et al. 2003). It is
proposed that small interfering RNAs (siRNAs) are generated from centromeric
double-stranded RNAs by the RNAi machinery. These siRNAs induce the formation of heterochromatin in the centromeric regions by targeting repetitive DNA
and directing its methylation.

The conservation of features including size, structure and multilayered organization have led to the suggestion that the fission yeast centromere represents the
basic modular structure of complex centromeric DNA in higher eukaryotes
(Kniola et al. 2001; Appelgren et al. 2003). These common features and the involvement of the RNAi components (essential for heterochromatin formation in
vertebrate cells) are inevitably making fission yeast a valuable model for eukaryotic chromatin remodelling and centromere function.

2.3 Subtelomeric regions

Approximately 50-60 kb of the region immediately proximal to the telomeric repeats of all four of the sequenced subtelomeric regions is highly similar (~99%
sequence identity for most of the regions). This is consistent with the observation
that the telomeres of fission yeast and other eukaryotes are known to cluster at
meiotic prophase (Chikashige et al. 1994; Scherthan et al. 1994), because telomere
clustering may promote the more frequent exchange of genetic information which
appears to occur in these regions (reviewed in Scherthan 2001).

One striking feature is a large (6.3 Kb) open reading frame (ORF),
SPAC212.11, with homology to ReqQ helicases present at the ends of the two
fully sequenced chromosome arms immediately proximal to the degenerate telomeric repeats. This helicase has recently been shown to be highly expressed in rare
survivors of crisis in telomerase mutants (Mandell et al. 2004; Mandell et al.
2005). There are also 19 highly conserved, telomere associated, Y' elements in *S.
cerevisiae*, containing a predicted helicase domain which have similarly been implicated in the maintenance of telomeres in telomerase defective populations of *S.
cerevisiae* (Louis and Haber 1998; Yamanda et al. 1998; Maxwell et al. 2004).

This is the only example of a conserved protein function and genomic location between the two yeasts. The *S. pombe* RecQ helicase appears to be partially transcriptionally regulated by RNAi, suggesting that this mechanism also operates at the telomeres (Mandell et al. 2004).

The subtelomeric regions of *S. pombe* appear to contain an increased density of species-specific predicted cell-surface glycoprotein families relative to the whole genome (Wood et al. 2002). Similarly, the *S. cerevisiae* Seripauperin and TIP or PAU family (26 members) and COS/DUP family (24 members), and flocculin family (6 members) which are also cell-surface molecules of unknown function, are typically telomerically encoded (Goffeau et al. 1996). It is possible that the subtelomeric regions of both yeasts may favour duplication and that this may result in the generation of novel, organism specific genes important for cell identity (Wood et al. 2002; Kellis et al. 2003). One feature of telomeric regions, which may be significant in providing a potential reservoir for surface variation, is that these regions are usually transcriptionally silent (Nimmo et al. 1998). A novel form of epigenetic regulation at the telomeres has recently been identified in an *S. cerevisiae* strain where only *FLO11* of the glycosylphosphatidylinositol (GPI) anchored flocculin family is normally expressed. In some mutants, the loss of Sir2 induced transcriptional silencing increases switching frequency and turns on silenced proteins (Halme et al. 2004). The observed redundancy may therefore not exist solely to provide protection against mutation, but instead, to provide a reservoir of contingency genes whose advantageous features can be positively selected for in response to novel or rare environmental conditions. Such a positional preference is already well documented for contingency genes involved in immune evasion of parasitic protozoan (reviewed in Barry et al. 2003). Under these circumstances it would be beneficial for essential housekeeping genes to concentrate away from the highly plastic subtelomeric regions. Intriguingly such a positional preference has also already been reported for *C. elegans* based on correlations between chromosome location and lethality, and chromosome location and sequence similarity (Kamath et al. 2003; The *C. elegans* Sequencing Consortium 1998).

Genome wide expression studies in *S. pombe* have identified the telomeres as chromosomal regions enriched for meiotic genes induced in response to nitrogen starvation leading to the suggestion that spatial arrangement has a role in the activation of genes required for this process (Mata et al. 2002; See also section 4.2). More recently Hansen and colleagues assayed the global effects of the silencing mutants in histone deacetylases (Clr3 and Clr6) and the histone methyltransferase (Clr4), using microarrays (Hansen et al. 2005). Many genes repressed by the Clr proteins cluster in extended regions close to the telomeres and these are largely overlapping with those shown previously to be expressed in response to nitrogen starvation (Mata et al. 2002). Hansen and colleagues also observed that the telomeric regions contained genes, including transporters, whose expression in response to nutrient depletion may facilitate survival. A similar histone dependent repression of environmental responsive genes in subtelomeric regions is observed in *S. cerevisiae* (Robyr et al. 2002).

Finally, Kellis and colleagues reported that the majority of the 18 species-specific genes which were present in *S. cerevisiae* but absent from syntenic positions in the closely related *Saccharomyces (sensu stricto)* strains were all at subtelomeric locations (Kellis et al. 2003). Therefore, although subtelomeric duplicated ORFs are highly similar within a species; between species they appear to be rapidly diverging.

In *S. pombe* functional categories of genes implicated in adaptations to environmental stresses appear to be frequently overrepresented for subtelomerically encoded genes. The observed changes in the expression of these genes when silencing factors are mutated, coupled with their frequent duplication and rapid divergence, suggest that sub-telomeric regions may provide the ideal genomic environment to create, test and select for novel genes which could be applicable to all fungi, or even eukaryotes in general. Future studies using refined datasets and annotations will allow this hypothesis to be tested fully.

2.4 Gene density, GC composition and gene structure

Protein coding gene density is similar for chromosomes I and II, with one gene every 2462 and 2495 base pairs respectively, but lower for chromosome III which has one gene every 2766 base pairs. The reason for the substantially lower gene density on chromosome III is not known, but is not due to a difference in average gene length which is similar for all three chromosomes (1405-1444 base pairs). There are other notable differences between chromosome III and the other two chromosomes, including the maintenance of the tandem rDNA repeats and the more repetitive structure of its centromere. Chromosome III has also been shown to harbour an increased density of the remnants of transposable elements (Bowen et al. 2003; see also 2.11). It is possible that all of these observations are due to the different physical environment of some regions of this chromosome which may contribute to an enhanced capacity for the retention of duplicated sequence and indirectly, the lower gene density.

Protein coding genes are absent from the centromeres and gene density is lower than average at the telomeres. Overall gene density is one gene every 2528 base pairs compared with only one gene every 2088 base pairs for *S. cerevisiae*. This may reflect more complex regulatory structures, as average gene length (excluding introns) is approximately equivalent (1424/1460) but *S. pombe* intergenic regions are correspondingly larger (Wood et al. 2002).

Protein coding sequence accounts for 57% of the *S. pombe* genome, compared to 70.5% for *S. cerevisiae*. The overall GC composition is very similar for the two yeasts (36% and 38.3% for *S. pombe* and *S. cerevisiae,* respectively), and for the protein coding portion it is identical at 39.6%.

Introns are present in 2260 (46%) of fission yeast protein coding genes, and a total of 4722 have so far been identified. Intron length varies from 28 to 819 nucleotides with a mean of 82 nucleotides and the largest number found within a single gene is 15. Introns are much rarer in *S. cerevisiae* with only 301 identified in 5% of protein coding genes, although curiously, the mean length of *S. cerevisiae* introns is substantially longer at 216 base pairs (Dolinski et al. 2002, ftp://ftp.yeastgenome.org/yeast 12th July 2002).

Most *S. pombe* introns have GT donors and AG acceptors (only three confirmed introns have a GC donor). The branch site is also well defined, with 95% of introns having a consensus YTRAY. Four additional branch sites, related to the consensus, are experimentally confirmed but used with decreased frequency. Fewer than 50 confirmed or predicted introns do not one have a verified branch site within 6-34 bases of the acceptor. At publication, 638 introns were experimentally confirmed by mRNA and EST data. This number has now increased to 722, although many more are supported by the absence of gaps across splice sites when aligned with related proteins.

For genes with one to six introns, a 5' bias has been observed based on values expected if introns were evenly distributed within genes (Wood et al. 2002). A similar bias was observed previously in *S. cerevisiae* where it was hypothesised to be due to *in vivo* reverse transcription generating cDNAs which then replaced the original chromosomal gene (Fink 1987). Because cDNAs are extended from their 3' ends, 5' introns would have a reduced tendency to be removed. In addition, the number of genes with a specified number of introns decreases exponentially as intron number increases from two to six (614 have two introns, 324 have three introns,148 have four introns, 70 have five introns and 40 have six introns; Wood et al. 2002). Both of these observations may be of relevance to the speculation concerning the mechanism of intron removal.

The substantially larger intron number in fission yeast may provide a greater potential for post transcriptional regulation of biological processes via the controlled regulation of intron processing. It has also been proposed that the splicing machinery in *S. pombe* is closer to higher eukaryotes in both similarity and content (Käufer and Potashkin 2000; Kuhn and Käufer 2003). In support of this, recent studies have shown that some components of the splicing machinery are conserved from fission yeast to humans but absent from *S. cerevisiae* and that these appear to play a role in the splicing of particular subsets of genes (Webb and Wise 2004).

2.5 Proteome complement

A central goal of biological research is to describe fully the information encoded in a genome and how this is integrated into the orchestrated collections of processes and functions which combine to produce living cells. Towards this goal, continual refinement of the gene structures and gene complements of sequenced genomes is necessary to provide the most accurate 'parts list' possible. Such a list is a prerequisite for a summary of an organism's functional capabilities, to partition the non-coding portion of the genome, and for accurate orthology mappings. Gene prediction in the relatively densely packed genomes of single celled fungi is substantially easier than for higher eukaryotes. However, the presence of splicing, and the difficulty in distinguishing short genes from short spurious ORFs means that even the basic statistic of gene number is not trivial to obtain. Gene structures are revised primarily by the incorporation of new information from both similarity searches and experimental data. Gene complement is refined by; (i). the identification of new genes; (ii). 'partitioning' of dubious ORFs which are unlikely to be protein coding; (iii). detection of distant orthologs or other signals which provide evidence for the biological significance of a predicted translation; and (iv). experimental verification.

The publication of the *S. cerevisiae* genome in 1996 reported 6275 ORFs but estimated that only around 5800 were likely to be coding, based on the predicted number of small but spurious ORFs which would be included by chance due to the 100 amino acid cut-off threshold (Goffeau et al. 1996). Early efforts to establish absolute protein coding gene complement were thwarted by the absence of; (i). homologous sequences in the public databases; (ii). adequate tools for gene discovery and (iii). available experimental data. Subsequent re-analyses based on additional data and the annotation methods implemented for *S. pombe,* and comparisons with the partial shotgun sequence from 13 hemiascomycetes predicted similar protein complements (a maximum of 5570 'real' ORFs over 100 codons, and a minimum of 5600 'real' ORFs including ORFs under 100 codons respectively; Wood et al. 2001; Blandin et al. 2000). Both of these studies also provided improved gene coordinates and status calls for individual ORFs.

Recently, detailed comparisons with the genomes of four closely related syntenic *Saccharomyces* species (*sensu stricto*) and the slightly more distantly related filamentous ascomycete *Ashbya gossypii* have provided an increasingly refined gene complement (Cliften et al. 2003; Kellis et al. 2003; Brachat et al. 2003). Modifications included improved gene structure coordinates, the identification of small genes and improved distinction between dubious and verified ORFs using reading frame conservation. The changes reported by these and other analyses (affecting approximately 10% of the genome) have been reviewed and incorporated into the Saccharomyces Genome Database (SGD). This database currently reports a total of 6606 protein coding genes, 829 of which are dubious (using numerous criteria) giving a likely total of 5777 which includes 309 coding sequences under 100 amino acids (SGD http://www.yeastgenome.org/ 9th Nov 2004).

Publication of the fission yeast genome in 2002 recorded an upper estimate of 4940 protein coding sequences (including 11 mitochondrial proteins and 116 dubious ORFs), the smallest number for a sequenced free-living eukaryote at publication. This number has since increased to 4973 through the addition of 22 genes in sequenced gaps, and 14 genes which were missed during first pass annotation because they were either below the threshold size of 100 codons, or highly spliced. These are documented at http://www.genedb.org/genedb/ pombe/newgenes.jsp. Due to stricter annotation criteria, only 90 genes are now reported as dubious, so the present protein coding gene count is 4883.

The initial gene predictions were performed using GeneFinder trained on experimentally verified *S. pombe* genes (Green and Hillier, unpublished software). These preliminary gene structures were refined by multiple rounds of manual inspection within the Artemis analysis and annotation tool (Rutherford et al. 2000). When applicable the results of sequence similarity searches using BLAST, FASTA and Genewise against the UniProt (formerly Swissprot and TrEMBL), EMBL and Pfam databases were incorporated to extend gene predictions and to correct intron/exon boundaries (Altschul et al. 1990; Pearson and Lipman 1988;

Birney et al. 1996; Apweiler et al. 2004; Kulikova et al. 2004; Bateman et al. 2004). Intron boundaries were refined using EST data mapped onto the genome data using EST_GENOME, and by a Hidden Markov Model trained on *S. pombe* intron sequences using HMMER (Mott 1997; Hughey and Krogh 1996). All splice sites were manually inspected for ungapped homology across intron/exon boundaries and the presence of a consensus branch site, and adjusted when necessary. These integrated methods, coupled with manual intervention, have provided highly accurate gene structures for the fission yeast.

Since publication, additional experimental data (sequenced mRNAs), and homology, have resulted in updated structures for 31 of the original gene predictions. These are documented at http://www.genedb.org/genedb/pombe/coord-Changes.jsp. The changes include; (i). the addition of small N- or C-terminal exons; (ii). changes to the N-terminal methionine (sequence extended or reduced); (iii). replaced N-terminal exons; (iv). alteration of intron boundaries; (v). gene splits (two); (vi). additional in-frame splice (one); (vii). single base deletion (one). It is likely that 5777 and 4883 are close to the actual protein coding totals for both genomes although undoubtedly further small or highly spliced genes remain to be discovered.

2.6 Non coding RNA complement

Non coding RNAs (ncRNAs) include all RNAs other than mRNA and are central to a wide range of biological processes including transcription, translation, gene regulation and splicing. The number of known ncRNAs is expanding, but in the absence of an obviously detectable signal for many ncRNAs, especially those present in low copy number, their computational identification is still difficult (Eddy 2002).

In addition to the ~5000 protein coding genes there are ~600 known or predicted genes for various cellular RNAs (more than 10% of the gene content). At present, 170 transfer RNAs (tRNAs; 195 including mitochondrially encoded tRNAs) are reported, compared to 288 in *S. cerevisiae*. This is likely to encompass the complete tRNA complement for the *S. pombe* genome and reflects the relative ease and accuracy with which tRNAs can be predicted by tRNAscan-SE (Lowe and Eddy 1997).

The 5.8S, 18S and 26S ribosomal RNAs (rRNAs) are present in tandem arrays of which there are an estimated 100-120 copies (Schaak et al. 1982; Barnitz et al. 1982). The genome sequence has a couple of representative copies of this repeat from the beginning of each tandem array. The 5S rRNAs are present in 32 copies dispersed throughout the genome in contrast to the 100-200 present in the *S. cerevisiae* rDNA repeats (Mao et al. 1982; Aarstad and Oyen 1975).

The spliceosomal RNAs (U1-U6), together with 34 small nucleolar RNAs (snoRNAs), are dispersed throughout the genome[4]. The snoRNAs cannot be detected by similarity alone and are difficult to predict computationally, although

---

[4] U3 has 2 copies in *S. pombe,* U5 has 2 copies in *S. cerevisiae*.

there have been advances in methods for their detection (Lowe and Eddy 1999). Based on the number of snoRNAs identified in *S. cerevisiae* to date (68), at least 30 additional snoRNAs are likely to be present in *S. pombe* (T. Lowe, personal communication).

Besides the major classes of RNA, 8 ncRNAs have been identified experimentally: RNase P K-RNA (Krupp et al. 1986), *sme2*-meiRNA (Watanabe and Yamomoto 1994), 7SL-RNA (Ribes et al. 1998), *meu3, meu11, meu16, meu19* and *meu20* (Watanabe et al. 2001). An additional 124 loci have been annotated as potential RNA genes (for example transcripts with no detectable open reading frame; Watanabe et al. 2002). It is likely that some of the 68 uncharacterised *prl* loci (which correspond to cDNAs lacking apparently long open reading frames, and often overlap with previously identified transcripts), and *tos1-3* which are antisense to *rec7,* have regulatory roles (Watanabe et al. 2002; Molnar et al. 2001). Inevitably, many more unidentified RNA genes (antisense, structural and catalytic) will play important roles in fission yeast and other organisms. The complete RNA complement can be accessed from http://www.genedb.org/shortcuts.jsp.[5].

2.7 Intergenic regions

Intergenic regions are larger, on average, between divergent genes containing two promoters (1341 bp) than between convergent genes containing two downstream regions, and therefore promoterless (558 bp), while intergenic regions between tandem genes containing one promoter, and one downstream region, show an intermediate length distribution (955 bp; Wood et al. 2002)[6]. All mean intergenic distances for *S. pombe* are larger than the corresponding mean distances for *S. cerevisiae*, although the difference for divergent genes is larger and the difference for convergent genes is smaller. Intergene regions for *S. pombe* have a mean of 952 bp, compared to the *S. cerevisiae* mean of 515 bp. Several explanations can account for this observation. The untranslated regions (UTRs) may be systematically longer in *S. pombe* than in *S. cerevisiae*. Mean lengths of identified 5' UTRs are 178 nucleotides and 95 nucleotides, and 3' UTRs 225 and 180 nucleotides for *S. pombe* and *S. cerevisiae* respectively. (*S. pombe* data, ftp://ftp.sanger.ac.uk/pub/yeast/pombe/UTRs; *S. cerevisiae* data, E. Hurowitz, personal communication)[7]. Although *S. pombe* UTRs are apparently longer, this difference would not account for the species differences for intergenic length. In addition, the 5' > 3' bias can also not be attributed to longer 5' UTRs as 3' UTRs appear to be on average, longer. The promoter regions may be more complex and therefore longer in *S. pombe*, although there is no evidence to support this at pre-

---

[5] The numbers reported here exclude the small complementary microRNAs for centromeric function (Volpe et al. 2002).

[6] Intergene distance is calculated from the stop and/or start codons between adjacent genes.

[7] The *S. cerevisiae* average sizes were obtained from RACE-PCR experiments which have higher success rates for genes with shorter UTRs, so the average reported here may be lower than the true genome average.

sent. However, there is evidence that classes of promoter proximal mammalian transcription activation domain, which are non functional in *S. cerevisiae,* are functional in a proximal promoter context in *S. pombe* suggesting there may be a closer relationship with higher eukaryotic promoters (Remacle et al. 1997). Replication origins are known to be more extended in *S. pombe* than in *S. cerevisiae* (see section 2.8 below). There are also annotated examples of extended low complexity gene free regions in *S. pombe* (around 10 per chromosome) which, at 4-8kb fall outside the normal distribution of lengths associated with average intergenic regions (Wood et al. 2002). These gene free tracts are usually flanked by divergently oriented genes and exhibit a (G-C) / (G+C) base compositional bias which switches strand in the centre of the gene free region. One such region in cosmid c4G8 corresponds to a prominent meiotic DNA break site (Young et al. 2002). No such gene free regions have been identified in *S. cerevisiae.* Intergenic regions are also more AT rich (69.4%) than the genome average (64%; Dai et al. 2005).

Publicly available EST data and mRNAs in the EMBL database have been mapped on to the genome sequence using EST_GENOME (Morimyo et al.1997; Kulikova et al. 2004; Mott 1997). When sequence quality was sufficient to determine transcriptional start or end, these have been manually curated to create features for untranslated regions. This dataset provides 370 5' UTRs and 742 3' UTRs which are available to download from http://www.sanger.ac.uk/Projects/ S_pombe/DNA_download.shtml. These features provide a preliminary dataset of truly coding regions for a subset of genes by providing delimiters between gene boundaries and truly intergenic regions.

2.8 Replication origins

DNA replication origins (ORIs) are specific sites within a DNA molecule where DNA replication is initiated. Researchers would usually include in this definition any '*cis* acting' sequences which affect origin function by binding the machinery that initiates and regulates replication (Masakuta et al. 2003). Replication origins have been identified in a variety of organisms including mammals, but are best studied in the two yeasts. Replication origins in *S. cerevisiae* are as short as 75 base pairs with an 11 base pair consensus and a number of partially redundant elements with varying distribution (Broach et al. 1983; Theis and Newlon 1997; Theis and Newlon 2001). Recent approaches based on chromatin immunoprecipitation and density labelling have predicted the distribution of 400 putative ORIs in *S. cerevisiae* (Wyrick et al. 2001; Raghuraman et al. 2001). In comparison, *S. pombe* replication origins are substantially larger and have a modular structure, possibly because more protein-DNA interactions are involved in replication initiation (Dubey et al. 1996). They require a minimum length of 0.5-1 kb and have no recognisable consensus, although they do contain asymmetric and non-asymmetric A-T stretches (Maundrell et al. 1988; Clyne and Kelly 1995). Like mammalian replication origins, they appear to be located preferentially upstream of RNA Polymerase II promoters (Gomez and Antequera 1999).

The first genome wide survey of potential replication origins in fission yeast showed that 90% of A+T rich islands colocalised with active ORIs (Segurado et al. 2003). The mean genomic frequency of the 384 A+T rich islands is one every 33 kb, and these all map to intergenic regions. A bias was also observed for their location in divergent transcription units, although this may be due to the larger size of these regions (see section 2.7 above). A similar number and distribution has also been observed using microarrays (C. Heichinger, personal communication). There are significant clusters of 'replication origin associated' AT rich islands in the centromeres, and in the subtelomeric regions of chromosomes I and II and the mating-type locus (fourfold higher than the genome average), although the significance of this is not known.

It was recently reported by Dai and colleagues that the relative origin activity of an intergene in *S. pombe* is a function of its length and AT content rather than a specific nucleotide sequence requirement, and that sequence properties ascribed to origins are therefore general characteristics of intergenic regions (Dai et al. 2005). It is proposed that the intergenes which function as origins are likely to form a broad continuum, and demonstrated that any intergenic region over ~900 kb in length and greater than 70% AT (close to the intergene average) is likely to have origin activity. A stochastic model is proposed, where the binding affinity of the origin recognition complex (ORC) subunit Orc4 is dependent on both AT content and length, in a departure from the classical model which predicts binding to a small number of sites with high specificity. This model explains the observation that the origins studied so far in *S. pombe* are not used in every cell cycle (because the number of potential origins greatly exceeds the number of ORC molecules), and may also explain some features of origins in metazoans.

Although the number of predicted ORIs in *S. pombe* (385) and *S. cerevisiae* (400) are very similar, they do not appear to be similar in composition. *S. pombe* ORIs are more similar to mammalian ORIs in their lack of consensus sequences, presence of multiple dispersed partially redundant elements, and preference for association with promoter regions. These preliminary global analyses of replication will provide a framework to study the contribution of replication origin structure and function to replication dynamics and for the dissection of organismal similarities and differences.

2.9 Mitochondrial genome

The mitochondrial genome of fission yeast is considerably smaller than that of budding yeast (20 kb versus 85.8 kb) and contains a smaller number of protein coding genes (11 versus 28; Lang et al. 1987; Foury et al. 1998). However, in *S. cerevisiae*, 9 of these appear to be complete orphan genes of small size (<134 amino acids) and are likely to be spurious ORFs. The remainder of the non-conserved genes are involved in intron metabolism and are absent from some close relatives of *S. cerevisiae.* Therefore, the 'ancient' coding portion of the mitochondrial genome is almost identical between the two yeasts.

2.10 Pseudogenes

The incidence of pseudogenes is relatively low for both yeasts. The fission yeast genome database (GeneDB *S. pombe* http://www.genedb.org) reports a total of 47 pseudogenes (9 of which are transposon or wtf related) compared to 22 pseudogenes reported by SGD for S88 strain of *S. cerevisiae* (http://www.yeastgenome.org/ 14th July 2004). In *S. pombe,* the majority of genes designated as pseudogenes have more than one frameshift, some are extremely degraded and were only identified as former coding sequences by BLASTX sequence similarity searches. It is not presently possible to identify genes which may be pseudogenes due to inactivated promoters. It is also possible that some genes reported as pseudogenes may in fact be sequencing errors resulting from spontaneous mutations in the clone libraries. Apparently frameshifted genes (for example *spa1* in *S. pombe*) may also have valid translations due to ribosomal frameshifting mechanisms (Ivanov et al. 1998; Zhu et al. 2000). A number of *S. pombe* annotated pseudogenes have been shown to be transcribed (Mata et al. 2002; Chen et al. 2003), and in human the RNA of an expressed pseudogene has been shown to have a regulatory function (Hirotsune et al. 2003). The current inventories of pseudogenes for both species should therefore be evaluated with caution.

2.11 Transposable elements

LTR (long terminal repeat) retrotransposons and endogenous retroviruses constitute variable proportions of their host genomes, and genome sequencing has revealed a diverse range of organismal transposon content. The availability of the complete fission yeast genome sequence has provided the opportunity to perform a comprehensive analysis of the entire complement of transposable elements with respect to their chromosomal distribution, insertion site preferences and evolution (Bowen et al. 2003). Only two families of transposons (Tf1 and Tf2) belonging to the Ty3/Gypsy group were known to exist in *S. pombe* (Levin et al. 1990; Levin 1995). Homology based methods confirmed that the *S. pombe* sequenced strain contained only 13 full length copies of a single family of active transposon (Tf2) and that there were no Tf1 elements in the laboratory strain. The transposon complement is therefore substantially lower than the 50 LTR-retrotransposons reported for budding yeast (Kim et al. 1998). It has been speculated that this difference may be due to the loss of the RNAi machinery from *S. cerevisiae* because of the involvement of RNAi in the removal of duplicated sequence (Aravind et al. 2000). In addition, 274 intact and 75 fragmented (<200 base pairs) solo LTRs and five transposon fragments, marking the site of former transposition events, were identified. The intact LTRs were classified into at least three large groups; (i). those closely related to Tf2 (35; ii). those closely related to Tf1 (28), and (iii). many more distantly related small families (111). Some of these more distant lineages were identical or highly similar to each other. Close examination revealed that these were all subtelomerically located and that their similarity was a result of telomeric duplications. This is consistent with the increased sequence similarity at these locations (see section 2.3). In total, transposon derived sequences account for ~133,000 base pairs or 1.1% of the sequenced portion of the genome compared to 2.4% for *S. cerevisiae.*

Experimental studies of insertion site preference in *S. pombe* have shown that the Tf1 element has a significant preference for insertion into intergenic sequence within 300 nucleotides of the 5' end of a coding sequence (CDS; Behrens et al. 2000; Singleton and Levin 2002). Bowen and colleagues provide complementary studies using a bioinformatics approach to support the previous experimental data for integration site preference (Bowen et al. 2003). Analysis of the 186 intact transposons and LTRs revealed that all insertions were exclusively intergenic. The frequency of insertion into intergenic regions proximal to CDS in tandem, divergent or convergent orientation was analysed. A positive correlation was detected between the number of expected transposon insertions and the number of expected RNA polymerase II promoters, in different spatial contexts. Insertions into intergenic regions between convergent genes containing no promoters were found to be statistically under-represented (incorporating corrections for size differences). Furthermore, the distance between each insertion and the end of the nearest ORF was significantly biased for insertions associated with the 5' end of genes, the majority clustering between 100 and 400 base pairs of the 5' end of the neighbouring CDS. Therefore, in contrast to *S. cerevisiae,* where transposons appear to target upstream of RNA polymerase III transcribed genes by specifically interacting with a component of the RNA polymerase III transcription machinery (Chalker and Sand Meyer 1992; Yieh et al. 2000); *S. pombe* transposon insertion sites appear to show an increased preference for RNA polymerase II promoters. *S. cerevisiae* is reported to contain 344 transpositions derived insertions or their remnants (Kim et al. 1998), so the overall numbers of transposons, or transposon footprints are similar for these two yeasts.

During sequencing and annotation, a novel species specific high number copy family was identified (Wood et al. 2002). They were named *wtf* (for with Tf) because many members of this family were flanked by Tf2-type LTRS. There are 25 sequences related to the *wtf* family, which was identified as the largest family of *S. pombe* specific genes in an analysis of lineage specific gene expansions (Lespinet et al. 2002). The only experimental data available has shown these genes to be upregulated up to 100 fold during meiosis (Watanabe et al. 2001; Mata et al. 2002). Surprisingly, 23 of the 25 copies were located on chromosome III.

Bowen et al. also analysed the genome wide distribution of insertion elements and showed that chromosome III contained almost twice as many insertions as the other two chromosomes. Further investigation revealed the association of *wtfs* with LTRs was responsible for 80% of the over-representation of LTRs on this chromosome. The nature of the mechanism of expansion of the *wtf* family is currently unclear but it now appears that the targeted integration of Tf elements and subsequent duplications have contributed to their association with LTRs. It is interesting to speculate whether the higher transcription level of the *wtfs* may have contributed to the accumulation of nearby Tf insertions and is analogous to the reported preference of HIV-1 integrations for actively transcribed genes (Bowen et al. 2003; Schroder et al. 2002). Furthermore, the integrase of Tf1 and Tf2 contains a chromodomain, which is implicated in chromatin remodeling via its interactions with histones (Malik and Eickbush 1999). It is therefore possible that the insertional preference of Tf insertion into actively transcribed genes is mediated by this chromodomain (Bowen et al. 2003). It appears that, despite the low abundance of transposable elements, the study of transposition mechanisms and insertion site preference in *S. pombe* will continue to be informative regarding the contribution of transposition to the shaping of genome content.

2.12 Genome features summary

A summary of the genome features and contents described here are presented in Table 1. Data is accessible via the GeneDB database (http://www.genedb.org/S_pombe/; Hertz-Fowler et al. 2004), or the *S. pombe* project page at the Wellcome Trust Sanger Institute (http://www.sanger.ac.uk/Projects/S_pombe/; WTSI).

3 Genome and proteome sequence comparisons

3.1 Introduction

Genome and proteome sequence comparisons provide insights into the functional similarities and differences, and evolutionary relationships, between the species compared. To fully elucidate the events operating on evolutionary timescales, it is necessary to compare sequences with different degrees of evolutionary relatedness. Distantly related genomes reveal ancient events and relatively slow changes, whereas more closely related genomes reveal recent and more rapid changes. Comparison of genomes identifies genes and other functional elements, regions of genome duplication and syntenic regions with other organisms. *S. pombe* is too divergent from currently available fungal genomes for direct genome comparisons to be informative in terms of genome rearrangements or content. However, the availability of the predicted proteomes of these two eukaryotic models has allowed the comparison of their protein complements to assess the similarities and differences in both size and content. Preliminary proteome comparisons, using pairwise sequence similarity, provide an overview of the potential conserved and species specific components of an organism. More specific classification of proteins, according to their potential evolutionary relationships, provides a natural framework for comparative genomics, functional annotation and evolutionary analysis. In this section, a summary of the initial global genome and proteome comparisons, and an overview of the more granular classification of orthologs, is presented.

Table 1. Comparative genome features and contents of the *S. pombe* and *S. cerevisiae* genomes

	S. pombe	*S. cerevisiae*
Genome size (sequenced/total)	12.5 Mb (~14.1 Mb)	12.1 Mb (~13.0 Mb)
Chromosomes number	3	16
Chromosome size range	3.5-5.7 Mb	0.2-1.5 Mb
Centromere size	35-110 kb	~0.15 kb
Gene density (average bp/gene)	~2,530 bp	2,090 bp
Average gene length	~1,430 bp	1,460 bp
Overall GC content	36.0%	38.3%
GC content in protein coding sequence	39.6%	39.6%
GC content in intergenic sequence	30.6%	-
Intron number	~4,730	~272
Genes with introns	2260 (46%)	257 (5%)
Average intron length	82 bp (29 bp-819 bp)	216 bp
Maximum number of introns/gene	15	3
Gene number (protein coding)	4,973	6606
Gene number (ex dubious)	4883	5777
tRNA genes	195	288
5.8S, 18S, 26S rRNA genes	100-120 tandem repeats (2 arrays)	~150 tandem repeats (1 array)
5S rRNA genes	32 dispersed genes	1-200 in rDNA repeats
small nuclear RNA genes(snRNAs)	7	7
small nucleolar RNA genes (snoRNAs)	34	68
Other RNA encoding genes	8	4
Inter-gene regions (mean/median)	952 bp/423 bp	515 bp/200 bp
Mean distance between divergent genes	1341 bp	570 bp
Mean distance between tandem genes	955 bp	586 bp
Mean distance between convergent genes	558 bp	339 bp
UTR length 3'	225	180
UTR length 5'	178	95
Replication origins	~400	~400
Replication origin sizes	0.5-1 kb	75-150 bp
Mitochondrial genome	20 kb (11 genes)	85.8 kb (28 genes)
Pseudogenes (excluding wtf)	39	22
Tf type transposons	13 /2 pseudo	59
Long terminal repeats (LTRs) solo intact	274	268
wtf elements (with tf2 type LTRs)	25/9 pseudo	0

3.2 Genome sequence comparisons

Numerous tracts of co-linear duplicated genes are detected in *S. cerevisiae* and were proposed to be the remnants of a whole genome duplication event in its evolutionary history (Wolfe and Shields 1997). The availability of the genomes of syntenic species, which diverged both before and after the proposed split, has provided irrefutable evidence for this event (Wong et al. 2002 ; Kellis et al. 2004; Dietrich et al. 2004). Similar searches for tracts of conserved gene order did not reveal evidence for large scale genome duplications in *S. pombe* (Keogh et al. 1998; Wood et al. 2002). Synteny is not detectable between fission yeast and any other available fungal genomes at the time of writing; any relationships have been obscured by chromosomal rearrangements, gene duplications and losses. However, a small number of segmental duplications are detectable in *S. pombe*, as blocks of intra genome conserved gene order at the sequenced subtelomeric regions of chromosomes I and II (see section 2.3). Thirty two tandemly repeated genes are also recorded.

3.3 Proteome sequence comparisons

Preliminary proteome comparisons between *S. pombe*, *S. cerevisiae* and *C. elegans* indicated that around 4050 (83%) *S. pombe* genes were common between the two yeasts (3281/67% of these also common to *C. elegans*; Wood et al. 2002). A small number (145/3%) were reported as present in *C. elegans* but not *S. cerevisiae*, and 681 (14%) were unique to *S. pombe*. Reciprocal comparisons revealed a larger number (4523) of *S. cerevisiae* proteins were conserved in *S. pombe* (3605 also in *C. elegans*) and 1104 were unique to *S. cerevisiae*. The number of genes conserved only between the two fungal species was greater in *S. cerevisiae* than in *S. pombe* (918 versus 769). These differences can only be explained by a greater number of duplicated genes being present in *S. cerevisiae*. The number of unique genes was greater for *S. cerevisiae* than *S. pombe* (1104 versus 681). This difference is primarily due to an increased number of duplicates in *S. cerevisiae*. However, it is possible that an increased number of newly evolved genes not generated by a duplication event, or a larger number of horizontally transferred genes are also contributory factors[8].

Further analysis based on protein clustering estimated the numbers of multi-member families versus singletons in both yeasts. This showed that *S. cerevisiae* has around 716 protein coding genes belonging to multi-member families but *S. pombe* has only around 361, supporting the conclusion that more duplicated genes are present in *S. cerevisiae*. These observations, and the absence of any co-linear duplicated segments indicate that *S. pombe* is unlikely to have undergone any whole genome duplication events since it separated from the *Saccharomyces* lineage, an estimated 300-400 My ago.

---

[8] 'Unique' is used here only with respect to the two species compared.

3.4 Orthologous groups

The concept of homologs (genes descended from a common evolutionary ancestor), and the implicit inference of evolutionary history which accompanies this concept, originated from the seminal work of Ohno in the 1970's (Ohno 1970). Homologs are further classified as orthologs (direct evolutionary counterparts by vertical descent i.e. the same gene in a different species) and paralogs (genes which have arisen by duplication events within a genome after a speciation event; Fitch 1970). These concepts are now routinely used in global genome comparisons and annotation protocols (Tatusov et al. 1997; Chervitz et al. 1998).

The identification of candidate orthologs, and orthologous groups between species, is a prerequisite for the rigorous evaluation of the nature and frequency of the events in their evolution affecting protein number and type; specifically gene duplications, lineage specific gene loss gene divergence and horizontal transfer. Accurate orthology mapping between *S. pombe* and *S. cerevisiae* provide a framework for the reconstruction of the evolutionary events giving rise to these two species.

Preliminary global analysis using BLAST with a threshold cut-off provided initial estimates of the level of protein conservation and redundancy between the two yeasts. Analyses of this type provide a useful overview but are unsuitable for the transfer of functional information because of a failure to detect many similarities (false negatives) and the inability to distinguish spurious matches (false positives). Functional transfer based on top scoring BLAST hits is only suitable for a proportion of any proteome, even when the alignment appears to be significant, and should be applied with extreme caution in any annotation pipeline. For robust functional annotation, orthologous relationships should ideally be identified by phylogenetic analysis of entire families but evolutionary inferences of orthology can usually be made without phylogenetic methods. A number of resources for automated ortholog detection are available. The most commonly used are COGS/KOGS, Inparanoid and OrthoMCL (Tatusov et al. 2003; Koonin et al. 2004; Remm et al. 2001; Li et al. 2003). These are based on initial candidate ortholog identification using pairwise BLAST comparisons followed by different methods (clustering or reciprocal best hit identification) to generate orthologous groups. Differing output and coverage indicate that these methods are currently sub-optimal (Li et al. 2003).

Most algorithms are ultimately dependent on reciprocal best hits which provide a good approximation of orthology. However, not all orthologs are reciprocal best hits, or even best hits. Extremely divergent proteins with lower levels of sequence conservation can often generate spurious matches, and obscure truly homologous relationships. The large number (30%) of reported KOGS orthologous clusters with unexpected phyletic patterns may be artificially large as a result of this restriction. Lineage specific gene losses can also complicate ortholog determination by generating spurious false positives. Finally, a global threshold cut off for candidate ortholog identification will impose an arbitrary restriction whereby extremely divergent orthologs will not be detected.

3.4.1 Establishing orthology

An orthology mapping between *S. cerevisiae* and *S. pombe* has been created based on manual inspection of pairwise alignments, multiple alignments and protein clusters, using alignments seeds from numerous algorithms including BLAST, PSI-BLAST, FASTA, Pfam-B/Domainer; Alschtul et al. 1990; Alschtul et al. 1997; Pearson and Lipman 1988; Sonnhammer et al. 1997). Ambiguous relationships are inspected after clustering using CLUSTAL W (Thompson et al. 1994) and identified orthologs are corroborated by experimental evidence where available. This has a number of advantages over automated methods including increased accuracy, increased specificity, greater coverage and the ability to combine data from multiple resources, including ortholog identification software (Wood et al. manuscript in preparation).

Firstly, accuracy is increased by manual curation through improved discrimination for multi domain proteins by the inspection of domain organization. In addition, 'fusion proteins' (a protein in one organism which maps independently to two unrelated proteins in another organism), can be identified. For example, *S. pombe* Pdf1 is a fusion between palmitoyl-protein thioesterase (PPT) and dolichyl pyrophosphate (Dol-P-P) phosphatase which is proteolytically cleaved after translation. The two mature proteins are functionally connected but the domain combination is not observed in other organisms, possibly indicating a recent fusion event. The PPT is the functional homolog of the neuronal ceroid lipofuscinosis (Battens disease) protein in humans and is absent from *S. cerevisiae*, although the Dol-P-P is present. These complex patterns of conservation are difficult to unravel with automated methods which usually rely on arbitrary thresholds for the length of the similarity hit and sequence identity when identifying candidates.

Not all similarities are due to homology, and unrelated proteins can sometimes generate reciprocal best hits. Manual inspection and experimental data can be used to distinguish non-orthologous sequences and increase accuracy. Granularity can be increased by detecting orthologous pairs within cluster members. Independent orthologs can also be detected for related proteins with promiscuous domains particularly the WD, TPR, HEAT and LRR families of repeat containing proteins. For example, KOG0266 includes three *S. cerevisiae* and three *S. pombe* proteins. This cluster can form independent orthologous groups between *S. cerevisiae* Cps30 and *S. pombe* Swd3 and between *S. cerevisiae* Tup1 and *S. pombe* Tup1 and Tup11. Uncharacterised *S. cerevisiae* YGL004C is more distantly related to all of the other cluster members. The discrimination of independent orthologs is crucial for accurate functional transfer based on sequence similarity.

Most importantly, increased coverage can be obtained by distant ortholog detection. Orthologous proteins show a broad distribution of sequence similarity (evolutionary rate). Not all orthologs are significantly similar, and the inspection of individual pairwise or multiple alignments can often result in the detection of truly homologous relationships which are not necessarily best hits. For example, the *S. pombe/S. cerevisiae* orthologous pairs Orc6/Orc6p, Rpa34/Rpa34p, Pcp1/Spc110p, Ker1/Rpa14p, and Swi5/Sae3p are not BLAST reciprocal best hits, and are not detected by KOGS or Inparanoid.

For statistically insignificant short motifs, confidence in ortholog assignments can be increased by the consideration of:

- i.   conserved residue type; residue properties likely to have functional significance, for example rare or charged amino acids, especially when conserved in all cluster members.
- ii.   spatial context of alignments; correspondence of the positions of the conserved region(s) in the protein i.e. co-linear high scoring pairs (HSPs)
- iii.   spatial context and conservation of other protein features; transmembrane domains, signal sequences, predicted posttranslational-modification sites
- iv.   correspondence of protein length
- v.   phylogenetic distribution and copy number; especially if conserved in a single copy in all sequenced eukaryotes
- vi.   functional context; supporting experimental data, for example similar knockout phenotype, or missing member of conserved stoichiometric complex

Directed searches of the orphan (non conserved) protein set can be performed to detect candidate orthologs for less conserved proteins. In most cases, multiple lines of evidence will be used to support such a prediction. The *S. cerevisiae* ortholog of *S. pombe* DNA recombination/repair protein Swi5 was identified by a directed search and proposed adjustment to the *S. cerevisiae* gene prediction for SAE3 (Akamatsu et al. 2003; Young et al. 2004). The orthology prediction was supported by the conserved residue type and context, length, single copy distribution in sequenced eukaryotes, and a conserved recombination defective phenotype. *S. pombe* Sad1 spindle pole body component is predicted to be the ortholog of *S. cerevisiae* Mps3 on the basis of a reciprocal BLAST hit with low significance, but is supported by co-linear HSPs, a transmembrane region in similar sequence context, and similar cellular localization. Orthologs have also been detected for members of conserved complexes by targeted searches which identified small and highly spliced genes missed by the first pass annotation. These include potential orthologs for *S. cerevisiae* Pop8, Ost4, Sen15, Dad3 and Sus1.

The annotation procedures outlined here remove the biologically artificial restriction of genome wide cut-off threshold for sequence similarity and match length, and the dependence on a single algorithm. Orthology assignments can be incorporated from multiple sources (both software and experimental results). For example, recent comparison of the remaining orphan set against KOGs identified predicted orthologs for 7 sequences[9]. Manual inspection determined three other KOG predictions for SPAC1687.10/YOR058C, SPAP8A3.13/YGR066C/YBR105C, and SPAC1A6.07/YLR330W to be false positives based on additional evidence. For example, *S. cerevisiae* YOR058C is a microtubule associated protein and its predicted ortholog is SPAPB1A10.09 (Pfam family PF03999).

A total of 3636 *S. pombe* proteins and 3842 *S. cerevisiae* proteins have curated orthologs in the other yeast (summarised in Table 2). The remaining 1235 *S. pombe* proteins and 1704 *S. cerevisiae* proteins have no predicted ortholog in the

---

[9] SPAC6F12.08c, SPCC1620.07c, SPCC736.12c, SPAC553.06, SPAC25B8.02, SPCC1289.09 and SPBC24C6.08.

other yeast at present. However, a number of these have homologs in other species (498 and 346 for *S. pombe* and *S. cerevisiae* respectively; see section 3.5)[10]. A number of proteins in both organisms (68 and 307 for *S. pombe* and *S. cerevisiae* respectively), have conserved domains, but their respective orthologs cannot be distinguished as multiple duplications and gene losses have obscured their evolutionary relationships. The majority of these are regulatory proteins, and include a high proportion of transcription factors and proteins with RNA binding motifs. Further work and additional sequenced genomes will allow the relationships between these to be resolved.

3.4.2 Orthologous relationship type and function

Ortholog identification is complicated by duplication events and can only be described accurately by multiple relationship type mappings (Table 2). Most orthologous relationships have a 'one to one' mapping (2396) where a single *S. pombe* gene maps to a single *S. cerevisiae* gene and *vice versa*. One to one mappings are usually functionally equivalent, especially when universally conserved in a single copy in most or all eukaryotic genomes. These are predominantly core 'informational' proteins (those involved in processes related to genome stability and maintenance, transcription, translation and biosynthetic metabolism). The remaining mappings represent instances of duplication in either one or both organisms and will be discussed in this context.

Recently duplicated genes are likely to have the same function and the most likely fate is rapid loss of one duplicate. However, duplicate genes which are retained usually have one of two fates; i). one copy will retain the original function and the other copy will evolve (often undergoing accelerated evolution) to gain a novel function or specificity (derived function, or neofunctionalization) or ii). the existing function is partitioned between the duplicate copies often by differential expression or compartmentation (subfunctionalization; reviewed in Prince and Pickett 2002).

S. cerevisiae genes which can be mapped to the most recent polyploidization event can be assumed to have formed simultaneously. Approximately 16% of the *S. cerevisiae* gene complement (~500 pairs) is estimated to be part of a duplicate pair dating from this whole genome duplication (reviewed in Wolfe 2004). The observation that *S. pombe* has 912 duplicated gene products in the conserved set (and additional duplicates in the non-conserved set) implies that both yeasts have been consistently prolific in generating and retaining duplicates since their divergence[11].

---

[10] The number of likely protein coding *S. cerevisiae* sequences reported here is 5546. This is 231 less than the SGD current total. Some of these discrepancies are due to gene merges not reported in SGD. The remainder are all under 100 amino acids, and some appear to be spurious as they are not reported in syntenic regions of the closely related yeasts (unpublished observation).

[11] It should be noted that this does not represent the absolute total for duplicates for these species, as many members of the non conserved set are also duplicated.

Table 2. Distribution of the relationship types of conserved and non conserved proteins between *S. pombe* and *S. cerevisiae*. 'Species specific' comprises sequence orphans, or duplicated in only one species, and characterised genes with no identifiable ortholog. 'Ortholog cannot be distinguished' refers to proteins with an identifiable domain but which cannot be assigned an ortholog in the other species. 'Conserved but not in *S. cerevisiae/S. pombe*' refers to lineage specific losses in the respective organisms. 'One to One', 'One to many' and 'many to many' refer to the numbers of orthologous proteins mapped from and to in the respective species.

	S. pombe	*S. cerevisiae*
Non conserved set		
Species specific	669	1051
Ortholog cannot be distinguished	68	307
Conserved but not in *S. cerevisiae/S. pombe*	498	346
Subtotal	1235	1704
Conserved Set		
Orthlog relationship type		
One to one	2396	2396
One *S. pombe* to many *S. cerevisiae*	328	731
One S. cerevisiae to many *S. pombe*	429	202
Many to many	483	513
Total with orthologs	3636	3842
Total predicted protein (ex dubious)	4871	5546

The number of proteins which map from a single copy in one yeast to more than one copy in the other yeast (one to many mappings) is higher for *S. pombe* than *S. cerevisiae* (328 versus 202). However, *S. cerevisiae* has a larger number of proteins mapped to, which is consistent with the previously observed increased number of duplicates (*S. pombe* 429 *S. cerevisiae* 731). The duplicated proteins in the 'one to many' set often have related, or overlapping functions. In some cases, subfunctionalization has occurred, either by altered expression, localization or specificity.

Parallel duplications (those which appear to have duplicated independently since divergence in both lineages) account for 483 *S. pombe* proteins and 513 *S. cerevisiae* proteins belonging to 193 orthologous clusters, (compared to 202 duplicated in *S. pombe* only, and 328 duplicated in *S. cerevisiae* only). These 'many to many' duplicates are predominantly involved in monitoring or responding to nutrients or specific environmental stresses, or are signalling pathway components. Specifically, cell surface glycoproteins implicated in the assimilation and catabolism of nutrients (proteases, glycosyl transferases amylases etc.) and membrane transporters are the most highly represented. Although the functions of the parallel duplicates are usually related, they are sometimes involved in different processes. For example, members of the expanded glycosyl transferase 48 family in *S. pombe* are variously required for normal growth and sporulation. Some expanded families appear more likely to be reutilised in different contexts by species specific adaptations. Annotation transfer should therefore be more conservative when mappings are multiple.

After the removal of ribosomal proteins (44 clusters), histones (3 clusters) and translation elongation and initiation factors (5 clusters), informational proteins are almost wholly absent from this set of 193 parallely duplicated clusters. In addition, the informational duplicates are frequently highly similar, or even identical, whereas non-informational duplicates tend to be more divergent. The frequency of occurrence of duplicates, and lack of divergence for these particular gene products in most genomes implies that mechanisms exist for the maintenance of copy number and similarity. Several lines of experimental evidence have been presented and mechanisms proposed to support this (Koszul et al. 2004; Prado et al. 2005; Pyne et al. 2005).

The *S. pombe* /*S. cerevisiae* curated ortholog mapping described here provides an inventory of potential orthologous sequences between these two species. By using a combination of methods, approximating to a natural classification, greater sensitivity and selectivity for the detection of orthologs and paralogs can be achieved to provide a rigorous and comprehensive inventory based on evolutionary relatedness. The nature of ortholog detection for divergent pairs (biological knowledge, multiple software and methods) make automation difficult. However, novel protein families identified during ortholog detection are submitted to the Pfam protein family database, and the Hidden Markov Models (HMMs) created for these divergent gene families will be useful for the detection of candidate orthologs (in combination with other methods) in other genomes.

This dataset will continue to be refined and extended by the identification of further distant orthologs and refined by the inclusion of intermediate species as they become available. However, it is already providing a rigorous dataset for applications including annotation by functional transfer, comparative analysis, evolutionary analysis and hypothesis development. [12] Future analysis of the nature of the evolutionary events shaping these two genomes will determine more fully how the biological capabilities of these two organisms are manifested in their respective protein complements.

3.5 Lineage Specific Gene Loss

Comparative analysis of *S. pombe* and *S. cerevisiae* identified lineage specific gene losses as a major contributor to the shaping of eukaryotic genome content (Aravind et al. 2000). This analysis identified approximately 300 genes which were either lost from, or diverged beyond expectation in *S. cerevisiae* but present in *S. pombe*. A large number of these genes were also conserved in other nonfungal eukaryotes. Co-elimination of functionally connected groups in *S. cerevisiae,* including some subunits of the signalosome and the spliceosome and all components of the RNAi machinery, were recorded. Some of the proposed gene losses reported by Aravind and colleagues, including *S. cerevisiae MEC3* and

---

[12] The pre-publication ortholog table is available on request.

DDC1, which are the functional orthologs of *S. pombe* Hus1 and Rad9 respectively, would be more appropriately described as diverged beyond expectation (Sunnerhagen 2002). Manual inspection of alignments and protein clusters has since identified 498 protein coding genes which are absent from *S. cerevisiae* but conserved in *S. pombe* and other species and 346 protein coding genes conserved in *S. cerevisiae* but absent from *S. pombe.* Sequences absent from *S. pombe* but present in *S. cerevisiae* are more often fungally conserved (and may therefore have evolved since the divergence of the two yeasts, or be rapidly evolving) while those absent from *S. cerevisiae* are frequently universally eukaryotically conserved [13].

3.6 Orphan and species-specific sequences

One of the most unexpected findings of the *S. cerevisiae* genome project was the sheer number of completely unstudied genes. Only 40%-50% of identified genes could be assigned a preliminary process or function from similarity or experimentation. A staggering ~30% of the gene set had remained elusive to genetic or biochemical techniques in *S. cerevisiae* and appeared to have no homolog in any other sequenced species and became known as orphans (Oliver et al. 1992; Goffeau et al. 1996).

The existence of orphans can only be attributed to; i). spurious ORFs which are not protein coding ii). the acquisition of novel species specific functions by the generation of *de novo* genes and proteins iii). rapidly evolving proteins for which the sequence similarity between the available species is obscured, and species closely related enough to detect orthologs have yet to be sequenced (Wood et al. 2001).

Over the past decade the number of orphans in all species has decreased rapidly, either through experimentation, the detection of distant orthologs or the sequencing of additional species. This is illustrated by the sequencing of the *Saccharomyces (sensu stricto)* quartet which resulted in only 18 genes being identified as *Saccharomyces cerevisiae* specific (Kellis et al. 2003). However, many of these now form numerous largely *Saccharomyces* and *hemiascomycete* specific families.

There is an accumulating body of empirical and conjectural evidence that many apparent orphans or phylogenetically restricted genes are more rapidly evolving than broadly conserved genes (Copley et al. 2003). These genes also appear to be frequently implicated in processes which involve interacting with and monitoring of external environmental signals. The sequence of the close *S. cerevisiae* relative *K. lactis* identified orthologs for previous orphans and showed that sequence similarity was, on average, lower for these than for more ubiquitously conserved genes (Ozier-Kalogeropoulos et al. 1998). Gaillardin and colleagues indicated that hemiascomycete specific proteins are highly represented in the functional classes of cell wall organization, extracellular and secreted proteins and transcriptional regu-

---

[13] These lists are accessible via GeneDB http://www.genedb.org/shortcuts.jsp.

lators, suggesting that these functional groups diverge more rapidly than other classes of protein (Gaillardin et al. 2000). Reports of rapid divergence of genes involved in taxon specific processes are not confined to fungi. Since the divergence of the mosquito *Anopheles gambiae* and the fruit fly *Drosophila melanogaster*, proteins involved in environmental defenses and signal transduction, have evolved faster on average than those involved in catalysis and maintenance of cellular structural integrity (Zdobnov et al. 2002; Domazet-Loso and Tautz 2003). Similarly, in comparisons between pufferfish and human, genes related to immunity and gametogenesis were identified as rapidly evolving (Aparicio et al. 2002).

There are now fewer than 500 complete orphans (less than 10% of the protein complement) remaining in *S. pombe* where experimentation has provided no clues about the process or function, and similarity has identified no orthologs or conserved domains. The majority of these are potential plasma membrane or cell surface molecules (based on sequence analysis of potential transmembrane domains, GPI anchors, N-terminal signal sequences and glycosylation sites) often identified as frequently rapidly evolving, and often involved in specific environmental adaptations.

Although many orphans are likely to be taxon specific adaptations, the detection of distant similarities between *S. pombe* and *S. cerevisiae* continues to reduce the *S. pombe* orphan set by identifying gene families, which although very divergent, are universally conserved 'core' genes. These include orthologous clusters containing *S. pombe* Rec10, Hop1, Spc24, Spc25, Sgo1, Nse1 (Lorenz et al. 2004; Asakawa et al. 2005; Kitajima 2004; Fujioka et al. 2002). Often, these divergent orthologs are part of the large proteinaceous complexes, for example those involved in chromosome synapsis and segregation. It is possible that the absence of interactions with invariable organic compounds (macromolecules, cofactors, substrates) reduces the selective pressures resulting in sequence conservation, because sequences can evolve via complementary mutations in interacting partners. Many components of these large complexes do not appear to be conserved. Detection of distant orthologs will almost certainly further reduce the orphan set to reveal the truly genus specific components of these species.

4 Comparative and functional genomics

4.1 Gene expression studies

The development of microarray technologies enabling the analysis of thousands of expression probes in parallel, has provided a mechanism to derive and test broad hypotheses on a genome wide basis, through the study of global expression profiles for defined developmental or lifecycle stages or under specific environmental conditions (DeRisi et al. 1997). The effect of perturbations to these systems (either natural or induced) can also be evaluated. Moreover, the integrated analysis of microarray expression data not only provides insights into global transcription pattern but it may also provide insights to function as co-expressed genes are likely to be involved in similar processes (Eisen et al. 1998).

S. pombe microarray data and analyses are now available for a number of biological processes fundamental to cell survival. These include sexual development and meiosis (Mata et al. 2002), stress responses (Chen et al. 2003), and the mitotic cell cycle (Rustici et al. 2004). The availability of complementary microarray datasets for *S. cerevisiae,* and a curated inventory of orthologous pairs, also allows the comparative analysis of these transcriptional programs.

During the fission yeast transcriptional program for sexual development, almost 2000 genes were significantly up-regulated in four temporal classes corresponding to the four main stages of sexual differentiation (Mata et al. 2002). Five chromosomal regions were highly enriched for meiotically induced genes. Significantly, four of these regions were close to the usually transcriptionally inactive regions at the telomeres. This raises the possibility that spatial arrangement has a role in the activation of clusters of genes in this process. Of all conditions studied, genes upregulated during sexual development show a lower proportion conserved between the two yeasts (Mata and Bähler 2003; see also sections 3 and 4.3.3). Both of these results are consistent with the observation that the subtelomeric regions of *S. pombe,* and other eukaryotes, harbour an increased density of apparently species specific families (see also section 2.3). The observed up-regulation of telomerically encoded and species-specific genes at meiosis may therefore also be significantly correlated.

The evaluation of transcriptional responses to environmental stress defined a core environmental stress response (CESR) in *S. pombe* common to all, or most stresses (Chen et al. 2003). A substantial overlap between these, and the CESR genes of budding yeast was demonstrated showing that many stress induced changes are evolutionarily conserved. Finally, comparisons of global expression data for the cell-cycle control of transcription have revealed conservation of transcription factors between fission yeast and budding yeast, yet major differences in regulatory circuits (Rustici et al. 2004). Periodic transcription appeared not to be conserved, except for a core set of ~40 genes expected to be critical for cell cycle control.

Transcriptional control may be the primary mechanism for gene regulation but this operates at multiple levels from the sequence level (i.e. recognition and binding of transcription factors), to the chromatin level (i.e. histone modification status) and the nuclear (level based on the 3D compartmentation of the genome in the nucleus; reviewed in van Driel et al. 2003). Gene expression is also controlled at additional levels: transcripts are regulated by their localization, processing and decay. Microarrays are being successfully exploited to evaluate various aspects of regulation by extensions to the original technology including chromatin immuno-precipitation (ChIP)-on-chip for the identification of binding sites for transcription factors and other DNA binding proteins (reviewed in Pollack and Iyer 2002). Other innovations include the analysis of polysome bound mRNA to determine global translation rates (Pradet-Balade et al. 2001), and combinatorial approaches to data analysis. The first *S. pombe* experiments correlating spatial genome expression patterns with specific chromatin modifiers have identified telomeric clustering of some of the target genes (Hansen et al. 2005). Experiments using ChIP-on-chip and polysome bound RNA also underway (personal communication, J Bähler) and will provide a wealth of data for the reconstruction of regulatory networks.

Microarray experiments using the yeast models will undoubtedly continue to be informative in terms of the biology of unicellular eukaryotes, and in providing a framework for evaluating what can be successfully achieved using microarray analysis for the understanding of the gene expression programmes of more complex organisms.

4.2 Regulatory sequences

The complete understanding of an organism's functional capabilities will depend not only on the analysis of individual gene products and their interactions, but also on the concurrent identification of shared regulatory motifs in the genome. Although the prediction of regulatory motifs is substantially more difficult than gene prediction, pattern discovery methods have been used with some success to identify potential regulatory patterns in the *S. cerevisiae* genome (Brazma et al. 1998; Ettwiller et al. 2003). Comparative genomics approaches relying on synteny using closely related yeasts, have also been successful for the *Saccharomyces* genus (Cliften 2001; Kellis et al. 2003). However, the lack of any sequenced yeast displaying synteny with fission yeast precludes analyses of this type at present.

There are currently around 55 experimentally verified transcription factor binding site motifs reported for *S. cerevisiae* (Kellis et al. 2003). However, fewer than a dozen transcription factor binding site motifs are so far experimentally identified in fission yeast (K. Kivinen, personal communication). Despite a similar genome size, the intergenic regions are significantly larger for *S. pombe* than for *S. cerevisiae,* which may be indicative of more complex regulatory mechanisms (see section 2.7). The availability of the genome sequences of these two yeast species provides an opportunity to assess the similarities and differences by the comparison of pattern discovery methods and assessment of the number and type of motifs found by applying the same procedures to evolutionarily distant yeasts.

Additional information can be extracted from microarray data by targeted pattern discovery; based on the assumption that genes involved in the same biological processes, and genes with similar expression patterns are more likely to share regulatory mechanisms. *S. pombe* and *S. cerevisiae* data clustered by sequence similarity, co-annotation or co-expression, coupled with evaluation of pattern significance, were evaluated for over-represented motifs (K Kivinen, PhD thesis; http://www.sanger.ac.uk/Info/theses/). Initial comparisons confirmed expectations that the two yeasts were too divergent for comparative genomics approaches using pairwise alignments of predicted regulatory regions of orthologous sequences. However, analyses based on the comparison of functionally connected genes and co-expressed genes have provided a comprehensive set of sequence patterns, many of which are likely to have regulatory roles in one, or both yeasts. Firstly, analysis of co-annotated clusters of genes identified all but two of the known regulatory sites in fission yeast, and novel regulatory sites (both upstream and down-stream) were identified for both yeasts (Kivinen et al. manuscript in preparation). Secondly, analysis of co-expressed clusters from microarray data during meiotic differentiation, stress response and mitotic cell cycle were studied using the published datasets for both organisms (Chu et al. 1998; Mata et al. 2002; Chen et al. 2003; Gasch et al. 2000; Rustici et al. 2004; Spellman et al. 1998). This approach also identified most known regulatory sites including the patterns common to the two yeasts and many novel potential regulatory motifs (Chen et al. 2003; Rustici et al. 2004). Additional observations from these studies include:

i.   The identification of unstudied, but shared motifs (K Kivinen, PhD thesis; http://www.sanger.ac.uk/Info/theses/; Kivinen et al. manuscript in preparation).

ii.   An extended functional role for the FLEX site which is conserved from yeast to man (previously identified as meiosis specific) through a likely involvement in both meiotic and mitotic cell cycles (Rustici et al. 2004).

iii.   Approximately 50% of known budding yeast and fission yeast regulatory sites show a spatial bias relative to translation start sites (K Kivinen, PhD thesis; http://www.sanger.ac.uk/Info/theses/; Kivinen et al. manuscript in preparation).

iv.   A set of genes containing a downstream motif in the 3' UTR were identified. This motif, an AU-rich element (ARE), is involved in the mRNA stability of interferons, cytokines and proto-oncogenes (reviewed in Chen and Shyu 1995). The same element has recently been implicated in the stability of the periodically abundant cyclin dependent kinase (CDK) inhibitor *rum1* mRNA in fission yeast (Daga et al. 2003).

v.   Sets of genes containing novel downstream motifs which appear to have a functional role (Groocock et al. manuscript in preparation; K Kivinen, PhD thesis; http://www.sanger.ac.uk/Info/theses/; Kivinen et al. manuscript in preparation).

It is known that some regulatory patterns have survived millions of years of evolution with no apparent change, for example, the shared regulatory sites MCB (Lowndes et al. 1992), and ATF/CRE (Jones and Jones 1989). For others, sequence patterns have diverged but the functional role has been retained. However, in the majority of cases, these two model yeasts have diverged so far from each other that their regulatory regions appear to be unrelated (K. Kivinen, personal communication). Extension of the complementary approaches of co-expression and co-annotation, for the identification of regulatory regions, will have enormous potential as the expression datasets increase in coverage; annotation increases in specificity; and analysis tools for identifying similarities in expression pattern and sequence improve. Future developments will undoubtedly support the use of pattern discovery as a predictive tool for suggesting functional links between groups of genes.

4.3 Integrative comparative studies

Complete genomes and their associated data are providing the opportunity to systematically examine the connections between the determinants of evolutionary history and other quantifiable characteristics of genes and proteins. Global correlations between different types of data; either genome wide experimental observations, computationally derived data, or genome wide functional annotations, can be assessed. Preliminary comparisons are beginning to provide insights into the relative contributions of these quantifiable characteristics to the biological constraints and selective pressures which determine genome content.

4.3.1 Dispensibility and divergence

A pilot gene deletion project was undertaken to estimate the percentage of essential genes in fission yeast, investigating 100 contiguous CDS (Decottignies et al. 2003). The percentage of essential genes was found to be 17.5%, almost identical to the 17.8% that are essential for *S. cerevisiae* growth on a rich medium (Garrels 2002). Amongst the 81 *S. pombe* genes with a predicted homolog in *S. cerevisiae*, 88% (71 genes) showed the same deletion phenotype in both yeasts. Of the 15 essential fission yeast genes, only 10 (67%) are also essential for budding yeast growth. Therefore, despite the absolute percentage of essential genes being almost identical between the two yeasts, only two-thirds of these appear to overlap. This did not appear to be due to gene duplication and functional redundancy for any of the genes studied. A correlation was observed between the likely time of origin of a gene and dispensability, leading to the conclusion that more ancient genes (maintained in all eukaryotic, or all eukaryotic and prokaryotic species sequenced) are more likely to be essential, and yeast specific genes are less likely to be essential. Previous analyses of both *C. elegans* and *S. cerevisiae* revealed similar conclusions (Fraser et al. 2000; Garrels 2002).

A relationship between evolutionary rate and fitness has proven difficult to detect, but Hirsh and Fraser demonstrated that there is a highly significant correlation between protein dispensability and evolutionary rate (based on the number of substitutions per amino acid site using *S. cerevisiae*) which is not always detectable from categorical comparisons of essential and non essential proteins (Hirsh and Fraser 2001). The relationship is apparently obscured because proteins with small but measurable fitness effects can be considered essential in evolutionary terms. It is likely that many highly conserved proteins involved in central processes are not lethal because biological systems make extensive use of 'fail-safe' mechanisms.

4.3.2 Correlations with gene loss

Krylov and colleagues explored the connection between the propensity of a gene to be lost in evolution, protein sequence divergence, dispensability, the number of protein-protein interactions and expression level for genes in clustered orthologous groups for seven fully sequenced eukaryotic genomes including *S. pombe*

(Krylov et al. 2003). Significant correlations were detected between the potential for a gene to be lost and all other categories. Genes with a lower propensity to be lost accumulate fewer changes, and tend to be essential, highly expressed and have many interaction partners. However, in this analysis no appreciable correlation was found between evolution rate and dispensability.

4.3.3 Correlations with expression level

Fission yeast gene expression levels were compared to the degree of species conservation, by integrating expression data with core eukaryotic genes (present in worm, budding yeast and fission yeast), yeast specific genes (present in budding and fission yeast) and *S. pombe* specific genes (Mata and Bähler 2003). In vegetatively growing cells, *S. pombe* specific genes tended to be expressed at a lower level and a disproportionate number of core conserved genes were highly expressed. These results support the hypothesis that core genes carry out basic functions, and are globally expressed in all conditions. Conversely, in sexually differentiating cells, although many core genes were still expressed, the bias was weaker, and many *S. pombe* specific genes became highly expressed. This enrichment of expression of *S. pombe* specific genes supports the hypothesis that organism-specific genes function in specialised processes (see also section 3.4.2 and 3.6). Organism specific genes were over-represented at all stages of sexual differentiation but the trend was most prevalent for genes in the cluster involved in chromosome pairing and recombination (meiotic prophase). This is consistent with observations that meiotic structural proteins are poorly conserved across eukaryotes (Villeneuve and Hillers 2001; see also section 3.6). It is speculated that differences in the chromosome pairing machinery may help to prevent fruitful meiosis between closely related organisms and drive the separation between species.

4.3.4 Conservation level and interaction number

Theoretical arguments propose that proteins evolve more slowly if they participate in many interactions. In addition, structural analysis has shown that amino acid residues at protein interfaces are generally more conserved than the average for all proteins (reviewed in Teichmann 2002). In order to investigate globally the constraints protein-protein interactions place on sequence variation, the sequence similarities of *S. cerevisiae* proteins were compared to their *S. pombe* orthologs and evaluated with respect to interaction type (Teichmann 2002). The large variation in sequence conservation between orthologs (>20-<90% identity) was used to demonstrate that stable complexes were, on average, more conserved than proteins involved in transient interactions. However, the trend for complexes to be more highly conserved than transient interactions, which are in turn more conserved than monomers, was found to be independent of whether a protein is involved in informational activities (transcription, translation, and replication) or not. This trend was also independent of protein dispensability. In contrast, Jordan et al. identified only a weak relationship between the number of protein interactions and evolutionary rate (estimating evolutionary rate from *S. cerevisiae/ S. pombe* comparisons, and using *S. cerevisiae* interaction data), and concluded that only the most prolific interactors showed a reduction in evolutionary rate (Jordan et al. 2003). Two further studies have subsequently identified a significant positive correlation between the number of protein-protein interactions in *S. cerevisiae* and evolutionary distance to other organisms including *S. pombe* (Fraser et al. 2003; Pagel et al. 2004). A preference for interacting proteins to be conserved together was also identified, but no bias was detected with respect to functional roles (Pagel et al. 2004). Inevitably studies of this type will be difficult to perform and interpret with current interaction datasets which are biased for well studied genes, error prone and incomplete.

4.3.5 Dispensability, distribution and interaction number

The availability of global datasets for genetic and physical interactions has made biology amenable to the application of techniques and theories governing the formation, behaviour, and development of networks. It has been proposed that most biological networks have a scale-free "small world" topology (Jeong et al. 2000). That is, most 'nodes' have a small number of connections, but a few highly connected nodes (or hubs) hold the network together. It was subsequently shown that centrality in the network (i.e. highly connected proteins) correlated positively with lethality (Jeong et al. 2001). Kunin and colleagues used *S. cerevisiae* protein interaction data to trace the origin of proteins in the interaction network and to evaluate the evolution of a scale-free topology (Kunin et al. 2004). They did not detect a direct correlation between connectivity and age as expected by the 'preferential attachment model' whereby older nodes should display higher connectivity, and proposed that this is due to the functional heterogeneity of the protein interaction network. Instead, it was found that proteins which evolved after the split which lead to the fungi, and those which evolved after the split from fission yeast, displayed on average, reduced connectivity. Surprisingly however, the proteins of oldest origin did not show the highest connectivity. The majority of the most highly connected proteins are found to have emerged during the eukaryotic radiation which seems to reflect the emergence of many highly connected proteins involved in eukaryotic cellular organization, such as cytoskeleton components, transcription complexes and the nuclear pore. They observed that different functional classes display different average connectivity. Specifically, proteins involved in cell wall organization and biogenesis appear to be the least connected, followed by proteins involved in transport, binding and metabolism. Proteins of unknown function also have lower levels of connectivity. Conversely, proteins involved in transcription, replication, cellular processes and regulatory functions have, on average, almost twice as many binding partners. The age of a protein also correlates well with what is known about its function; far fewer ancient proteins are uncharacterised, which is expected because phylogenetically extended families tend to be well studied. It is proposed that protein function determines the types of binding partner, degree of connectivity and the time of emergence in the network.

4.4 Section summary

Further genome wide studies integrating information from function, physical interactions, lethality, sequence conservation, duplication and phylogenetic distribution will continue to define factors affecting the evolution and characteristics of a eukaryotic cell, and to assess their relative contributions to genome content. These will become more accurate as annotation, curation and methods for comparison improve, providing the potential to propose and test numerous evolutionary hypotheses on a genome-wide scale. An accurate and comprehensive model will also be a powerful predictive tool to determine which genes are likely to be involved in core eukaryotic processes, or species-specific adaptations based on phylogenetic distribution, copy number, evolutionary rate and network position.

5 Curation

The accumulation of biological data produced by genome-scale biology has required a revolution in the approaches used to describe, integrate and retrieve this huge volume of diverse information. Numerous attributes of gene products can be recorded during annotation or literature curation but the molecular activity (function), biological process and cellular localization (component) are generally considered the most immediately useful information to describe an organism's biology. Any robust system to capture these features of gene products has the following necessary or desirable requirements:

1. The ability to describe gene products consistently and unambiguously so that similar characteristics are grouped, (including the grouping of gene products for which no functional data is available)
2. To support the inherent pleiotropy of the data, recognising that gene products may have multiple functions, and locations, and participate in multiple processes
3. The ability to describe gene products using different levels of granularity (different levels of detail) depending how much is known or can be inferred (hierarchical)
4. Mechanisms to qualify annotations with different levels of confidence, and to support the annotations with a method or citation
5. Sophisticated consistency checks to maintain the integrity of the data
6. Be readily and rapidly extensible to incorporate new biological concepts
7. To support negative annotations
8. Be species independent, to support inter-organism queries
9. To enable researchers to retrieve specified groups of genes or to identify candidate gene products for specific functions

The annotation standards provided by controlled vocabularies, and more sophisticated 'ontologies' are now crucial to the annotation process for most genomes. These define 'terms' to describe aspects of a gene product's biology, which can be interpreted identically both within, and between organisms, by both biologists and computers. The most vital resource for maintaining consistent annotation of genes and gene products is provided by the Gene Ontology (GO) Consortium, which fulfils all of the nine requirements above, and is the annotation system of choice for the majority of model organism databases (MODs; http://www.geneontology.org; The Gene Ontology Consortium 2004). The GO Consortium is a collaborative open source project to develop shared controlled vocabularies, which are continually refined and expanded to reflect accumulating biological knowledge (Ashburner et al. 2000). The GO provides three ontologies to describe the orthogonal biological domains of biological process, cellular component and molecular function, in a species-independent manner.

5.1 Gene Ontology structure

Gene Ontology terms are arranged so that broader parents give rise to more specific children. The relationships are represented in the form of a directed acyclic graph (DAG), which is similar to a hierarchy, except that it captures biological relationships more realistically by allowing individual child terms to have many parent terms. At present, two types of relationship are implemented in GO ('is_a' and 'part_of'), although it is conceivable that other relationship types will be added in the future. For example, the cellular component term 'nuclear pore (GO:0005643)', has two parents, it is 'part_of' 'nuclear membrane (GO:0005635)' , and 'is_a' 'pore complex (GO:0046930)' (Figure 1 shows a screenshot of the cellular component term 'nuclear pore' in the 'Amigo' GO browser. This view shows the term 'nuclear pore' with all its parent terms, together with the numbers of *S. pombe* gene products associated with each term).
Every GO term must obey the 'true path rule'; this means every possible path from any term back to the root (most general term) must be biologically accurate. When a gene product is annotated to a term, it is therefore automatically annotated to all of the parent terms. For example, a gene product annotated to 'inner plaque of spindle pole body (GO:0005822)' is 'part_of' the 'spindle pole body (GO:0005816)' which is 'part_of' the 'spindle pole (GO:0000922)' and so forth back to the root node 'cellular component (GO:0005575)'. If a path back to the root node is incorrect for a valid annotation a 'true path violation' occurs and the ontology must be revised. This structure allows curators to assign properties at different levels of granularity depending how much is known, or can be inferred, about a gene product. Multiple associations (ontology terms) can be applied to a single gene product, reflecting the fact that a gene product may have several functions, be present in different locations, participate in different processes and interact with numerous other proteins.

AmiGO

nuclear pore

Accession: GO:0005643
Aspect: cellular_component
Synonyms:
    nuclear pore membrane protein
    GO:0005644
    NPC
    nuclear pore complex
Definition:
    Any of the numerous similar discrete openings in the nuclear envelope of a eukaryotic cell, where the inner and outer nuclear membranes are joined.

Term Lineage

Graphical View

```
all : all ( 4229 )
    GO:0005575 : cellular_component ( 3052 )
        GO:0005623 : cell ( 3038 )
            GO:0005622 : intracellular ( 2443 )
                GO:0043229 : intracellular organelle ( 2185 )
                    GO:0043231 : intracellular membrane-bound organelle ( 1837 )
                        GO:0005634 : nucleus ( 1034 )
                            GO:0005635 : nuclear membrane ( 69 )
                                GO:0005643 : nuclear pore ( 47 )
                GO:0005634 : nucleus ( 1034 )
                    GO:0005635 : nuclear membrane ( 69 )
                        GO:0005643 : nuclear pore ( 47 )
            GO:0016020 : membrane ( 1048 )
                GO:0012505 : endomembrane system ( 179 )
                    GO:0005635 : nuclear membrane ( 69 )
                        GO:0005643 : nuclear pore ( 47 )
                GO:0016021 : integral to membrane ( 545 )
                    GO:0046930 : pore complex ( 47 )
                        GO:0005643 : nuclear pore ( 47 )
        GO:0043226 : organelle ( 2185 )
            GO:0043229 : intracellular organelle ( 2185 )
                GO:0043231 : intracellular membrane-bound organelle ( 1837 )
                    GO:0005634 : nucleus ( 1034 )
                        GO:0005635 : nuclear membrane ( 69 )
                            GO:0005643 : nuclear pore ( 47 )
            GO:0043227 : membrane-bound organelle ( 1837 )
                GO:0043231 : intracellular membrane-bound organelle ( 1837 )
                    GO:0005634 : nucleus ( 1034 )
                        GO:0005635 : nuclear membrane ( 69 )
                            GO:0005643 : nuclear pore ( 47 )
```

Fig. 1. A screenshot of the cellular component term nuclear pore and its parent terms in the 'Amigo' GO browser (B. Marshall and S. Lewis, unpublished software). The numbers in parentheses show the number of *S. pombe* gene products associated to each term.

5.2 Gene Ontology implementation

GO collaborators use the GO schema to annotate individual gene products. These annotations are maintained in a common file format (the gene association file: see http://www.geneontology.org/GO.annotation.shtml?all#file), which is incorporated into the contributing database (GeneDB http://www.genedb.org/ in the case of *S. pombe*) and submitted to the Gene Ontology consortium.

A comprehensive set of GO annotations (gene associations) are provided for *S. pombe* within GeneDB. These associations are derived from a number of non-redundant sources, and are continually refined and updated. There are currently 4889 manual gene associations for 1300 gene products. Of these, 2013 are derived

Table 3. Sources of the non-redundant GO data for *S. pombe* in GeneDB, by evidence code. Key: [1] GOC:pombekw2GO, [2] GOC:ec2go, [3] GOA:interpro, [4] GOA:spkw, [5] GOA:spec.

Evidence code	GeneDB	GOA/Uniprot
IEA	7732 [1]	4162 [3]
	364 [2]	1530 [4]
		46 [5]
IMP	571	12
IDA	408	23
IEP	5	0
IGI	149	1
IPI	249	4
ISS	2876	6
IC	294	9
NAS	16	9
TAS	321	1
	12985	5738

IEA = inferred from electronic annotation, IMP = inferred from mutant phenotype, IDA = inferred from direct assay, IEP = inferred from expression profile, IGI = inferred from genetic interaction, IPI = inferred from physical interaction, ISS = inferred from sequence similarity, IC= inferred by curator, NAS = non traceable author statement, TAS = traceable author statement. Also see the online GO Evidence documentation http://www.geneontology.org/GO.evidence.shtml?all

from experimental data via literature curation (1008 publications) and are supported by the appropriate evidence code for the type of experiment (see Table 3 legend for a list of evidence codes). A further 2876 are 'inferred from sequence similarity' (ISS evidence code) based on manual inspection of sequence alignments to characterised proteins.

The manual gene associations are supplemented by electronically inferred annotations (IEA evidence code) from *S. pombe* primary annotation, the Gene Ontology Annotation (GOA) database (Camon 2004) and UniProt, (Apweiler 2004). These include:

i.   A keyword mapping from the primary *S. pombe* annotation to GO terms (GOC:pombekw2GO)

ii.   A mapping of enzyme commission (EC) numbers assigned to GeneDB entries (GOC:ec2go) and Uniprot entries (GOA:spec) to GO terms

iii.   A mapping of Interpro families and domains to GO terms (GOA:interpro)

iv.   A mapping of Uniprot keywords to GO (GOA:spkw)

Within the GeneDB database, redundancy of GO mappings is prevented by presenting IEA mappings only when they are more granular than manual associations. These provide 13834 additional associations giving a total of 18723 non redundant associations. The sources of these associations and their distribution between the various evidence codes are summarised in Table 3.

Presenting automated mappings and integrating bioinformatics predictions with the manual annotations, provides greater annotation coverage to the *S. pombe* research community in the absence of manual curation. For instance, a researcher looking for a specific activity like dolichyl-phosphate beta glucosyltransferase activity (GO:0004581) may retrieve all genes which are inferred by sequence similarity or electronic annotation to have the activity of the broader parent terns UDP-glucosyltransferase activity (GO:0035251) or glucosyltransferase activity (GO:0046527). Inferred annotations not only allow researchers to identify groups of candidate genes, but are also beneficial to the curation process, as they alert the curator to relevant terms which may have been overlooked, or to missing relationships in the ontologies. Moreover, assessing the output of global mapping resources from the perspective of individual gene products identifies erroneous mappings and allows them to be corrected, which can radically reduce false positive mappings for the automated annotation of other organisms. One future aim of the curation strategy for *S. pombe* is to process the literature backlog, in order to convert IEA associations to experimentally supported evidence codes where applicable, or ISS codes supported by a manually assessed alignment to a characterised protein or protein family if direct experimental results are not available.

Three qualifiers (NOT, contributes_to and colocalizes_with) are available within GO to modify the interpretation of the annotation. The 'NOT' qualifier is used to support negative annotations. This would normally be used if experimental evidence has shown a particular assignment not to be true, but where an association might otherwise be made based on other evidence. The 'contributes_to' qualifier is used when a complex has an activity but the individual subunits do not, for example the subunits of RNA polymerases. The 'colocalizes_with' qualifier is used when gene products are associated transiently or peripherally with a cellular component, or where the resolution is inconclusive.

Within *S. pombe* GeneDB, additional qualifiers are applied to GO associations to increase their informational content further. Examples include 'phase' qualifiers' used to specify the life cycle or cell cycle stage when a particular localization is observed, or process occurs, which is especially useful for pleiotropic gene products. A selection of qualifiers are also used in conjunction with the 'inferred from genetic interaction' (IGI) evidence code to establish the type of genetic interaction (epistasis, localization_dependency, acts_upstream_of, parallel_pathway etc). These qualifiers provide information about the position in the genetic hierarchy, the directionality of the interaction or whether the gene product is in the same, or a different, pathway and will be pertinent to the reconstruction of genetic networks.

5.3 Dynamic aspects of the Gene Ontology and the associated annotations

The Gene Ontology is a dynamic resource. Changes to the ontologies are frequently made to correct legacy terms and relationships, to improve consistency, and to add new terms and relationships as advances are made in biology. Literature curation is not a passive process and necessarily includes contributing to the development of the GO by identifying missing relationships, extending vocabularies, refining existing term definitions and identifying new terms. New terms added

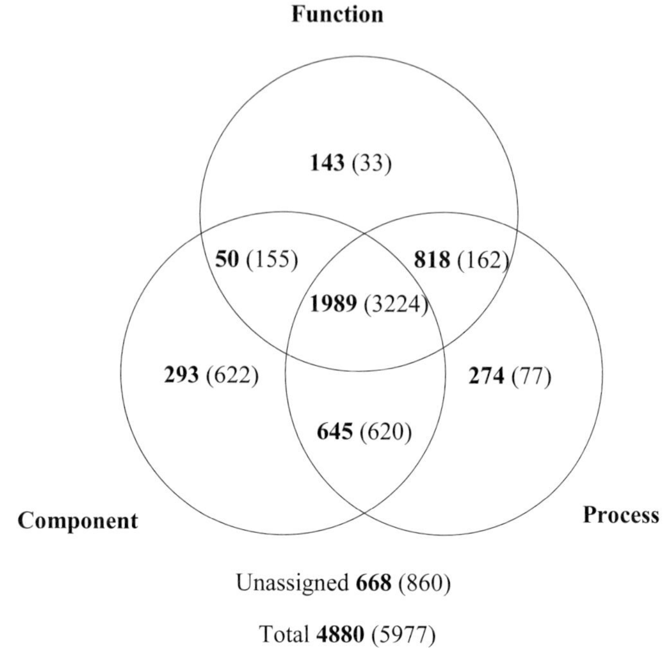

Fig. 2. Gene Ontology association coverage for *S. pombe* showing the number of gene products with at least one association to each of the three ontologies; molecular function, biological process and cellular component. The corresponding figures for *S. cerevisiae* are shown in parentheses.

recently to describe biological phenomena studied in *S. pombe* include the cellular component terms 'medial ring (GO:0031097)' and 'linear element (GO:0030998)', the biological process terms 'sister chromatid biorientation (GO:0031134)' and 'horsetail movement (GO:0030989)' and the molecular function terms 'ornithine N5-monooxegenase activity (GO: 0031172)' and 'glucan endo-1,3-alpha glucosidase activity (GO:0051118)'.

5.4 *S. pombe* gene associations, coverage and comparison with *S. cerevisiae*

Of the 4880 known and predicted protein coding genes 4215 are assigned to at least one GO term (Figure 2). This includes 3726 with at least one biological process term, 2977 with at least one cellular component term and 3000 with at least one molecular function term. Only 668 genes considered likely to be protein coding, have no known or predicted component process or function. I n contrast,

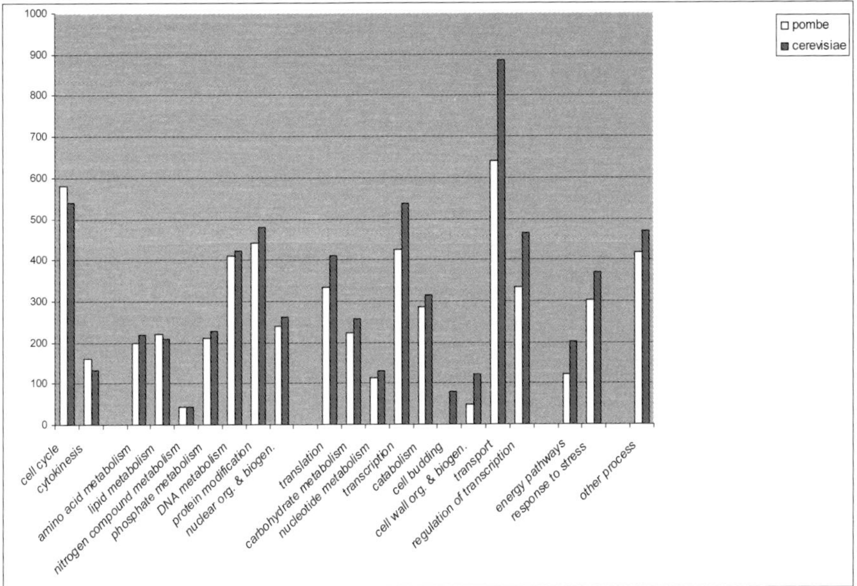

Fig. 3. A comparative overview of the distribution of 'high level' GO biological process annotations for *S. pombe vs. S. cerevisiae*. The *S. cerevisiae* totals are derived from SGD manual annotations supplemented by GOA mappings. Annotations to 'unknown' process function and component terms, and annotations for non-protein coding genes have been filtered. The terms are not mutually exclusive as terms may belong to more than one category, or, in the case of transcription and transcriptional regulation, one may be a complete subset (child) of another.

S. cerevisiae has more gene products assigned to at least one term in all three ontologies (3224), but also, a greater number of genes with unknown function process or component (860)[14].

A comparative overview of the distribution of 'high level' annotations for the process ontology annotations of the two yeasts is presented in Figure 3. This distribution of annotations corresponds with what is known about the broad biology of these two organisms based on the accumulation of literature and comparative analysis reviewed earlier in this chapter (see section 3).

The GO terms which have annotations in approximately equal numbers in both yeasts are biased towards universally conserved proteins involved in informational processes. The GO terms which have annotations in increased numbers for S. cerevisiae have a larger number of species specific genes (those without an apparent

---

[14] *S. cerevisiae* figures were derived from a non redundant set combining SGD manual assignments and IEA annotations derived from GOA as described for *S. pombe* in the text. *S. cerevisiae* annotations to noncoding RNAs were removed, as these have not yet been implemented for *S. pombe*. Annotations to 'cellular_component_unknown', 'biological_process_unknown' and 'molecular_function_unknown' were removed for both organisms.

ortholog) and genes which are commonly duplicated (many to many), and are processes implicated more often in interactions with the environment (nutrient acquisition, toxicity modulating and aspects of regulation; see also section 3).

5.5 Searching and accessing GO

The *S. pombe* gene associations can be accessed via the Gene Ontology consortium website and the GeneDB website using the Amigo Gene Ontology browser (http://www.godatabase.org/cgi-bin/amigo/go.cgi; http://www.genedb.org/ amigo/perl/go.cgi; B. Marshall and S. Lewis, unpublished software). Amigo allows the browsing of GO terms and the relationships between them and the retrieval of gene products associated with those terms, or all the terms associated with specific gene products. It also allows searching of the ontology by term name and of the annotations by gene name, sequence, evidence code or species.

GeneDB also supports a query facility with Boolean capability, this allows the results of queries to any GO term to be combined using AND or OR. These queries can also be combined with other biological attributes including protein domains, keywords, protein length and Mass, presence of transmembrane regions, signal peptides, exon number, and chromosomal location. The results can be saved to a query history, combined with previous queries (added, subtracted and intersected) and downloaded in a number of formats (for example, as gene names, description, protein or nucleotide sequence; http://www.genedb.org/gusapp/serv-let?page=boolq; Hertz-Fowler et al. 2004).

The gene association files are available for download from the GO consortium and the Wellcome Trust Sanger Institute (WTSI) websites (http://www.geneontology.org/GO.current.annotations.shtml; ftp://ftp.sanger. ac.uk/ pub/yeast/pombe/Gene_ontology. The WTSI files also include the non-redundant associations from other sources.

5.6 Curation summary

The curation process is improved by GO, far beyond the provision of controlled vocabularies to consistently describe biological phenomenon. The GO also provides a framework for quality control, of both data input, and of the subsequent revisions or extensions to biological knowledge which affect the description and implementation of concepts within the ontologies. In addition, it provides a mechanism for the identification of relevant terms; either by the application of curated mappings (i.e. Interpro to GO, or EC to GO); or by the consideration of commonly co-annotated terms from orthogonal ontologies. The resulting associations, in turn, provide robust datasets for inter-species comparisons, and facilitate uniform queries based on shared biological roles. Increasingly, GO is being used by biologists to identify interesting gene products, and has the potential to identify areas and genes which are relatively unstudied. All of these applications become increasingly powerful as the annotations are refined ontologies become more complete.

6 Future prospects

The availability of the genome sequence has revolutionised experimental research for *S. pombe*. When genome sequencing began, the number of studied genes was around 200; around 1400 genes now have some degree of published experimental characterization. Fission yeast is therefore no longer only the bastion of cell cycle research. Its efficacy as a general eukaryotic model is now promoting research in areas of cell biology that were traditionally more confined to *S. cerevisiae*. Despite the advances made by the *S. pombe* research community and the enormous potential of *S. pombe* as a eukaryotic model organism, the published genome-wide functional interrogations are currently limited to microarray analyses. Genome wide datasets for deletion and localization are in progress are therefore eagerly anticipated.

Functional and comparative genomics initiatives and the emerging field of systems biology are intimately dependent on accurate parts list and continued primary sequence analysis is therefore paramount. The availability of close *S. cerevisiae* relatives has been instrumental in refining gene structures and identifying missing genes for this organism, resulting in alterations to more than 10% of the gene complement (Kellis et al. 2003; Brachat et al. 2003; Cliften et al. 2001). *S. pombe* will benefit similarly from the availability of the genomes of *S. japonica*, *S. octosporus* and *S. kambucha* which have recently been approved for sequencing as part of the Whitehead Institute Fungal Genomes Initiative (http://www.broad.mit.edu/annotation/fungi/fgi/candidates.html).

Whole genome comparisons are now central to the development and testing of hypotheses relating to the mechanisms of evolution. Accurate inventories of orthologs and partitioning of conserved, non-conserved and dubious proteins will provide accurate functional transfer, but will also benefit integrative studies to provide a framework for the dissection of species similarities and differences. The identification of further factors that determine, or correlate strongly with the rate of duplication, divergence and loss of proteins will continue to reveal the prevailing trends in protein evolution. Additional data partitions based on biology (i.e. metabolic versus non-metabolic, nuclear versus cytoplasmic) are likely to reveal more subtle correlations and evolutionary constraints.

Sequencing projects need a commitment to consistent curation to make meaningful computational comparisons based on functional roles a realistic prospect. However, data curation remains a major bottleneck for comparative analysis. The Gene Ontology (GO) schema provides a workable framework to make accurate and consistent curation a feasible goal. As the annotation becomes more complete, and GO is refined and extended in coverage, the possibilities for *in silico* research will increase in parallel. Integration of bioinformatics predictions with experimental data will in turn provide testable hypotheses and models for bench scientists.

Ultimately the parts lists provided by genome sequencing and curation, and the data generated by functional genomics experiments are creating a platform for systems biology and network approaches for the elucidation of biological function. Systems biology aims to describe the global organization of genes and proteins in the control and maintenance of cells and organisms. Ultimately, systems approaches will go far beyond the mere description of a network's connectivity and its global dynamics. However, to explore fully the nature of the relationships within and between the identified modules will require new approaches for obtaining organizing and analysing data (Nurse 2003). Integration of functional genomics datasets is paramount; integration will corroborate statistically significant data and improve functional predictions. Fraser and Marcotte have recently outlined some considerations for these systems and begun to assess how this might be achieved (Fraser and Marcotte 2004). A complete description of cellular networks is a realistic goal for the post-genomics era, and *S. pombe* is an exemplary organism to pioneer systems level research.

One of the challenges of biology is to identify the fundamental requirements for a functioning eukaryotic cell. This will be achieved by the integrated efforts of individual bench scientists and genome wide studies. Despite a general correlation in proteome size between fission yeast and budding it appears that fission yeast is more similar in protein complement to higher eukaryotes than any single celled organism sequenced so far. To quote Mitsohiro Yanagida in the review '*S. pombe* the model eukaryotic organism'; "Researchers who are seriously interested in the evolution and establishment of eukaryotic organisms must consider fission yeast as a premier organism for study" (Yanagida 2002).

Acknowledgements

The author would like to thank M Aslett, J Bähler and M Harris for proofreading comments. M Aslett and the GeneDB programmers for technical support, L Groocock for mitochondrial proteome reannotation, and the staff at SGD and the GO editorial office.

References

Aarstad K, Oyen TB (1975) On the distribution of 5s RNA cistrons on the genome of *Saccharomyces cerevisiae*. FEBS Lett 51:227-231

Akamatsu Y, Dziadkowiec D, Ikeguchi M, Shinagawa H, Iwasaki H (2003) Two different Swi5-containing protein complexes are involved in mating-type switching and recombination repair in fission yeast. Proc Natl Acad Sci 100:15770-15775

Alschtul SF, Gish W, Miller W, Myers EW, Lipman DJ (1990) Basic local alignment search tool. J Mol Biol 215:403-410

Alschtul SF, Madden TL, Schaffer AA, Zhang J, Zhang Z, Miller W, Lipman DJ (1997) gapped BLAST and PSI-BLAST a new generation of database search programs. Nucleic Acids Res 1:3389-3402

Aparicio S, Chapman J, Stupka E, Putnam N, Chia JM, Dehal P, Christoffels A, Rash S, Hoon S, Smit A, Gelpke MD, Roach J, Oh T, Ho IY, Wong M, Detter C, Verhoef F, Predki P, Tay A, Lucas S, Richardson P, Smith SF, Clark MS, Edwards YJ, Doggett N, Zharkikh A, Tavtigian SV, Pruss D, Barnstead M, Evans C, Baden H, Powell J, Glusman G, Rowen L, Hood L, Tan YH, Elgar G, Hawkins T, Venkatesh B, Rokhsar D, Brenner S (2002) Whole-genome shotgun assembly and analysis of the genome of *Fugu ribicans*. Science 297:1301-1310

Appelgren H, Kniola B, Ekwall K (2003) Distinct centromere domain structures with separate functions demonstrated in live fission yeast cells. J Cell Sci 116:4035-4042

Apweiler R, Bairoch A, Wu CH, Barker WC, Boeckmann B, Ferro S, Gasteiger E, Huang H, Lopez R, Magrane M, Martin MJ, Natale DA, O'Donovan C, Redaschi N, Yeh LS (2004) UniProt: The universal protein knowledgebase. Nucleic Acids Res 32:D138-D141

Aravind L, Watanabe H, Lipman DJ, Koonin EV (2000) Lineage-specific gene loss and divergence of functionally linked genes in eukaryotes. Proc Natl Acad Sci 97:11319-11324

Asakawa H, Hayashi A, Haraguchi T, Hiraoka Y (2005) Dissociation of the Nuf2-Ndc80 complex releases centromeres from the spindle-pole body during meiotic prophase in fission yeast. Mol Biol Cell 16:2325-2538

Ashburner M, Ball CA, Blake JA, Botstein D, Butler H, Cherry JM, Davis AP, Dolinski K, Dwight SS, Eppig JT, Harris MA, Hill DP, Issel-Tarver L, Kasarskis A, Lewis S, Matese JC, Richardson JE, Ringwald M, Rubin GM, Sherlock G (2000) Gene Ontology: tool for the unification of biology. Nat Genet 25:25-29

Bähler J, Wyler T, Loidl J, Kohli J (1993) Unusual nuclear structures in meiotic prophase of fission yeast: a cytological analysis. J Cell Biol 121:241-256

Barry JD, Ginger ML, Burton P, McCulloch R (2003) Why are parasitic contingency genes often associated with telomeres? Int J Parasitol 33:29-45

Barnitz JT, Cramer JH, Rownd RH, Cooley L, Soll D (1982) Arrangement of the ribosomal RNA genes in *Schizosaccharomyces pombe*. FEBS Lett 143:129-132

Bateman A, Coin L, Durbin R, Finn RD, Hollich V, Griffiths-Jones S, Khanna A, Marshall M, Moxon S, Sonnhammer EL, Studholme DJ, Yeats C, Eddy SR (2004) The Pfam protein families database. Nucleic Acids Res 32:D138-D141

Baum M, Ngan VK, Clarke L (1994) The centromeric K-type repeat and the central core are together sufficient to establish a *Schizosaccharomyces pombe* centromere. Mol Biol Cell 5:747-761

Behrens R, Hayles J, Nurse P (2000) Fission yeast retrotransposons Tf1 integration is targeted to the 5' ends of open reading frames. Nucleic Acids Res 28:4709-4716

Berbee ML, Taylor JW (1993) Dating the evolutionary radiations of the true fungi. Can J Bot 71:1114-1127

Birney E, Thompson JD, Gibson TJ (1996) PairWise and SearchWise: finding the optimal alignment in a simultaneous comparison of a protein profile against all DNA translation frames. Nucleic Acids Res 24:2730-2739

Blandin G, Durrens P, Tekaia F, Aigle M, Bolotin-Fukuhara M, Bon E, Casaregola S, de Montigny J, Gaillardin C, Lepingle A, Llorente B, Malpertuy A, Neuveglise C, Ozier-Kalogeropoulos O, Perrin A, Potier S, Souciet J, Talla E, Toffano-Nioche C, Wesolowski-Louvel M, Marck C, Dujon B (2000) The genome of *Saccharomyces cerevisiae* revisited. FEBS Lett 487:31-36

Bowen NJ, Jordan IK, Epstein JA, Wood V, Levin HL (2003) Retrotransposons and their recognition of pol II promoters: A comprehensive survey of the transposable elements from the complete genome sequence of *Schizosaccharomyces pombe*. Genome Res 13:1984-1997

Brachat S, Dietrich FS, Voegeli S, Zhang Z, Stuart L, Lerch A, Gates K, Gaffney T, Philippsen P (2003) Reinvestigation of the *Saccharomyces cerevisiae* genome annotation by comparison to the genome of a related fungus: *Ashbya gossypii*. Genome Biol 4:R45

Brazma A, Jonassen I, Vilo J, Ukkonen E (1998) Predicting gene regulatory elements in silico on a genomic scale. Genome Res 8:1202-1215

Broach JR, Li YY, Feldman J, Jayaram M, Abraham J, Nasmyth KA, Hicks JB (1983) Localization and sequence analysis of yeast origins of DNA replication. Cold Spring Harb Symp Quant Biol 47 Pt2:1165-1173

Camon E, Magrane M, Barrell D, Binns D, Fleischmann W, Kersey P, Mulder N, Oinn T, Maslen J, Cox A, Apweiler R (2004) The Gene Ontology Annotation (GOA) database: sharing knowledge in Uniprot with gene ontology. Nucleic Acids Res 32:D262-66

Chalker DL, Sandmeyer SB (1992) Ty3 integrates within the region of RNA polymerase III transcription initiation. Genes Dev 6:117-128

Chen CY, Shyu AB (1995) AU-rich elements: characterization and importance in mRNA degradation. Trends Biochem Sci 20:465-470

Chen D, Toone WM, Mata J, Lyne R, Burns G, Kivinen K, Brazma A, Jones N, Bähler J (2003) Global responses of fission yeast to environmental stress. Mol Biol Cell 14:214-229

Chervitz SA, Aravind L, Sherlock G, Ball CA, Koonin EV, Dwight SS, Harris MA, Dolinski K, Mohr S, Smith T, Weng S, Cherry JM, Botstein D (1998) Comparison of the complete protein sets of worm and yeast: Orthology and divergence. Science 282:2022-2028

Chikashige Y, Kinoshita N, Nakaseko Y, Matsumoto T, Murakami S, Niwa O, Yanagida M (1989) Composite motifs and repeat symmetry in *S. pombe* centromeres: Direct analysis by integration of NotI restriction sites. Cell 57:739-751

Chikashige Y, Ding DQ, Funabiki H, Haraguchi T, Mashiko S, Yanagida M, Hiraoka Y (1994) Telomere-led premeiotic chromosome movement in fission yeast. Science 264:270-273

Chu S, DeRisi J, Eisen M, Mulholland J, Botstein D, Brown PO, Herskowitz I (1998) The transcriptional program of sporulation in budding yeast. Science 282:699-705

Clarke L, Baum MP (1990) Functional analysis of a centromere from fission yeast: a role for centromere-specific repeated DNA sequences. Mol Cell Biol 10:1863-1872

Cliften PF, Hillier LW, Fulton L, Graves T, Miner T, Gish WR, Waterston RH, Johnston M (2001) Surveying *Saccharomyces* genomes to identify functional elements by comparative DNA sequence analysis. Genome Res 11:1175-1186

Cliften P, Sudarsanam P, Desikan A, Fulton L, Fulton B, Majors J, Waterston R, Cohen BA, Johnston M (2003) Finding functional features in *Saccharomyces* genomes by phylogenetic footprinting. Science 301:71-76

Clyne RY, Kelly TJ (1999) Genetic analysis of an ARS element from the fission yeast *Schizosaccharomyces pombe*. EMBO J 14:6348-6357

Copley R, Goodstadt L, Ponting C (2003) Eukaryotic domain evolution inferred from genome comparisons. Curr Opin Genet Dev 13:623-628

Dai J, Chuang R-Y, Kelly T (2005) DNA replication origins in the *Schizosaccharomyces pombe* genome. PNAS 102:337-342

Daga RR, Bolanos P, Moreno S (2003) Regulated mRNA stability of the Cdk inhibitor Rum1 links nutrient status to cell cycle progression. Curr Biol 13:2015-2024

Davis JC Petrov DA (2004) Preferential duplication of conserved proteins in eukaryotic genomes. PLoS 2:E55

Decottignies A, Sanchez-Perez I, Nurse P (2003) *Schizosaccharomyces pombe* essential genes: A pilot study. Genome Res 13:399-406

DeRisi JL, Iyer VR, Brown PO (1997) Exploring the metabolic and genetic control of gene expression on a genomic scale. Science 278:680-686

Dietrich FS, Voegeli S, Brachat S, Lerch A, Gates K, Steiner S, Mohr C, Pohlmann R, Luedi P, Choi S, Wing RA, Flavier A, Gaffney TD, Philippsen P (2004) The *Ashbya gossypii* genome as a tool for mapping the ancient *Saccharomyces cerevisiae* genome. Science 304:304-307

Doe CL, Wang G, Chow C, Fricker MD, Singh PB, Mellor EJ (1998) The fission yeast chromodomain encoding gene chp1(+) is required for chromosome segregation and shows a genetic interaction with alpha-tubulin. Nucleic Acids Res 26:4222-4229

Dolinski K, Balakrishnan R, Christie KR, Costanzo MC, Dwight SS, Engel SR, Fisk DG, Hirschman JE, Hong EL, Nash R, Oughtred R, Theesfeld CL, Binkley G, Lane C, Schroeder M, Sethuraman A, Dong S, Weng S, Miyasato S, Andrada R, Botstein D, Cherry JM "*Saccharomyces* Genome Database" http://www.yeastgenome.org/

Domazet-Loso T, Tautz D (2003) An evolutionary analysis of orphan genes in *Drosophila*. Genome Res 13:2213-2219

Dubey DD, Kim SM, Todorov IT, Huberman JA (1996) Large, complex modular structure of a fission yeast DNA replication origin. Curr Biol 6:467-473

Eddy SR (2002) Computational genomics of noncoding RNA genes. Cell 109:137-140

Ekwall K, Javerzat JP, Lorentz A, Schmidt H, Cranston G Allshire R (1995) The chromodomain protein Swi6: a key component of fission yeast centromeres. Science 269:1429-1431

Eisen MB, Spellman PT, Brown PO, Botstein D (1998) Cluster analysis and display of genome-wide expression patterns. Proc Natl Acad Sci 95:14863-14864

Ettwiller LM, Rung J, Birney E (2003) Discovering novel *cis*-regulatory motifs using functional networks. Genome Res 13:883-895

Fan JB, Chikashige Y, Smith CL, Niwa O, Yanagida M, Cantor CR (1988) Construction of a *Not* I restriction map of the fission yeast *Schizosaccharomyces pombe*. Nucleic Acids Res 17:2801-2818

Fink GR (1987) Pseudogenes in yeast? Cell 49:5-6

Fitch WM (1970) Distinguishing homologs from analogous proteins. Berlin-Heidelberg-New York, Springer-Verlag

Fitzgerald-Hayes, Clarke L, Carbon J (1982) Nucleotide sequence comparisons and functional analysis of yeast centromere DNAs. Cell 29:235-244

Forsburg SL (1999) The best yeast? Trends Genet 15:340-344

Foury F, Roganti T, Lecrenier N, Purnelle B (1998) The complete sequence of the mitochondrial genome of *Saccharomyces cerevisiae*. FEBS Lett 440:325

Fraser AG, Kamath RS, Zipperlen P, Martinez-Campos M, Sohrmann M, Ahringer J (2000) Functional genomic analysis of *C. elegans* chromosome I by systematic RNA interference. Nature 408:325-330

Fraser AG, Marcotte EM (2004) A probabilistic view of gene function. Nat Genet 36:559-564

Fraser HB, Wall DP, Hirsh AE (2003) A simple dependence between protein evolution rate and the number of protein-protein interactions. BMC Evol Biol 3:11

Fujioka Y, Kimata Y, Nomaguchi K, Watanabe K, Kohno K (2002) Identification of a novel non-structural maintenance of chromosomes (SMC) componet of the SMC5-SMC6 complex involved in DNA repair. J Biol Chem 277:21585-21591

Gaillardin C, Duchateau-Nguyen G, Tekaia F, Llorente B, Casaregola S, Toffano-Nioche C, Aigle M, Artiguenave F, Blandin G, Bolotin-Fukuhara M, Bon E, Brottier P, de Montigny J, Dujon B, Durrens P, Lepingle A, Malpertuy A, Neuveglise C, Ozier-Kalogeropoulos O, Potier S, Saurin W, Termier M, Wesolowski-Louvel M, Wincker P, Souciet J, Weissenbach J (2000) Genomic exploration of the hemiascomycetous yeasts: 21 Comparative functional classification of genes. FEBS Lett 487:134-149

Garrels JI (2002) Yeast genomic databases and the challenge of the post-genomic era. Funct Integr Genomics 2:212-237

Gasch AP, Spellman PT, Kao CM, Carmel-Harel O, Eisen MB, Storz G, Botstein D, Brown PO (2000) Genomic expression programs in the response of yeast cells to environmental changes. Mol Biol Cell 11:4241-4257

Goffeau A, Barrell BG, Bussey H, Davis RW, Dujon B, Feldmann H, Galibert F, Hoheisel JD, Jacq C, Johnston M, Louis EJ, Mewes HW, Murakami Y, Philippsen P, Tettelin H, Oliver SG (1996) Life with 6000 genes. Science 274:546-567

Gomez M, Antequera F (1999) Organization of DNA replication origins in the fission yeast genome. EMBO J 18:5683-5690

Halme A, Bumgarner S, Styles C, Fink GR (2004) Genetic and epigenetic regulation of the FLO gene family generates cell-surface variation in yeast. Cell 116:405-415

Hall IM, Shankaranarayana GD, Noma K, Ayoub N, Cohen A, Grewel SI (2002) Establishment and maintenance of a heterochromatin domain. Science 297:2215-2218

Hansen KR, Burns G, Mata J, Volpe TA, Martienssen RA, Bähler J, Thon G (2005) Global effects on gene expression in fission yeast by silencing and RNA interference machineries. Mol Cell Biol 25:590-601

Heckman DS, Geiser DM, Eidell BR, Stauffer RL, Kardos NL, Hedges SB (2001) Molecular evidence for the early colonization of land by fungi and plants. Science 293:1129-1133

Hertz-Fowler C, Peacock CS, Wood V, Aslett M, Kerhornou A, Mooney P, Tivey A, Berriman M, Hall N, Rutherford K, Parkhill J, Ivens AC, Rajandream MA, Barrell B (2004) GeneDB: a resource for prokaryotic and eukaryotic organisms Nucleic Acids Res 32:D339-D343

Hirsh A, Fraser HB (2001) Protein dispensability and rate of evolution. Nature 411:1046-1049

Hirotsune S, Yoshida N, Chen A, Garrett L, Sugiyama F, Takahashi S, Yagami K, Wynshaw-Boris A, Yoshiki A (2003) An expressed pseudogene regulates the messenger-RNA stability of its homologous coding gene. Nature 423:91-96

Hughey R, Krogh A (1996) Hidden Markov models for sequence analysis: extensions and analysis of the basic method. Comput Appl Biosci 12:95-107

Ivanov IP, Gesteland RF, Matsufuji S (1998) Programmed frameshifting in the synthesis of mammalian anitzyme is +1 in mammals predominantly +1 in fission yeast, but -2 in budding yeast. RNA 4:1230-1238

Jeong H, Tombor B, Albert R, Oltvai ZN, Barabasi AL (2000) The large scale organization of metabolic networks. Nature 407:651-654

Jeong H, Mason SP, Barabasi AL, Oltvai ZN (2001) Lethality and centrality in protein networks. Nature 411:41-42

Jones RH, Jones NC (1989) Mammalian cAMP-responsive element can activate transcription in yeast and binds a yeast factor(s) that resembles mammalian transcription factor ATF. Proc Natl Acad Sci 86:2176-2180

Jordan IK, Wolf YI, Koonin EV (2003) No simple dependence between protein evolution rate and the number of protein-protein interactions: only the most prolific interactors tend to evolve slowly. BMC Evol Biol 3:1

Kamath RS, Fraser AG, Dong Y, Poulin G, Durbin R, Gotta M, Kanapin A, Le Bot N, Moreno S, Sohrmann M, Welchman DP, Zipperlen P, Ahringer J (2003) Systematic functional analysis of the *Caenorhabditis elegans* genome using RNAi. Nature 421:231-237

Kanoh J, Ishikawa F (2003) Composition and conservation of the telomeric complex. Cell Mol Life Sci 60:2295-2302

Käufer NF, Potashkin J (2000) Analysis of the splicing machinery in fission yeast: a comparison with budding yeast and mammals. Nucleic Acids Res 28:3003-3010

Kellis M, Birren B, Lander ES (2004) Proof and evolutionary analysis of ancient genome duplication in yeast *Saccharomyces cerevisiae*. Nature 428:617-624

Kellis M, Patterson N, Endrizzi M, Birren B, Lander ES (2003) Sequencing and comparison of yeast species to identify genes and regulatory elements. Nature 423:241-254

Keogh RS, Seoighe C, Wolfe KH (1998) Evolution of gene order and chromosome number in *Saccharomyces, Kluyveromyces* and related fungi. Yeast 14:443-457

Kim JM, Vanguri S, Boeke JD, Gabriel A, Voytas DF (1998) Transposable elements and genome organization: A comprehensive survey of retrotransposons revealed by the complete *Saccharomyces cerevisiae* genome sequence. Genome Res 8:464-478

Kitajima TS, Kawashima SA, Watanabe Y (2004) The conserved kinetochore protein shugoshin protects centromeric cohesion during meiosis. Nature 427:510-517

Kniola B, O'Toole E, McIntosh JR, Mellone B, Allshire R, Mengarelli S, Hultenby K, Ekwall K (2001) The domain structure of centromeres is conserved from fission yeast to humans. Mol Biol Cell 12:2767-2775

Koonin EV, Fedorova ND, Jackson JD, Jacobs AR, Krylov DM, Makarova KS, Mazumder R, Mekhedov SL, Nikolskaya AN, Rao BS, Rogozin IB, Smirnov S, Sorokin AV, Sverdlov AV, Vasudevan S, Wolf YI, Yin JJ, Natale DA. (2004) A comprehensive evolutionary classification of proteins encoded in complete eukaryotic genomes. Genome Biol 5:R7

Koszul R, Caburet S, Dujon B, Fischer G (2004) Eucaryotic genome evolution through the spontaneous duplication of large chromosomal segments. EMBO J 23:234-243

Kunin V, Pereira-Leal JB, Ouzounis CA (2004) Functional evolution of the yeast protein interaction network. Mol Biol Evol 21:1711-1716

Krupp G, Cherayil B, Frendewey D, Nishikawa S, Soll D (1986) Two RNA species copurify with RNase P from the fission yeast *S. pombe*. EMBO J 5:1697-703

Krylov DM, Wolf YI, Rogozin IB, Koonin EV (2003) Gene loss, protein sequence divergence, gene dispensability, expression level, and interactivity are correlated in eukaryotic evolution. Genome Res 10:2229-2235

Kuhn AN, Käufer NF (2003) Pre-mRNA splicing in *Schizosaccharomyces pombe.* Curr Genet 42:241-251

Kulikova T, Aldebert P, Althorpe N, Baker W, Bates K, Browne P, van den Broek A, Cochrane G, Duggan K, Eberhardt R, Faruque N, Garcia-Pastor M, Harte N, Kanz C, Leinonen R, Lin Q, Lombard V, Lopez R, Mancuso R, McHale M, Nardone F, Silventoinen V, Stoehr P, Stoesser G, Tuli MA, Tzouvara K, Vaughan R, Wu D, Zhu W, Apweiler R (2004) The EMBL nucleotide sequence database. Nucleic Acids Res 32:D115-D119

Lang BF, Cedergren R, Gray MW (1987) The mitochondrial genome of the fission yeast, *Schizosaccharomyces pombe.* Sequence of the large-subunit ribosomal RNA gene, comparison of potential secondary structure in fungal mitochondrial large-subunit rRNAs and evolutionary considerations. Eur J Biochem 169:527-537

Langkjaer RB, Cliften P, Johnston M, Piskur J (2003) Yeast genome duplication was followed by asynchronous differentiation of duplicated genes. Nature 421:848-852

Lespinet O, Wolf YI, Koonin EV, Aravind L (2002) The role of lineage-specific gene family expansion in the evolution of eukaryotes. Genome Res 12:1048-1059

Levin HL (1995) A novel mechanism of self-primed reverse transcription defines a new family of retroelements. Mol Cell Biol 15:3310-3317

Levin H, Weaver DC, Boeke JD (1990) Two related families of retrotransposons from *Schizosaccharomyces pombe.* Mol Cell Biol 10:6791-6798

Li L, Stoeckert CJ, Roos DS (2003) OrthoMCL: Identification of ortholog groups for eukaryotic genomes. Genome Res 13:2178-2189

Lorentz A, Ostermann K, Fleck O (1994) Switching gene swi6, involved in repression of silent mating-type loci in fission yeast, encodes a homologue of chromatin-associated proteins from *Drosophila* and mammals. Gene 143:139-143

Lorenz A, Wells JL, Pryce DW, Novatchkova M, Eisenhaber F, McFarlane RJ, Loidl J (2004) *S. pombe* linear elements contain proteins related to synaptonemal complex components. J Cell Sci 117:3345-3351

Lowe TM, Eddy SR (1997) tRNAscan-SE: a program for improved detection of transfer RNA genes in genomic sequence. Nucleic Acids Res 25:955-964

Lowe TM, Eddy SR (1999) A computational screen for methylation guide snoRNAs in yeast. Science 283:1168-1171

Lowndes NF, McInerny CJ, Johnson AL, Fantes PA, Johnston LH (1992) Control of the DNA synthesis genes in fission yeast by the cell-cycle gene *cdc10+.* Nature 355:449-453

Lum PY, Edwards S, Wright R (1996) Molecular, functional and evolutionary characterization of the gene encoding HMG-CoA reductase in the fission yeast *Schizosaccharomyces pombe.* Yeast 12:1107-1124

Malik HS, Eikbush TH (1999) Modular evolution of the integrase domain in the Ty3/Gypsy class of LTR retrotransposons. J Virol 73:5186-5190

Mandell J, Goodrich KJ, Bähler J, Cech TR (2004) Expression of a RecQ helicase homolog affects progression through crisis in fission yeast lacking telomerase. J Biol Chem 280:5249-5257

Mandell JG, Bähler J, Volpe TA, Martienssen RA, Cech TR (2005) Global expression changes resulting from loss of telomeric DNA in fission yeast. Genome Biol 6:R1

Mao J, Appel B, Schaack J (1982) The 5S RNA genes of *Schizosaccharomyces pombe*. Nucleic Acids Res 10:487-500

Masakuto H, Huberman JA, Frattini MG, Kelly TJ (2004) DNA replication in *S. pombe*. In: The molecular biology of *Schizosaccharomyces pombe* (Egel R, Ed). Springer-Verlag Heidelberg, pp73-99

Mata J, Lyne R, Burns G, Bähler J (2002) The transcriptional program of meiosis and sporulation in fission yeast. Nat Genet 32:143-147

Mata J, Bähler J (2003) Corrlelations between gene expression and gene conservation in fission yeast. Genome Res 13:2686-2690

Maundrell K, Hutchison A, Shall S (1988) Sequence analysis of ARS elements in fission yeast. EMBO J 7:2203-2209

Maxwell PH, Coombes C, Kenny AE (2004) Ty1 mobilizes subtelomeric Y' elements in telomerase-negative *Saccharomyces cerevisiae* survivors. Mol Cell Biol. 24:9887-9898

Molnar M, Parisi S, Kakihara Y (2001) Characterization of rec7, an early meiotic recombination gene in *Schizosaccharomyces pombe*. Genetics 2:519-532

Morimyo M, Mita K, Hongo E, Higashi T, Sugaya K, Ajimura M, Yamauchi M, Tsuji S, Park W.-Y, Sasanuma S, Nohata J, Kimura T, Inoue H, Ishihara Y (1998) cDNA catalog of fission yeast (*Schizosaccharomyces pombe*) and its application for cloning of mammalian DNA repair gene. In: Biodefence mechanisms against environmental stress (Ozawa T, Hori T, Tatsumi K Eds), Springer Verlag Tokyo, Heidelberg, pp 115-123

Mott R (1997) EST_GENOME: a program to align spliced DNA sequences to unspliced genomic DNA. Comput Appl Biosci 4:477-478

Mundt KE, Porte J, Murray JM, Brikos C, Christensen PU, Caspari T, Hagan IM, Millar JB, Simanis V, Hofmann K, Carr AM (1999) The COP9/signalosome complex is conserved in fission yeast and has a role in S phase. Curr Biol 9:1427-1430

Murakami S, Matsumoto T, Niwa O, Yanagida M (1991) Structure of the fission yeast centromere cen3: direct analysis of the reiterated inverted region. Chromosoma 101:214-221

Nimmo ER, Pidoux AL, Perry PE, Allshire RC (1998) Defective meiosis in telomere silencing mutants of *Schizosaccharomyces pombe*. Nature 392:825-828

Nurse P (2000) A long twentieth century of the cell cycle and beyond. Cell 100:71-78

Nurse P (2003) Understanding cells. Nature 424:883

Ohno S (1970) Evolution by gene duplication. Springer-Verlag, Berlin-Heidelberg-New York

Oliver SG, van der Aart QJ, Agostoni-Carbone ML, Aigle M, Alberghina L, Alexandraki D, Antoine G, Anwar R, Ballesta JP, Benit P, et al. (1992) The complete DNA sequence of yeast chromosome III. Nature 357:38-46

Ozier-Kalogeropoulos O, Malpertuy A, Boyer J, Tekaia F, Dujon B (1998) Random exploration of the *K. lactis* genome and comparison to that of *S. cerevisiae*. Nucleic Acids Res 26:5511-5524

Pagel P, Mewes H-W, Frishman D (2004) Conservation of protein-protein interactions – lessons from *ascomycota*. Trends Genet 20:72-76

Pasero P, Marilley M (1993) Size variation of rDNA clusters in the yeasts *Saccharomyces cerevisiae* and *Schizosaccharomyces pombe*. Mol Gen Genet 236:448-452

Pearson W, Lipman DJ (1988) Improved tools for biological sequence comparison. Proc Natl Acad Sci 85:2444-2448

Piskur J (2001) Origin of the duplicated regions in the yeast genomes. Trends Genet 16:302-303

Pollack JR, Iyer VR (2002) Characterizing the physical genome. Nat Genet Suppl 32:515-521

Pradet-Balade B (2001) Translation control: bridging the gap between genomics and proteomics? Trends Biochem Sci 26:225-229

Prado F, and Aguilera A (2005) Partial depletion of histone H4 increases homologous recombination-mediated genetic instability. Mol Cell Biol 24:1526-1536

Prince VE, Pickett (2002) Splitting pairs: The diverging fates of duplicated genes. Nat Rev Genet 3:827-837

Pyne S, Skiena S, Futcher B (2005) Copy correction and concerted evolution in the conservation of yeast genes. PLoS Biol, in press

Raghuraman MK, Winzeler EA, Collingwood D, Hunt S, Wodicka L, Conway A, Lockhart DJ, Davis RW, Brewer BJ, Fangman WL (2001) Replication dynamics of the yeast genome. Science 294:115-121

Remacle JE, Albrecht G, Brys R, Braus GH, Huylebroeck D (1997) Three classes of mammalian transcription activation domain stimulate transcription in *Schizosaccharomyces pombe*. EMBO J 16:5722-5729

Remm M, Storm CE, Sonnhammer EL (2001) Automatic clustering of orthologs and in-paralogs from pairwise species comparisons. J Mol Biol 314:1041-1052

Ribes V, Dehoux P, Tollervey D (1988) 7SL RNA from *S. pombe* is encoded by a single copy essential gene. EMBO J 7:231-237

Robyr D, Suka Y, Xenarios I, Kurdisatani SK, Wang A, Suka N, Grunstein M (2002) Microarray deacetylation maps determine genome-wide functions for yeast histone deacetylases. Cell 1009:437-466

Rustici G, Mata J, Kivinen K, Lio P, Penkett CJ, Burns G, Hayles J, Brazma A, Nurse P, Bähler J (2004) Periodic gene expression program of the fission yeast cell cycle. Nat Genet 36:809-817

Rutherford K, Parkhill J, Crook J, Horsnell T, Rice P, Rajandream MA, Barrell B (2000) Artemis: sequence visualization and annotation. Bioinformatics 16:944-945

Schaak J, Mao J, Söll D (1982) The 5.8S RNA gene sequence and the ribosomal repeat of *S. pombe*. Nucleic Acids Res 10:2851-2864

Scherthan H, Bähler J, Kohli J (1994) Dynamics of chromosome organization and pairing during meiotic prophase in fission yeast. J Cell Biol 127:273-285

Scherthan H (2001) A bouquet makes ends meet. Nat Rev Mol Cell Biol 2:621-627

Schroder AR, Shinn P, Chen H, Berry C, Ecker JR, Bushman F (2002) HIV-1 integration in the human genome favors active genes and local hotspots. Cell 110:521-529

Segurado M, de Luis A, Antequera F (2003) Genome-wide distribution of DNA replication origins at A+T rich islands in *Schizosaccharomyces pombe*. EMBO reports 4:1048-1053

Singleton TL, Levin HL (2002) A long terminal repeat retrotransposon of fission yeast has strong preferences for specific sites of insertion. Eukaryot Cell 1:44-55

Sipiczki M (2001) Where does fission yeast sit on the tree of life? Genome Biol 1:1011.1-1011.4

Smith CL, Matsumoto T, Niwa O, Klco S, Fan JB, Yanagida M, Cantor CR (1987) An electrophoretic karyotype for *Schizosaccharomyces pombe* by pulsed field gel electrophoresis. Nucleic Acids Res 15:4481-4491

Sonnhammer EL, Eddy SR, Durbin R (1997) Pfam: a comprehensive database of protein domain families based on seed alignments Proteins 3:405-420

Spellman PT, Sherlock G, Zhang MQ, Iyer VR, Anders K, Eisen MB, Brown PO, Botstein D, Futcher B (1998) Comprehensive identification of cell-cycle regulated genes of the yeast *Saccharomyces cerevisiae* by microarray hybridization. Mol Biol Cell 9:3273-3297

Sunnerhagen P (2002) Prospects for functional genomics in *Schizosaccharomyces pombe*. Curr Genet 42:73-84

Takahashi K, Murakami S, Chikashige Y, Funabiki H, Niwa O, Yanagida M (1992) A low copy number central sequence with strict symmetry and unusual chromatin structure in the fission yeast centromere. Mol Biol Cell 3:819-835

Tatusov RL, Koonin EV, Lipman DJ (1997) A genomic perspective on global families. Science 278:631-637

Tatusov RL, Fedorova ND, Jackson JD, Jacobs AR, Kiryutin B, Koonin EV, Krylov DM, Mazumder R, Mekhedov SL, Nikolskaya AN, Rao BS, Smirnov S, Sverdlov AV, Vasudevan S, Wolf YI, Yin JJ, Natale DA (2003) The COG database: an updated version includes eukaryotes. BMC Bioinformatics 4:41

Teichmann SA (2002) The constraints protein-protein interactions place on sequence divergence. J Mol Biol 324:399-407

The *C. elegans* sequencing consortium (1998) Genome sequence of the nematode *C. elegans* : a platform for investigating biology. Science 282:2012-2018

The Gene Ontology Consortium (2004) The Gene Ontology (GO) database and informatics resource. Nucleic Acids Res 32:D258-D261

Theis JF, Newlon CS (1997) The ARS309 chromosomal replicator of *Schizosaccharomyces cerevisiae* depends on an exceptional ARS consensus sequence. Proc Natl Acad Sci USA 94:10786-10791

Theis JF, Newlon CS (2001) Two compound replication origins in *Saccharomyces cerevisiae* contain redundant origin complex binding sites. Mol Cell Biol 21:2790-2801

Thompson JD, Higgins DG, Gibson TJ (1994) CLUSTAL W: improving the sensitivity of progressive multiple sequence alignment through sequence weighting, position-specific gap penalties and weight matrix choice. Nucl Acids Res 22:4673-4680

van Driel R (2003) The eukaryotic genome: a system regulated at different hierarchical levels. J Cell Sci 116:4067-4075

Villeneuve AM, Hillers KJ (2001) Whence meiosis? Cell 106:647-650

Volpe TA, Kidner C, Hall IM, Teng G , Grewal SI, Martienssen RA (2002) Regulation of heterochromatin silencing and histone H3 lysine-9 methylation by RNAi. Science 297:1833-1837

Volpe T, Schramke V, Hamilton, White SA, Teng G, Martienssen RA, Allshire RC (2003) RNA interference is required for normal centromere function in fission yeast. Chromosome Res 11:137-146

Watanabe Y, Yamamoto M (1994) *S. pombe mei2$^+$* encodes an RNA-binding protein essential for premeiotic DNA synthesis and meiosis I, which cooperates with a novel RNA species meiRNA. Cell 78:487-498

Watanabe T, Miyashita K, Saito TT (2001) Comprehensive isolation of meiosis-specific genes identifies novel proteins and unusual non-coding transcripts in *Schizosaccharomyces pombe*. Nucleic Acids Res 29:327-337

Watanabe T, Miyashita K, Saito TT, Nabeshima K, Nojima H (2002) Abundant poly (A)-bearing RNAs that lack open reading frames in *S. pombe*. DNA Res 9:209-215

Webb CJ, Wise JA (2004) The splicing factor U2AF small subunit is functionally conserved between fission yeast and humans. Mol Cell Biol 10:4229-4240

Wood V, Rutherford K, Ivens A, Rajandream M-A, Barrell B (2001) A re-annotation of the *Saccharomyces cerevisiae* genome. Comp Funct Genom 2:143-154

Wood V, Gwilliam R, Rajandream MA, Lyne M, Lyne R, Stewart A, Sgouros J, Peat N, Hayles J, Baker S, Basham D, Bowman S, Brooks K, Brown D, Brown S, Chillingworth T, Churcher C, Collins M, Connor R, Cronin A, Davis P, Feltwell T, Fraser A, Gentles S, Goble A, Hamlin N, Harris D, Hidalgo J, Hodgson G, Holroyd S, Hornsby T, Howarth S, Huckle EJ, Hunt S, Jagels K, James K, Jones L, Jones M, Leather S, McDonald S, McLean J, Mooney P, Moule S, Mungall K, Murphy L, Niblett D, Odell C, Oliver K, O'Neil S, Pearson D, Quail MA, Rabbinowitsch E, Rutherford K, Rutter S, Saunders D, Seeger K, Sharp S, Skelton J, Simmonds M, Squares R, Squares S, Stevens K, Taylor K, Taylor RG, Tivey A, Walsh S, Warren T, Whitehead S, Woodward J, Volckaert G, Aert R, Robben J, Grymonprez B, Weltjens I, Vanstreels E, Rieger M, Schafer M, Muller-Auer S, Gabel C, Fuchs M, Fritzc C, Holzer E, Moestl D, Hilbert H, Borzym K, Langer I, Beck A, Lehrach H, Reinhardt R, Pohl TM, Eger P, Zimmermann W, Wedler H, Wambutt R, Purnelle B, Goffeau A, Cadieu E, Dreano S, Gloux S, Lelaure V, Mottier S, Galibert F, Aves SJ, Xiang Z, Hunt C, Moore K, Hurst SM, Lucas M, Rochet M, Gaillardin C, Tallada VA, Garzon A, Thode G, Daga RR, Cruzado L, Jimenez J, Sanchez M, del Rey F, Benito J, Dominguez A, Revuelta JL, Moreno S, Armstrong J, Forsburg SL, Cerrutti L, Lowe T, McCombie WR, Paulsen I, Potashkin J, Shpakovski GV, Ussery D, Barrell BG, Nurse P (2002) The genome sequence of *Schizosaccharomyces pombe*. Nature 415:871-880

Wolfe KH, Shields DC (1997) Molecular evidence for an ancient duplication of the entire yeast genome. Nature 387:708-713

Wolfe K (2004) Evolutionary genomics: Yeast accelerate beyond BLAST. Curr Biol 14: R392-R394

Wong S, Butler G, Wolfe KH (2002) Gene order evolution and paleopolyploidy in hemiascomycete yeasts. Proc Natl Acad Sci 14:9272-9277

Wyrick JJ, Aparicio JG, Chen T, Barnett JD, Jennings EG, Young RA, Bell SP, Aparicio OM (2001) Genome-wide distribution of ORC and NCN proteins in *S. cerevisiae*: high resolution mapping of replication origins. Science 294:2357-2360

Yamanda M, Hayatsu N, Matsuura A, Ishikawa F (1998) Y'-Help1, a DNA helicase encoded by the yeast subtelomeric Y' element, is induced in survivors defective for telomerase. J Biol Chem 273:33360-33366

Yanagida M (2002) The model unicellular eukaryote, *Schizosaccharomyces pombe*. Genome Biol 3:COMMENT2003.1-2003.4

Yieh L, Kassavetis G, Geiduscheck EP, Sandmeyer SB (2000) The Brf and TATA-binding proteins subunits of the RNA polymerase III transcription factor IIIB mediate position specific integration of the gypsy-like element, Ty3. J Biol Chem 275:29800-29807

Young JA, Schreckhise RW, Steiner WW, Smith GR (2002) Meiotic recombination remote from prominent break sites in *S. pombe*. Mol Cell 9:253-263

Young JA, Hyppa RW, Smith GR (2004) Swi5 acts in meiotic DNA joint molecule formation in *Schizosaccharomyces pombe*. Genetics 167:593-605

Zdobnov EM, von Mering C, Letunic I, Bork P (2002) Comparative genome and proteome analysis of *Anopheles gambiae* and *Drosophila melanogaster*. Science 298:149-159

Zhu C, Karplus K, Grate L, Coffino P (2000) A homolog of mammalian antizyme is present in fission yeast *Scizosaccharomyces pombe* but not detected in budding yeast *Saccharomyces cerevisiae.* Bioinformatics 16:478-481

Wood, Valerie
  Wellcome Trust Sanger Institute, Wellcome Trust Genome Campus, Hinxton, CB10 1SA, UK
  val@sanger.ac.uk

Index